Dynamic Modeling and Control of Engineering Systems

J. LOWEN SHEARER

Professor Emeritus
The Pennsylvania State University
University Park, Pa

BOHDAN T. KULAKOWSKI

Department of Mechanical Engineering
The Pennsylvania State University
University Park, Pa

Macmillan Publishing Company
NEW YORK

Collier Macmillan Publishers
LONDON

Editor: David Johnstone
Production Supervisor: J. Edward Neve
Production Manager: Richard C. Fischer
Cover Designer: Robert Freese
Illustrations: Wellington Studios

PRINTED IN THE UNITED STATES OF AMERICA

Macmillan Publishing Company
866 Third Avenue, New York, New York 10022

Collier Macmillan Canada, Inc.

LIBRARY OF CONGRESS CATALOGING IN PUBLICATION DATA

Shearer, J. Lowen.
Dynamic modeling and control of engineering systems / J. Lowen Shearer, Bohdan T. Kulakowski.
p. cm.
Includes index.
ISBN 0-02-409790-X
1. Engineering—Mathematical models. I. Kulakowski, Bohdan T. II. Title.
TA342.S54 1990
620′.0042—dc19 89-2513
CIP

Printing: 1 2 3 4 5 6 7 8 Year: 0 1 2 3 4 5 6 7 8 9

Preface

This text provides a comprehensive treatment of the fundamentals of mathematical modeling of engineering systems. Special attention is given to the roles of integration (with respect to time) and feedback: (a) natural, negative feedback existing in nature that accompanies inherent stability and (b) artificial feedback introduced deliberately in automatic control systems to improve system performance. The systems discussed are mechanical, electrical, thermal, fluid, and mixed, with emphasis on analogies among them.

The inherent lag with which a system responds to an input disturbance is shown to be the result of the same dynamic interactions among energy storage elements, energy dissipation elements, energy sources, and load systems, regardless of the type (mechanical, electrical, etc.) of system. The analogous behaviour of the different types of systems is first evident in the forms of their describing equations and second in their system characteristics, such as time constant, natural frequency, and damping ratio.

The use of state-variables in describing and modeling the dynamic performance of engineering systems has come to be routine, having evolved from the esoteric ''Modern Control Theory'' of an earlier era. State-space representation enhances the unified approach involving the analogies mentioned previously, facilitates programming on high-speed digital computers, and is used as the primary form of mathematical and computer model. Furthermore, the ready availability of high-speed digital computers to engineers makes possible the simulation of dynamic systems, formerly carried out with less accuracy but higher speed on analog computers. The ready availability enhances the role of the digital computer as a powerful tool for use by the system designer in predicting system performance and in designing systems to meet exacting requirements. Also, it is the authors' experience that students quickly learn state-space methods, especially when reinforced by ready access to high-speed digital computers.

Although state-space models will be emphasized here, older conventional system models, such as input-output differential equations, transfer functions, and frequency response graphs, are also employed. Use of the older models assures continuity of the educational process by making a smooth transition from earlier courses in design, vibrations, and instrumentation where state-space methods have been rarely employed. Also these earlier forms of system models are very useful in communication with scientists in related fields. Some comments need to be emphasized here with regard to the use of complex models.

- First, a general rule: **The simplest useful model is the best model.** Herein lies a problem: On one hand, excessive simplication may result in leaving out a significant feature of system performance. On the other hand, including minor features and parasitic effects leads to excessive complexity, wastes time and money, and may tend to obscure the more significant aspects being sought.
- Second, only lumped-parameter models are employed here, because they comprise such a large fraction of the systems that engineers are called upon to deal with. In addition, many distributed systems can be usefully modeled by lumped-parameter approximations.
- Third, the modeling of nonlinearities is very readily accomplished through the use of state-variable equations and digital simulation. In order to provide a bridge to the use of strictly linear methods, linearization techniques and their limitations are also discussed.

This text is intended primarily for use by students in mechanical engineering and related fields, who have already taken courses such as mechanical design, vibrations, and instrumentation. It is an outgrowth of a one-semester course taken by seniors in mechanical engineering at the Pennsylvania State University since 1980.

In Chapter 1 basic definitions and terminology associated with the field of dynamic systems are introduced. General types of system elements are defined according to their role in the storage and dissipation of system energy[1]. These elements, which are present in all kinds of dynamic systems, are the A-type energy-storing elements, the T-type energy storing elements, and the D-type energy-dissipating elements. Associated with these generalized elements are two types of generalized system variables: The A-type or ''across'' variable and the T-type or ''through'' variable. Analogous elements and variables are identified in mechanical, electrical, fluid, and thermal systems. The key role of integration in the performance of dynamic systems is emphasized. Linearization is also introduced in Chapter 1.

Chapter 2 deals with mechanical systems, including both translation and rotational components. Free-body diagrams are used in the derivation of differential equations based on Newton's laws of motion. Mechanical systems are introduced very early in this text in order to provide nontrivial examples in the following three chapters.

In Chapter 3 simulation block diagrams are introduced in the time domain. Originally developed for programming simulation studies on analog computers, now largely replaced by digital computers, these simulation block diagrams provide a ''road map'' showing the signal flows within the system; they also provide valuable insights when analyzing dynamic system performance.

[1]J. L. Shearer, A. T. Murphy, and H. H. Richardson, *Introduction to System Dynamics,* Addison-Wesley Publishing Co., Reading, MA, 1967.

In Chapter 4 basic forms of mathematical models of dynamic systems are presented. Although the traditional input-output form of the mathematical models will be reviewed, the main emphasis will be on state-variable models, which have become the dominant form employed in modern methods of system analysis and control.

Analytical methods for solving mathematical model equations are reviewed in Chapter 5. These methods, although limited in practice to linear, stationary, single input-single output equation models, are very useful in the analysis of low-order systems. They can also be used for approximate solutions needed for verification of the results obtained from computer simulations of more complex and nonlinear systems.

In Chapter 6 numerical methods for solving sets of first-order differential equations are reviewed, including the classical method introduced by Leonhard Euler in 1768. More advanced algorithms employing interpolation and automatic change of time step are then introduced and employed for the numerical integration of sets of state-variable equations.

In Chapters 7, 8, 9, and 10 the techniques for modeling and analysis developed in the earlier chapters are applied to electrical, thermal, fluid, and mixed systems. Special attention is given to electrical systems because of their widespread presence in measuring and control devices in all kinds of engineering systems. In the presentation of thermal, hydraulic, pneumatic, and mixed systems (mainly electromechanical and hydromechanical), the emphasis is again on simple, lumped models. Heat storage systems, hydraulic pumps driven by electric motors, hydraulic motor-driven mechanical systems, DC-motor and -generator drives, and valve-controlled fluid systems are analyzed in these chapters.

The next four chapters deal with basic methods of analyzing automatic control systems. In Chapter 11 the concept of system transfer function is introduced as an alternative form of mathematical model for linear systems. The approach employed here differs from that chosen by many authors who rely on Laplace transform theory to introduce the complex variable s. Here the development is based on the response of a linear system to the input e^{st}, where s is a complex number $\sigma + j\omega$,[2] with emphasis on the forced part (particular integral) of the response. This approach avoids the problem of assuring convergence of the Laplace integral and the tedious inversion procedure involved in the use of Laplace transforms to obtain the amplitude and phase of responses to sinusoidal inputs. For the reader who is more comfortable with the use of Laplace transform methods, the results obtained in this way are included in Appendix 2 for comparison. Chapter 12 then deals with methods of frequency response analysis, including sinusoidal transfer functions, Bode diagrams, and polar plots. In Chapters 13 and 14 two fundamental aspects of the performance of closed-loop control systems, namely stability and steady-state errors, are discussed. Both algebraic and graphical stability criteria are introduced for use with linear models. The classical dilemma arising from conflicting requirements for transient performance and steady-state response is considered.

In Chapters 15 and 16 introductory material on analysis, modeling, and control at discrete intervals of time in closed-loop systems is presented. The material will

[2]Ibid.

provide a base for the use of data sampling, required when digital computers are used in feedback control loops. This coverage of so-called "discrete-time systems" is brief and intended only to whet the appetite of those who might want to go more deeply into this topic[3,4].

J.L.S.
B.T.K.

[3]K. Ogata, *Discrete-Time Control Systems,* Prentice Hall, Inc., 1987.
[4]H. F. VanLandingham, *Introduction to Digital Control Systems,* Macmillan Publishing Company, New York, 1985.

Brief Contents

Contents

1

Introduction

1.1 Systems and System Models

The word "system" has become very popular in recent years. It is used not only in engineering but also in science, economics, sociology, and even in politics. In spite of its common use (or perhaps because of it), the exact meaning of the term is not always fully understood. A system is defined as a combination of components that act together and perform a certain objective. A little more philosophically, a system can be understood as a conceptually isolated part of the universe, which is of interest to us. Other parts of the universe that interact with the system compose the system environment, or neighboring systems.

All existing systems change with time, and when the rates of change are significant, they are referred to as dynamic systems. A car riding over a road can be considered as a dynamic system (especially on a crooked or bumpy road). The limits of the conceptual isolation determining a system are entirely arbitrary. Therefore any part of the car given as an example of a system—its engine, brakes, suspension, etc.—can also be considered a system (i.e., a subsystem). Similarly, two cars in passing maneuver or even all vehicles within a specified area can be considered as a major traffic system.

The isolation of a system from the environment is purely conceptual. Every system interacts with its environment through two groups of variables. The variables in the first group originate outside the system and are not directly dependent on what happens in the system. These variables are called input variables or simply inputs. The other group involves variables generated by the system as it interacts with its environment. Those dependent variables in this group that are of primary interest to us are called output variables or simply outputs.

In describing the system itself, a complete set of variables, called state variables is needed. The state variables constitute the minimum set of system variables

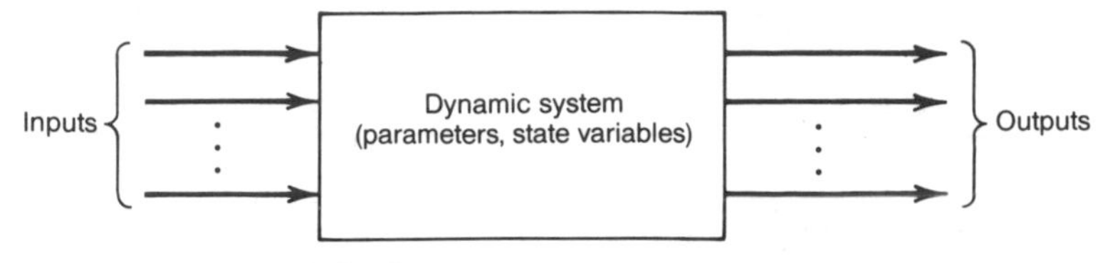

FIGURE 1.1 A dynamic system.

necessary to describe completely the state of the system at any given instant of time; and they are of great importance in the modeling and analysis of dynamic systems. Providing the initial state and the input variables have been all specified, the state variables then describe from instant-to-instant the behavior, or response, of the system. The concept of state of a dynamic system will be discussed in more detail in Chapter 4. In most cases the state variable equations employed in this text represent only a simplified model of the system, and their use leads only to an approximate prediction of system behavior.

Figure 1.1 shows a graphical presentation of a dynamic system. In addition to the state variables the system is also characterized by its parameters. In the example of a moving car, the input variables would include throttle position, position of the steering wheel, and road conditions such as slope and roughness. In the simplest model, the state variables would be the position and velocity of the vehicle as it travels along a straight path. The choice of the output variables is arbitrary, determined by the objectives of the analysis. The position, velocity, or acceleration of the car, or perhaps the average fuel flow rate or the engine temperature, can be selected as the outputs. Some of the system parameters would be the mass of the vehicle or the size of its engine. Note that the system parameters may change with time. For instance, the mass of the car will change as the amount of fuel in its tank decreases or when passengers are added. The change in mass may or may not be negligible for the performance of the car but would certainly be of critical importance in the analysis of dynamics of a ballistic missile.

Generally speaking, the main objective of the system analysis is to predict relevant performance characteristics of the system operating under specified environmental conditions. Without being able to predict this performance, a prototype system to experiment with would have to be built to obtain the necessary information. Although the data obtained that way would be very valuable, the cost, in time and money, of experimenting with an actual system is usually very high. Moreover, the flexibility of investigating alternative approaches by varying the system parameters or changing the structural interconnections is extremely limited. Therefore, one of the early major tasks in system analysis is to establish an adequate mathematical model of the system. Such a mathematical model is a description of relationships among the system variables in terms of mathematical equations. In many cases a final prototype system is eventually built to verify the adequacy of the model.

A model is a tool used in developing designs or control algorithms, and the major task in which it is to be used has basic implications on the choice of a particular form of a system model. In other words, if a model can be considered a tool, it is a specialized tool, developed specifically for a particular application. Constructing universal mathematical models, even for systems of moderate complexity, is impractical and uneconomic. Let us use a moving automobile as an example once again: The task of developing a model general enough to allow for studying ride quality, fuel economy, traction characteristics, passenger safety, and forces exerted on the road pavement (to name just a few problems typical for transportation systems) could be compared to the task of designing one vehicle to be used as a truck, for daily commuting to work in New York City, and as a racing car to compete in the Indianapolis 500. Moreover, even if such a supermodel was developed and made available to researchers (free), it is very likely that the cost of using it for most applications would be prohibitive.

Thus system models should be as simple as possible and should be developed with a specific application in mind. Of course, this approach may lead to different models being built for different uses of the same system. In the case of mathematical models, different types of equations may be used in describing the system in various applications.

Mathematical models can be grouped according to several different criteria. Table 1.1 shows classification of system models according to the four most common criteria: applicability of the principle of superposition, dependence on spatial coordinates as well as on time, variability of parameters in time, and continuity of independent variables. Based on these criteria, models of dynamic systems are classified as linear or nonlinear, lumped or distributed, stationary or time-varying, continuous or discrete, respectively. Each class of models is also characterized by

TABLE 1.1 Classification of System Models

Type of Model	Classification Criterion	Type of Model Equation
Nonlinear	Principle of superposition does not apply	Nonlinear differential equation
Linear	Principle of superposition applies	Linear differential equations
Distributed	Dependent variables are functions of spatial coordinates and time	Partial differential equations
Lumped	Dependent variables independent of spatial variables	Ordinary differential equations
Time-varying	Model parameters vary in time	Differential equations with time varying parameters
Stationary	Model parameters constant in time	Differential equations with constant parameters.
Continuous	Dependent variables defined over continuous range of independent variable	Differential equations
Discrete	Dependent variables defined only for distinct values of independent variables	Time difference equations

the type of mathematical equations employed in describing the system. All types of system models listed in Table 1.1 will be discussed in this book, although distributed models will only be given limited attention.

1.2 System Elements, Their Characteristics, and the Role of Integration

The modeling techniques to be developed in this text will focus initially on the use of a set of simple ideal system elements found in four main types of systems: mechanical, electrical, fluid, and thermal. Transducers, which enable the coupling of these types of system to create mixed system models, will be introduced later.

This set of ideal linear elements is shown in Table 1.2, which also provides their elemental equations and, in the case of energy storage elements, their energy

TABLE 1.2 Ideal System Elements (Linear)

System Type	Mechanical	Electrical	Fluid	Thermal
A-type element	Mass	Capacitor	Fluid capacitor	Thermal capacitor
Elemental equation	$F = m\frac{dv}{dt}$	$i = C\frac{de}{dt}$	$Q_f = C_f\frac{dP}{dt}$	$Q_h = C_t = \frac{dT}{dt}$
Energy stored	Kinetic	Electric field	Potential	Thermal
Energy equation	$\mathcal{E}_K = \frac{m}{2}v^2$	$\mathcal{E}_E = \frac{C}{2}e^2$	$\mathcal{E}_P = \frac{C_f}{2}P^2$	$\mathcal{E}_T = \frac{C_t}{2}T^2$
T-type element	Spring	Inductor	Inertor	None
Elemental equation	$v = \frac{1}{k}\frac{dF}{dt}$	$e = L\frac{di}{dt}$	$P = I\frac{dQ_f}{dt}$	
Energy stored	Potential	Magnetic field	Kinetic	
Energy equation	$\mathcal{E}_P = \frac{1}{2k}F^2$	$\mathcal{E}_M = \frac{L}{2}i^2$	$\mathcal{E}_K = \frac{I}{2}Q_f^2$	
D-type element	Damper	Resistor	Fluid resistor	Thermal resistor
Elemental equation	$F = bv$	$i = \left(\frac{1}{R}\right)e$	$Q_f = \left(\frac{1}{R_f}\right)P$	$Q_h = \left(\frac{1}{R_t}\right)T$
	$v = \frac{1}{b}F$	$v = Ri$	$P = R_fQ_f$	$T = R_tQ_h$
Energy dissipation rate	$\frac{d\mathcal{E}_D}{dt} = Fv$	$\frac{d\mathcal{E}_D}{dt} = ie$	$\frac{d\mathcal{E}_D}{dt} = Q_fP$	$\frac{d\mathcal{E}_D}{dt} = Q_hT$

Note: Each A-type variable represents a spatial difference across the element.

storage equations in simplified form. The variables, such as force F and velocity v used in mechanical systems and current i and voltage e in electrical systems, have also been classified as either T-type (through) variables, which act through the elements, or A-type (across) variables, which act across the elements. Thus force and current are called T variables, and velocity and voltage are called A variables. Note that this designation also corresponds to the type of instrument required to measure each variable in a physical system: force and current meters are used in series to measure what goes through the element, and velocity and voltage meters are connected in parallel to measure the difference across the element. Furthermore the energy storage elements are also classified as T-type or A-type elements, designated by the nature of their respective energy storage equations: mass stores kinetic energy, which is a function of the square of its velocity, an A variable; hence mass is A-type element, and so on.

The A-type elements are said to be analogous to each other; T-type elements are also analogs of each other. For a given kind (mechanical, electrical and so on) of system its A-type elements are called the duals of its T-type elements.

Because differentiation is seldom, if ever, encountered in nature, whereas integration is very commonly encountered, the essential dynamic character of each energy storage element is better expressed by converting its elemental equation from differential form to integral form. For instance, the elemental equation for the mass element may be integrated with respect to time, yielding

$$v(t) = v(0^-) + \left(\frac{1}{m}\right) \int_{0^-}^{t} F \, dt \tag{1.1}$$

This equation says that the velocity of a given mass m increases as the integral with respect to time of the net force applied to it, formally known as Newton's third law. It also implicitly says that, lacking a very, very large (infinite) force F, it takes time to change the velocity of mass m. It says that the kinetic energy $\mathscr{E}_K = (m/2)v^2$ of the mass is also accumulated with respect to time when the force F is finite, and it cannot be changed in zero time.

Similar elemental equations in integral form may be written for all the other energy storage elements, and similar conclusions can be drawn concerning the role of integration with respect to time and how it affects the accumulation of energy with respect to time. These two phenomena, integration and energy storage, are very important aspects of dynamic system performance, especially when energy storage elements interact and exchange energy with each other.

The energy dissipation elements or D elements, store no useful energy and have elemental equations that express instantaneous relationships between their A variables and their T variables, with no need to wait for time integration to take effect. For example, the force in a damper is instantaneously related to the velocity difference across it (i.e., no integration with respect to time is involved).

Furthermore these energy dissipators absorb energy from the system and perform a "negative feedback" effect (to be discussed in detail later), which provides damping and helps assure system stability.

1.3 Linearization

It should be emphasized that the arbitrary classifications presented in Table 1.1 and the ideal elements included in Table 1.2 are employed only as aids in system modeling. Real systems usually exhibit nonideal characteristics and/or combinations of characteristics that depart somewhat from the ideal models to be employed here. However, analysis of a system represented by nonlinear, partial differential equations with time-varying coefficients is extremely difficult and requires extraordinary computational resources to perform complex iterative solution procedures, an effort that can rarely be justified by the purpose of the analysis. Naturally, therefore, simplified descriptions of the actual system behavior are employed whenever possible. In particular, nonlinear system characteristics are approximated by linear models for which many powerful methods of analysis and design, involving non-iterative solution procedures, are available.

In general, one of the following three options can be taken in analysis of a nonlinear system:

(a) Replace nonlinear elements with "roughly equivalent" linear elements.
(b) Develop and solve a nonlinear model.
(c) Linearize the system equations for small perturbations.

Use of option (a) often leads to invalid models, to say the least. Approach (b) leads to the most accurate results, but the cost involved in nonlinear model analysis may be excessively high, often not justified by the benefits carried with a very accurate solution. Finally, option (c) represents the most rational approach, especially in preliminary stages of system analysis. Thus, option (c) will now be discussed in considerable detail.

Consider a nonlinear spring characterized by a relationship, $F_{NLS} = f_{NL}(x_{1r})$, describing a force exerted by the spring, F_{NLS}, when subjected to a change in length, x_{1r}, as shown in Figure 1.2. The value of x_{1r} is considered to be the

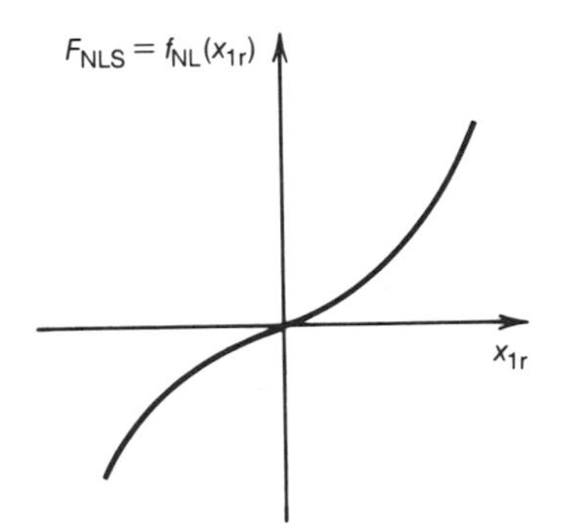

FIGURE 1.2 Nonlinear spring characteristic.

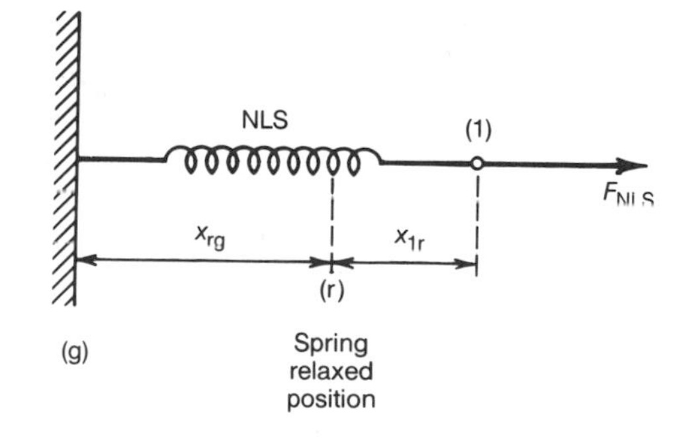

FIGURE 1.3 Elongation of spring subjected to force F_{NLS}.

extension of the spring from its relaxed length, x_{rg}, when no external forces are applied, as illustrated in Figure 1.3. The purpose of linearization is to replace a nonlinear characteristic with a linear approximation. In other words, linearizing the nonlinear function $f_{NL}(x_{1r})$ means replacing it locally with an approximating straight line. Such a formulation of the linearization process is not precise and may yield inaccurate results unless restrictions are placed on its use. In particular a limit must be somehow established for the small variation $\hat{x}_{1r}$ of the whole variable $x_{1r} = \bar{x}_{1r} + \hat{x}_{1r}$ from its normal operating point value $\bar{x}_{1r}$. This limit on the range of acceptable variation of the independent variable x_{1r} is influenced by the shape of the nonlinear function curve, and the location of the normal operating point on the curve. (In the case of two independent variables, of course, a surface would replace the role of a curve).

The term "normal operating point" here refers to the condition of a system when it is in a state of equilibrium with the input variables constant and equal to their mean values averaged over time. The variations of the inputs, to the system containing the nonlinear element, from these mean values must be small enough collectively, for the linearization error to be acceptable. In this case the maximum variation of x_{1r}, due to all the system inputs acting simultaneously, must be small enough so that the maximum resulting variation of the force F_{NLS} is adequately described by the straight-line approximation. At the normal operating point for the nonlinear spring the normal operating point force, $\bar{F}_{NLS}$, is related to the normal operating point displacement, $\bar{x}_{1r}$, by

$$\bar{F}_{NLS} = f_{NL}(\bar{x}_{1r}) \tag{1.2}$$

As variations occur from these normal operating point values,

$$x_{1r} = \bar{x}_{1r} + \hat{x}_{1r},\ F_{NLS} = \bar{F}_{NLS} + \hat{F}_{NLS}, \quad \text{and}$$

$$F_{NLS} = f_{NL}(\bar{x}_{1r} + \hat{x}_{1r}) = \bar{f}_{NL} + \hat{f}_{NL} \tag{1.3}$$

where $\bar{f}_{NL}$, the normal operating value of f_{NL}, is the first term of the Taylor's series expansion of a function near its operating point:

$$f_{NL}(\bar{x}_{1r} + \hat{x}_{1r}) = f_{NL}(\bar{x}_{1r}) + \hat{x}_{1r} \left.\frac{df_{NL}(x_{1r})}{dx_{1r}}\right|_{\bar{x}_{1r}} + \frac{\hat{x}_{1r}^2}{2!} \left.\frac{d^2 f_{NL}(x_{1r})}{dx_{1r}^2}\right|_{\bar{x}_{1r}} + \frac{\hat{x}_{1r}^3}{3!} \left.\frac{d^3 f_{NL}(x_{1r})}{dx_{1r}^3}\right|_{\bar{x}_{1r}} + \cdots$$

Hence,

$$\bar{f}_{NL} = f_{NL}(\bar{x}_{1r}) \tag{1.4}$$

And the incremental or "hat" value $\hat{f}_{NL}$ would represent the remaining terms of the Taylor's series expansion,

$$\hat{f}_{NL} = \hat{x}_{1r} \left.\frac{df_{NL}}{dx_{1r}}\right|_{\bar{x}_{1r}} + (\hat{x}_{1r}^2/2!) \left.\frac{d^2 f_{NL}}{dx_{1r}^2}\right|_{\bar{x}_{1r}} + \cdots \tag{1.5}$$

However imposition of the conditions discussed earlier, which justify linearization, make it possible to neglect the terms involving higher powers of $\hat{x}_{1r}$ in equation (1.5). Thus linearization for small perturbations of x_{1r} about the normal operating point employs the following approximation for $\hat{f}_{NL}$:

$$\hat{f}_{NL} \approx \hat{x}_{1r} \left.\frac{df_{NL}}{dx_{1r}}\right|_{\bar{x}_{1r}} \tag{1.6}$$

Hence the nonlinear spring force F_{NLS} may be approximated adequately in a small vicinity of the normal operating point where $x_{1r} = \bar{x}_{1r}$ by the following:

$$F_{NLS} \approx f_{NL}(\bar{x}_{1r}) + \hat{x}_{1r} \left.\frac{df_{NL}}{dx_{1r}}\right|_{\bar{x}_{1r}} \tag{1.7}$$

where

$f_{NL}(\bar{x}_{1r}) = \bar{F}_{NLS}$ and $\hat{x}_{1r}\, df_{NL}/dx_{1r}|_{\bar{x}_{1r}} = \hat{F}_{NLS}$.

The linearized equation for the nonlinear spring is represented by a new incremental coordinate system having its origin at the normal operating point $(\bar{x}_{1r}, \bar{F}_{NLS})$ as depicted in Figure 1.4. The linearized spring characteristic is represented by the straight line passing through the new origin with a slope having angle α. The slope itself is $\tan\alpha$ which is often referred to as the local incremental spring constant k_{inc}, where

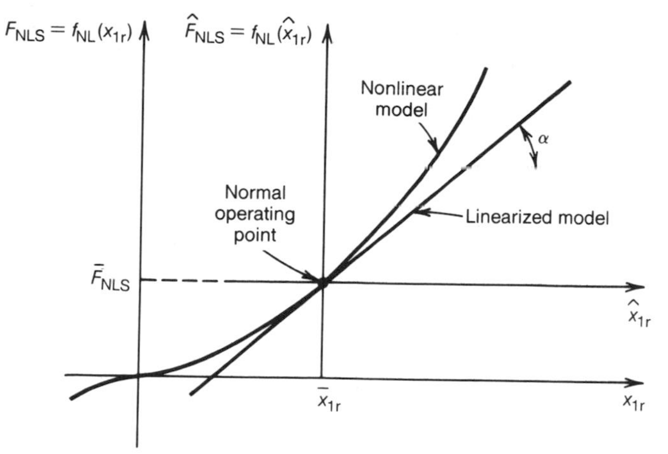

FIGURE 1.4 Incremental coordinate system.

$$k_{\text{inc}} = \tan\alpha = \left.\frac{dF_{\text{NLS}}}{dx_{1\text{r}}}\right|_{\bar{x}_{1\text{r}}}$$

$$= \left.\frac{df_{\text{NL}}}{dx_{1\text{r}}}\right|_{\bar{x}_{1\text{r}}} \tag{1.8}$$

The approximating linear formula can now be written in a simple form

$$F_{\text{NLS}} = \bar{F}_{\text{NLS}} + \hat{F}_{\text{NLS}} \approx \bar{F}_{\text{NLS}} + k_{\text{inc}}\,\hat{x}_{1\text{r}} \tag{1.9}$$

A similar procedure may be applied to a nonlinear element described by a function of two variables. Consider the example of a hydrokinetic type of fluid coupling (rotational damper) such as that shown in Figure 1.5.

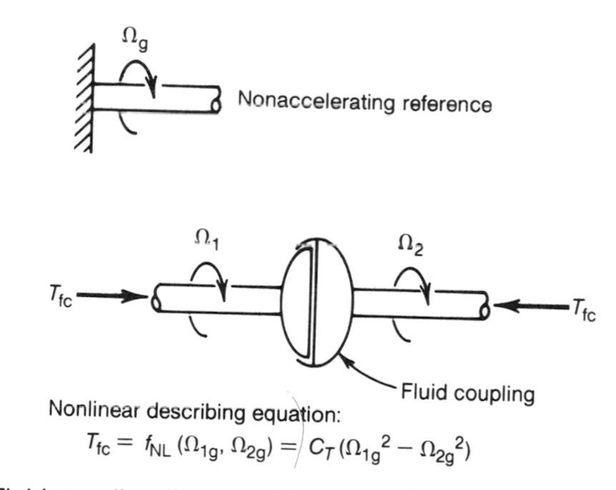

FIGURE 1.5 Fluid coupling described by a function of two shaft speeds Ω_{1g} and Ω_{2g}.

In this case the fluid coupling torque T_{fc} is a nonlinear function of two variables $f_{NL}(\Omega_{1g}, \Omega_{2g})$, having a Taylor's series expansion of the following form.

$$f_{NL}[(\overline{\Omega}_{1g} + \hat{\Omega}_{1g}), (\overline{\Omega}_{2g} + \hat{\Omega}_{2g})] = f_{NL}(\overline{\Omega}_{1g}, \overline{\Omega}_{2g}) + \hat{\Omega}_{1g} \left.\frac{\partial f_{NL}}{\partial \Omega_{1g}}\right|_{\overline{\Omega}_{1g}, \overline{\Omega}_{2g}} + \hat{\Omega}_{2g} \left.\frac{\partial f_{NL}}{\partial \Omega_{2g}}\right|_{\overline{\Omega}_{1g}, \overline{\Omega}_{2g}} + \frac{\hat{\Omega}_{1g}^2}{2!} \left.\frac{\partial^2 f_{NL}}{\partial^2 \Omega_{1g}^2}\right|_{\overline{\Omega}_{1g}, \overline{\Omega}_{2g}} + \frac{\hat{\Omega}_{2g}^2}{2!} \left.\frac{\partial^2 f_{NL}}{\partial^2 \Omega_{2g}}\right|_{\overline{\Omega}_{1g}, \overline{\Omega}_{2g}} + \cdots \tag{1.10}$$

Neglecting Taylor's series terms involving higher powers of $\hat{\Omega}_{1g}$ and $\hat{\Omega}_{2g}$,

$$f_{NL} = C_T(\Omega_{1g}^2 - \Omega_{2g}^2) \approx \overline{f}_{NL} + \hat{f}_{NL}$$

where

$$\overline{f}_{NL} = C_T(\overline{\Omega}_{1g}^2 - \overline{\Omega}_{2g}^2)$$

and

$$\hat{f}_{NL} = 2C_T(\overline{\Omega}_{1g}\hat{\Omega}_{1g} - \overline{\Omega}_{2g}\hat{\Omega}_{2g})^*$$

Thus we may write

$$\begin{aligned} T_{fc} &\approx \overline{f}_{NL} + \hat{f}_{NL} \\ &\approx C_T(\overline{\Omega}_{1g}^2 - \overline{\Omega}_{2g}^2) + 2C_T(\overline{\Omega}_{1g}\hat{\Omega}_{1g} - \overline{\Omega}_{2g}\hat{\Omega}_{2g}) \end{aligned} \tag{1.11}$$

Simplification of a system model obtained as a result of linearization is certainly a benefit but—as it usually is the case with benefits—it does not come free. The price that must be paid in this case represents the error involved in approximating the actual nonlinear characteristics by linear models, which can be called a linearization error. The magnitude of this error depends primarily on the particular type of nonlinearity being linearized and on the amplitude of deviations from a normal operating point experienced by the system. The effects of these two factors are illustrated in Figure 1.6a and b. In both figures the same nonlinear function, representing combined viscous and dry friction forces, is linearized. However, since different normal operating points are selected, different types of nonlinearities are locally approximated by the dashed lines. In Figure 1.6a the normal operating value of the velocity v is very close to zero where a very large discontinuity in the nonlinear function occurs. In this case a sudden change of the force F_{NLD} occurs at the dry friction discontinuity when the system deviates only a small amount from its normal operating state. On the other hand, if the normal operating point is far

*Instead of using the Taylor's series expansion, another approach here would be to simply multiply the whole variables and drop terms that are second-order small. In other words,

$$\begin{aligned} f_{NL} &= C_T[(\overline{\Omega}_{1g} + \hat{\Omega}_{1g})^2 - (\overline{\Omega}_{2g} + \hat{\Omega}_{2g})^2] \\ &= C_T[(\overline{\Omega}_{1g}^2 + 2\overline{\Omega}_{1g}\hat{\Omega}_{1g} + \hat{\Omega}_{1g}^2) - (\overline{\Omega}_{2g}^2 + 2\overline{\Omega}_{2g}\hat{\Omega}_{2g} + \hat{\Omega}_{2g}^2)] \\ &\approx C_T(\overline{\Omega}_{1g}^2 - \overline{\Omega}_{2g}^2) + 2C_T(\overline{\Omega}_{1g}\hat{\Omega}_{1g} - \overline{\Omega}_{2g}\hat{\Omega}_{2g}) \end{aligned}$$

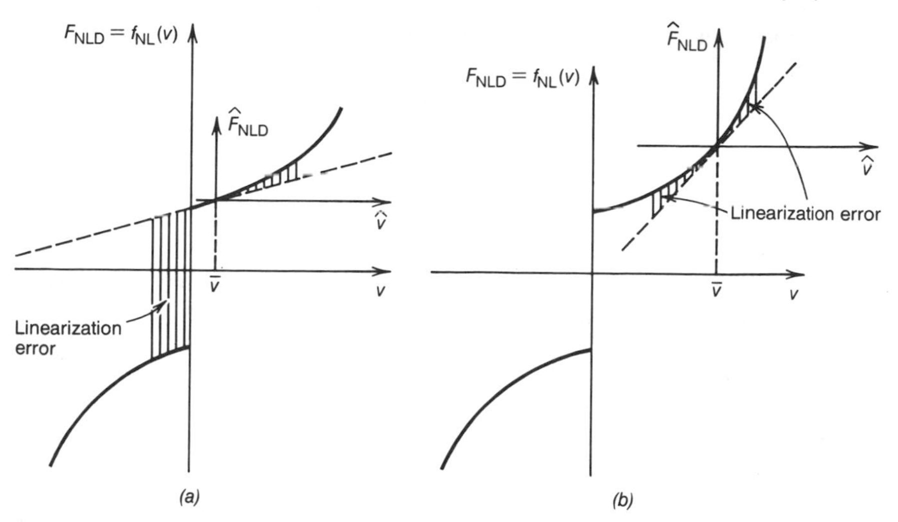

FIGURE 1.6 Effect of the type of nonlinearity on linearization error.

away from the discontinuity at zero velocity, the error resulting from linearization is reasonably small if the magnitude of the change in v is small.

1.4 Synopsis

The concept of a system within a given boundary interacting with its surroundings via input variables and output variables has been introduced, and the distinction between system variables and system characteristic parameters has been made. The technique of system modeling has been introduced as a tool to be used in predicting system behaviour or performance, with emphasis on the use of special-purpose simplified models that adequately approximate the operation of engineering systems in the early stages of design before expensive development and manaufacture has been carried out.

The general types of models to be employed have been classified and presented in Table 1.1.

The ideal linear system elements to be employed in modeling in this text have been introduced and classified into A-type, T-type, and D-type groups, the elements in each group being analogous to each other. These analogies are evident both in the similar nature of their describing equations and in the manner in which they store energy.

Finally, examples of nonlinear elements were introduced. Methods of linearization for small perturbations about a given operating point were demonstrated, and limitations on the use of this linearization were discussed. The underlying motivation of simplification to augment mathematical analysis has been balanced against the limitations imposed by linear approximation for a sufficiently small region near the selected normal operating point.

In summary, although linearized models should be used whenever possible, bear in mind that the results obtained with linearized models provide a simplified picture of the actual system behavior.

Problems

1.1 Employing an input-output block diagram, such as that shown in Figure 1.1, show what you consider to be the input variables and the output variables for an automobile engine, shown schematically in Figure P1.1.

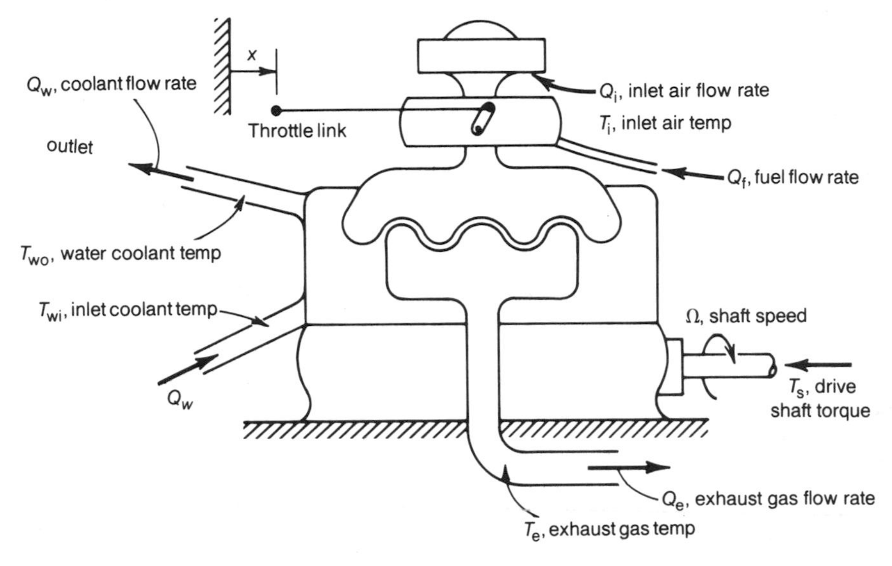

FIGURE P1.1

1.2 For the automotive alternator shown in Figure P1.2, prepare an input-output diagram showing what you consider to be inputs and what you consider to be outputs.

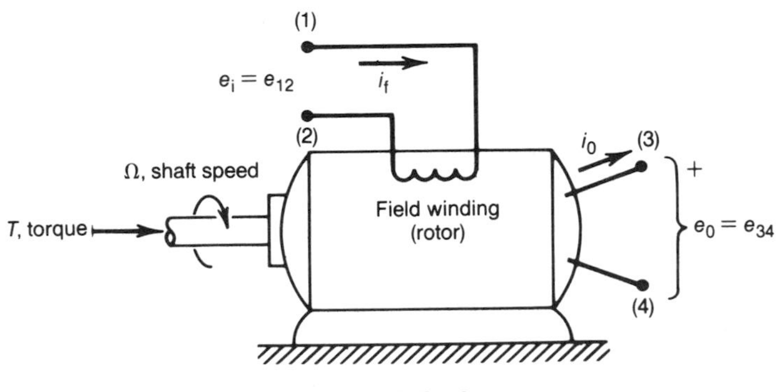

FIGURE P1.2

1.3 Prepare an input-output block diagram showing what you consider to be the inputs and the outputs for the domestic hot water furnace shown schematically in Figure P1.3.

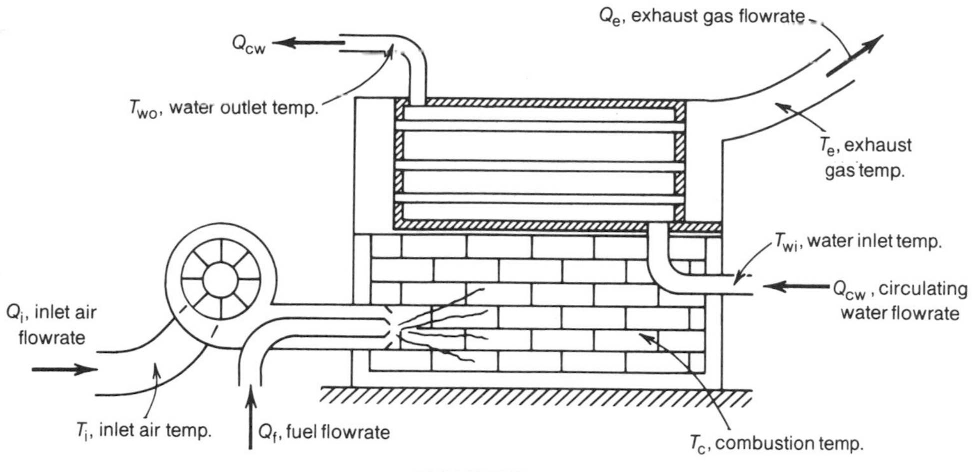

FIGURE P1.3

1.4 **(a)** For the jack-knife mechanism shown in Figure P1.4, develop the kinematic relationship between the input motion x and the output motion y.

(b) Linearize this relationship showing how $\hat{y}$ is related to $\hat{x}$ for small variations of x around $L/2$.

(c) Assuming no losses, i.e., "work in" equals "work out," find the relationship between $\hat{F}_o$ and $\hat{F}_i$.

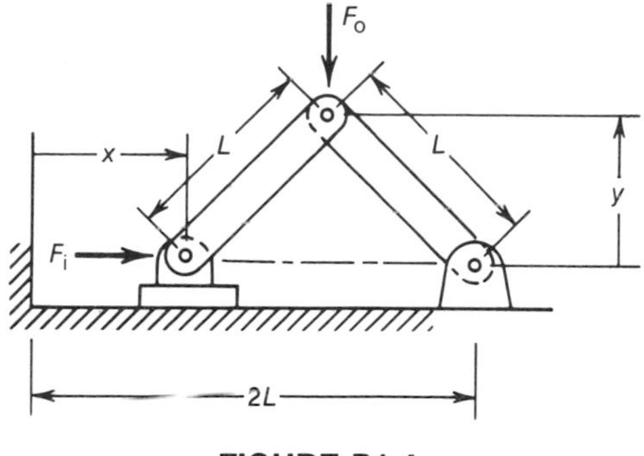

FIGURE P1.4

1.5 A nonlinear hydraulic shock absorber employing symmetrical orifice-type flow resistances, shown schematically in Figure P1.5, has the nonlinear force versus velocity characteristic shown in the accompanying graph.

(a) Find the operating point value $\overline{F}_{NLD}$ of the force F_{NLD} in terms of $\overline{v}_{1g}$.

(b) Linearize this characteristic showing how $\hat{F}_{NLD}$ is related to $\hat{v}_{1g}$ for small variations of v_{1g} around an operating point value of $\overline{v}$. In other words find the incremental damping coefficient $b_{inc} = \hat{F}_{NLD}/\hat{v}_{1g}$.

(c) Suggest what should be used as the approximate equivalent damping coefficient b_{eq} when large periodic variations occur through the large range, v_{R1}, about zero velocity as shown on the graph.

(d) Suggest what should be used for b_{eq} when a large periodic "one-sided" variation, v_{R2}, of v_{1g} occurs.

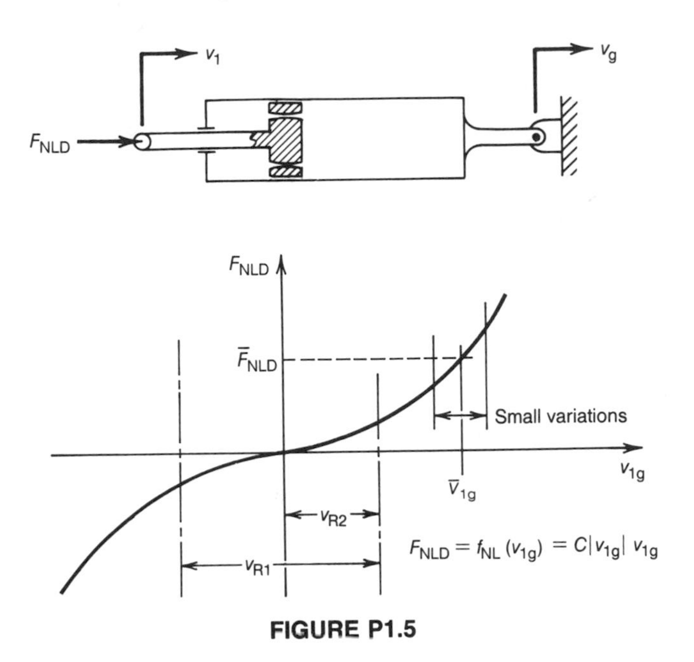

FIGURE P1.5

2

Mechanical Systems

2.1 Introduction

As indicated in Chapter 1, three basic ideal elements are available for modeling elementary mechanical systems: masses, springs, and dampers. Although each of these elements is itself a system with all the attributes of a system (inputs, parameters, state variables, and outputs), the use of the term system usually implies a combination of interacting elements. In this chapter, systems composed of only mechanical elements will be discussed. In addition to the translational elements (moving along a single axis) introduced in Chapter 1, a corresponding set of rotational elements (rotating about a single axis) will be introduced to deal with rotational mechanical systems and mixed (translational and rotational) systems.

Also, this chapter will deal only with so-called lumped-parameter models of real mechanical systems. In certain situations, such as when modeling a real spring having both mass and stiffness uniformly distributed from one end to the other, suitable lumped-parameter models can be conceived that will adequately describe the system under at least limited conditions of operation. For example, if a real spring is compressed very slowly, the acceleration of the distributed mass is very small so that all the force acting on one end is transmitted through it to the other end; under these conditions the spring may be modeled as an ideal spring. On the other hand, if the real spring is being driven in such a manner that the forces acting on it cause negligible deflection of its coils, it may be modeled as an ideal mass. When a real spring is driven so as to cause both significant acceleration of its mass and significant deflection of its coils, combinations of lumped ideal mass(es) and lumped ideal spring(s) are available to model the real spring adequately, depending on the type of vibration induced in it. And, if the real spring is being driven at frequencies well below the lowest frequency mode of vibration for that type of forcing, simple combinations of ideal masses and ideal springs will usually suffice to model it adequately. These combinations are shown for the corresponding forcing conditions in Figure 2.1.

As the frequencies of the forcing functions approach the lowest natural frequency

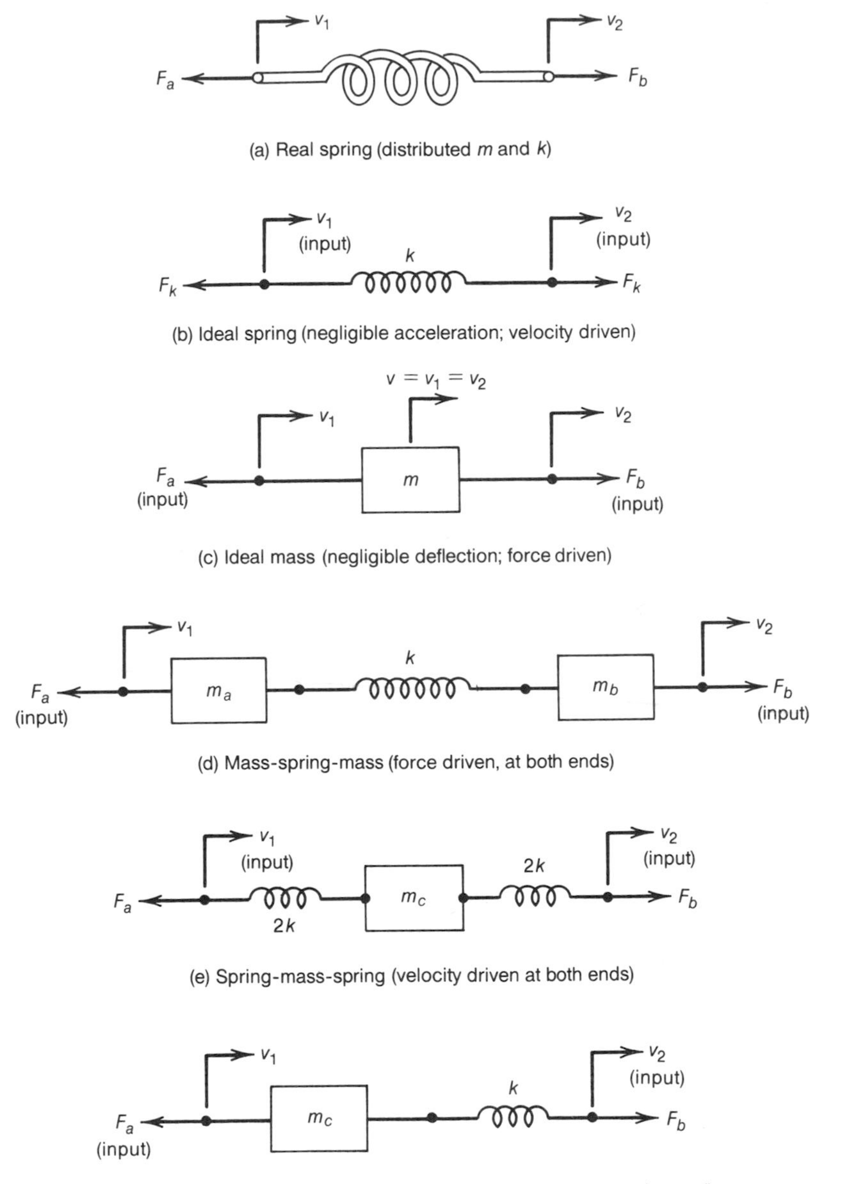

FIGURE 2.1 Several possible lumped-parameter models of a real spring. In (d), (e), and (f) the choice of values for m_a or m_b and m_c depends on relative amplitudes of the forcing functions, i.e., the inputs.

for that type of forcing, the use of lumped-parameter models becomes questionable; unless a many-element model is used (i.e., a finite-element model), the formulation of the partial differential equations for a distributed-parameter model is advisable.

2.2 Translational Mechanical Systems

2.2.1 Translational Masses

Analysis of mechanical systems is based on the principles embodied in Newton's laws of motion and the principle of compatibility (no gaps between connected elements). An ideal mass, depicted schematically in free-body diagram form in Figure 2.2, moves in relation to a nonaccelerating frame of reference, which is usually taken to be a fixed point on the earth (ground)—however, the frame of a nonaccelerating vehicle could be used instead.

The elemental equation for an ideal mass, m, based on Newton's second law[1] is

$$m\left(\frac{dv_{1g}}{dt}\right) = F_m \tag{2.1}$$

where v_{1g} is the velocity of the mass m relative to the ground reference point (g), and F_m is the net force (i.e., the sum of all the applied forces) acting upon the mass in the x-direction. Since $v_{1g} = dx_{1g}/dt$, the variation of the distance x_{1g} of the mass from the reference point is related to F_m by

$$m\left(\frac{d^2x_{1g}}{dt^2}\right) = F_m \tag{2.2a}$$

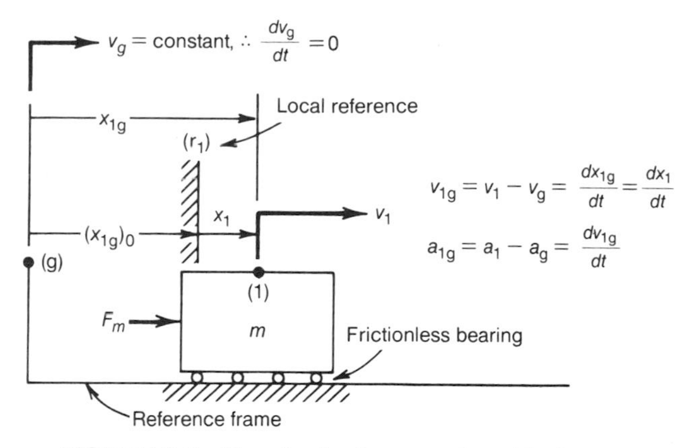

FIGURE 2.2 Free-body diagram of an ideal mass.

[1] Newton's second law expressed in more general form is

$$m\left(\frac{d^2x}{dt^2}\right) = \sum_{i=1}^{n} F_i$$

Or since $dx_{1g}/dt = dx_1/dt$,

$$m\left(\frac{d^2x_1}{dt^2}\right) = F_m \tag{2.2b}$$

EXAMPLE 2.1

Find the response (in terms of its acceleration, velocity, and position versus time) of a 3000-lb automobile to a force F_i, of 500 lb, which is suddenly applied by three members of the football squad (i.e., a step change in force occurring at $t = 0$), ignoring friction effects. See Figure 2.3. Assume that the football players are able to maintain the applied force of 500 lb. regardless of how fast the automobile moves. Reference r_1 is the local ground reference, in other words the starting point for vehicle motion.

Solution

Using Equation (2.1), that is, ignoring friction effects,

$$\frac{dv_{1g}}{dt} = \left(\frac{1}{m}\right) F_i(t) \tag{2.3}$$

Since the vehicle acceleration $a_{1g}(t) = dv_{1g}/dt$, we see that it undergoes a step change from 0 to (32.2)(500)/3000 at $t = 0$ and remains at that value until the applied force is removed. Next, we may separate variables in Equation (2.3) and integrate with respect to time to solve for $v_{1g}(t)$.

$$\int_{v_{1g}(0)}^{v_{1g}(t)} dv_{1g} = \left(\frac{1}{m}\right) \int_0^t F_i(t)\, dt \tag{2.4}$$

which yields

$$v_{1g}(t) - v_{1g}(0) = \frac{(32.2)}{6} t - 0$$

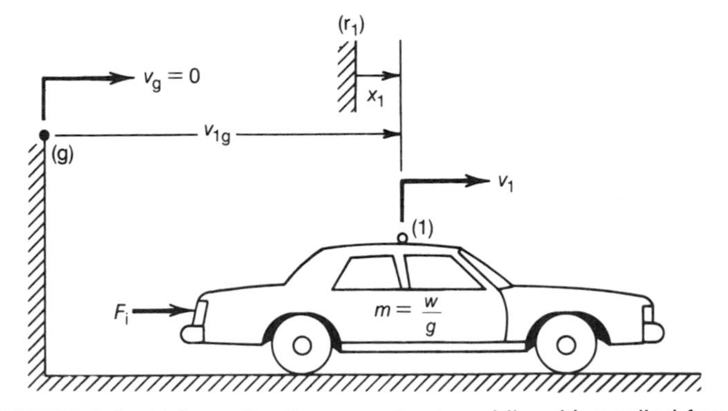

FIGURE 2.3 Schematic diagram of automobile with applied force.

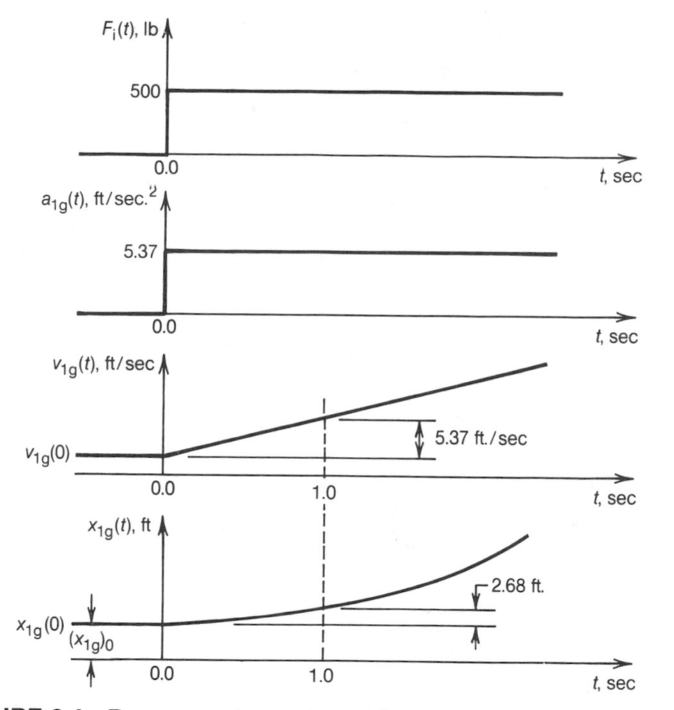

FIGURE 2.4 Response of an automobile to a suddenly applied force.

or

$$v_{1g}(t) = v_{1g}(0) + 5.37t \text{ ft/sec}; \quad \text{i.e., a ramp starting at } t = 0 \text{ having a slope of } 5.37 \text{ ft/sec}^2 \tag{2.5}$$

Similarly a second integration with respect to time yields

$$\int_{x_{1g}(0)}^{x_{1g}(t)} dx = 5.37 \int_0^t t\, dt$$

or

$$x_{1g}(t) = x_{1g}(0) + 2.68t^2 \text{ ft}; \quad \text{i.e., a parabola starting at } t = 0. \tag{2.6}$$

The results are portrayed as functions of time in Figure 2.4 along with the input force F_i. ■

Note that it takes time to build up changes in velocity and displacement because of the integrations involved. Example 2.1 displays one of the first indications of the role played by integration in determining the dynamic response of a system. From another point of view, the action of the applied force represents work being done upon the mass as it accelerates, increasing the kinetic energy stored in it as time goes by. The rate at which energy is stored in the system is equal to the rate at which work is expended upon it by the members of the football squad (the first law of thermodynamics)

$$\frac{d\mathscr{E}_K}{dt} = F_i v_{1g} \tag{2.7}$$

Separating variables and integrating with respect to time we find

$$\int_{\mathscr{E}_K(0)}^{\mathscr{E}_K(t)} d\mathscr{E}_K = F_i v_{1g}\, dt = \int_0^t m v_{1g} \left(\frac{dv_{1g}}{dt}\right) dt = m \int_{v_{1g}(0)}^{v_{1g}(t)} v_{1g}(t)\, dv_{1g}$$

so that

$$\mathscr{E}_K(t) = \mathscr{E}_K(0) + (m/2) v_{1g}(t)^2 \tag{2.8}$$

Thus the stored energy accumulates over time, proportional to the square of the velocity, as the work is being done on the system; and the mass is an A-type element storing energy which is a function of the square of its A-variable v_{1g}.

As this text proceeds to the analysis of more complex systems, the central role played by integration in shaping dynamic system response will become more and more evident.

It should be noted in passing that it would *not* be reasonable to try to impose on the automobile a step change in velocity, because this would be an impossible feat for three football players—or even for ten million football players! Such a feat would require a very, very great force as well as a very, very great source of power—infinite sources! Considering that one definition of infinity is that it is a number greater than the greatest possible imaginable number, would it be likely to find or devise an infinite force source and an infinite power source?

2.2.2 Translational Springs

An ideal translational spring that stores potential energy as it is deflected along its axis may also be depicted within the same frame of reference used for a mass. Figure 2.5 shows such a spring in two ways: (a) in mechanical drawing format in relaxed state with $F_k = 0$; and (b) in stylized schematic form with left end displaced relative to the right end due to the action of F_k shown acting at both ends in free-body diagram fashion. The references r_1 and r_2 are local ground references.

Note that since an ideal spring contains no mass, the force transmitted by it is undiminished during acceleration; therefore, the forces acting on its ends must always be equal and opposite (Newton's third law of motion). The elemental equation for such a spring derives from Hooke's law, namely,

$$F_k = k(x_{21} - (x_{21})_0) \tag{2.9}$$

where $(x_{21})_0$ is the free length of the spring, i.e., its length when $F_k = 0$. Differentiating Equation (2.9) with respect to time yields

$$\frac{dF_k}{dt} = k v_{21} \tag{2.10}$$

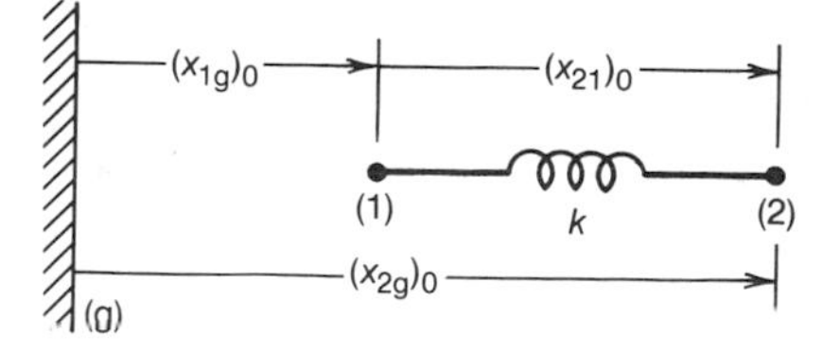

(a) Initial relaxed state, no force acting, no displacement

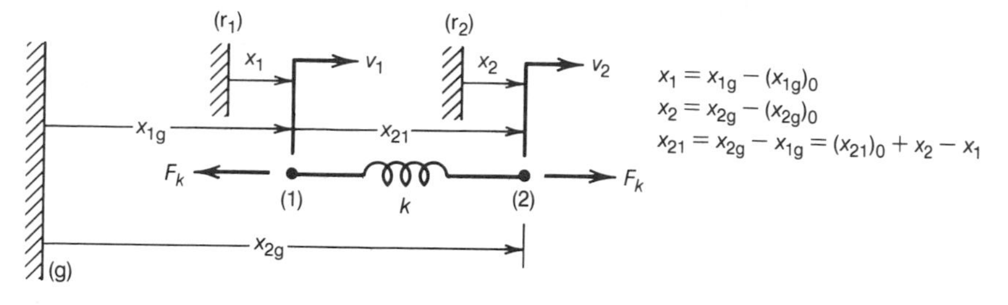

(b) Force acting, system displaced

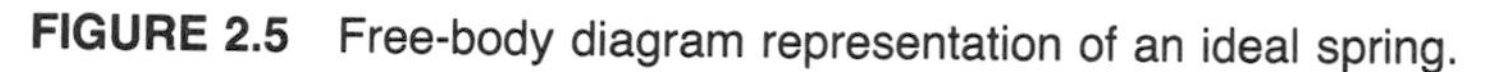
FIGURE 2.5 Free-body diagram representation of an ideal spring.

where $v_{21} = dx_{21}/dt$ is the velocity of the right-hand end of the spring, point (2), relative to the velocity of the left-hand end, point (1). Since $(x_{21} - (x_{21})_0) = x_2 - x_1$, Equation (2.9) is simplified to

$$F_k = k(x_2 - x_1) \tag{2.11}$$

where $(x_2 - x_1)$ is the deflection of the spring from its initial free length.

When the spring is nonlinear, it does not have truly constant stiffness k, and it is advisable to employ a nonlinear symbol designation and a nonlinear function to describe it as follows.

$$F_{\mathrm{NLS}} = f_{\mathrm{NL}}(x_2 - x_1)$$

x_1 x_2

F_{NLS} NLS F_{NLS}

Linearization of this equation for small perturbations will yield an incremental stiffness k_{inc} which is often adequate for use in a small region near the normal operating point for the spring, as discussed in Chapter 1.

EXAMPLE 2.2

Find the response (in terms of force F_k and deflection x_1) of the spring shown in Figure 2.6, having $k = 8000$ lb/in. when it is subjected to a 20 in./sec step change in input velocity from zero, starting from its free length at $t = 0$.

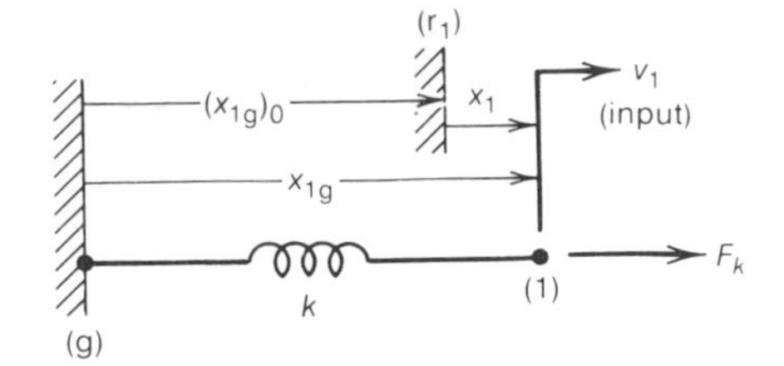

(a) Spring in tension, x_1 is positive when F_k is positive (x_{1g} is positive here)

Alternatively

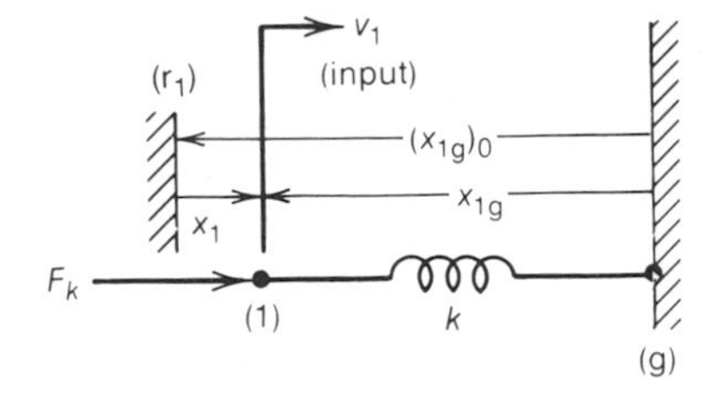

(b) Spring in compression, x_1 is positive when F_k is positive (x_{1g} is negative here)

FIGURE 2.6 An ideal spring subjected to a step input of velocity.

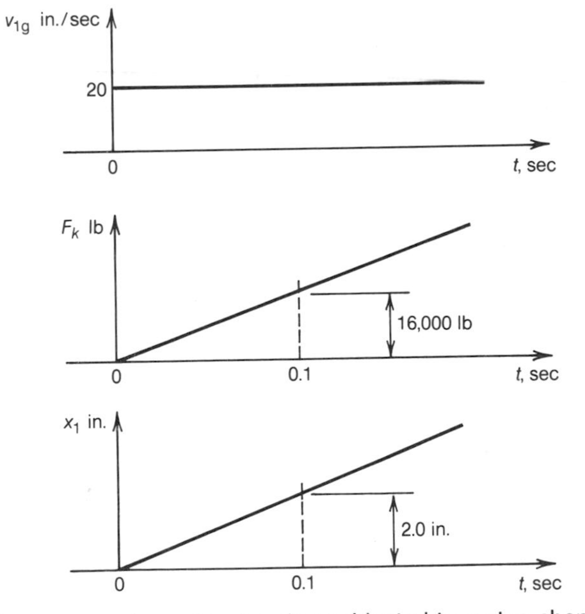

FIGURE 2.7 Responses for an ideal spring subjected to a step change in velocity.

Solution

Separating variables in Equation (2.10) and integrating with respect to time

$$\int_{F_k(0)}^{F_k(t)} dF_k = k\int_0^t v_{1g}\,dt = (8000)(20)\int_0^t dt$$

$$F_k(t) - 0 = (8000)(20)t - 0$$

or

$$F_k(t) = 160{,}000t \text{ lb} \tag{2.12}$$

Since $v_{1g} = dx_{1g}/dt = dx_1/dt + d(x_{1g})_0/dt$, and $d(x_{1g})_0/dt = 0$, we may use the definition

$$v_{1g} \equiv \frac{dx_1}{dt} \tag{2.13}$$

Separating variables in Equation (2.13) and integrating again with respect to time yields

$$\int_{x_1(0)}^{x_1(t)} dx_1 = \int_0^t v_{1g}\,dt = \int_0^t 20\,dt$$

or

$$x_1(t) - 0 = 20t - 0$$

or

$$x_1(t) = 20t \text{ in.} \tag{2.14}$$

The results are shown in Figure 2.7 ■

Again the essential role played by integration in finding the dynamic response of a system is evident. From an energy point of view, the rate at which potential energy is stored in the spring is equal to the rate at which the steadily increasing force does work on the spring as it deflects the spring.

$$\frac{d\mathscr{E}_P}{dt} = F_k v_{1g}$$

Separating variables and integrating yields

$$\int_{\mathscr{E}_P(0)}^{\mathscr{E}_P(t)} d\mathscr{E}_P = \int_0^t F_k v_{1g}\,dt = \left(\frac{1}{k}\right)\int_0^t F_k\left(\frac{dF_k}{dt}\right)dt = \left(\frac{1}{k}\right)\int_{F_k(0)}^{F_k(t)} F_k\,dF_k$$

so that

$$\mathscr{E}_P(t) - 0 = \frac{1}{2k}F_k^2 - 0$$

or

$$\mathscr{E}_P(t) = \left(\frac{1}{2k}\right) F_k^2 \tag{2.15}$$

Thus it takes time for the work input to add to the accumulated energy stored in the spring. A spring is a T-type element, storing energy as a function of the square of its T-variable F_k.

Again note as for changing velocity in Example 2.1, it would *not* be realistic here to try to apply a step change in *force* to a spring; such a force source would have to move at a very, very great velocity to deflect the spring suddenly, which would require a very, very great power source. *In general it can be stated that inputs that would suddenly add to the stored energy in a system are not realistic and cannot be achieved in the natural world.*

2.2.3 Translational Dampers

An ideal damper is shown in free-body diagram form in Figure 2.8. Since an ideal damper contains no mass, the force transmitted through it is undiminished during acceleration; therefore the forces acting at its ends must always be equal and opposite. The elemental equation for an ideal damper is given by

$$F_b = b(v_{2g} - v_{1g}) = bv_{21} \tag{2.16}$$

A damper is a D-type element which dissipates energy. With this element there is no storage of retrievable mechanical work (work being done by an applied force becomes dissipated as thermal internal energy), and the relationship between force and velocity is instantaneous. Thus it is realistic to apply step changes of either force or velocity to such an element. Damping plays a key role in influencing speed of response and stability of many systems.

Although ideal damping does exist, arising from viscous friction between well-lubricated moving mechanical parts of a system, nonideal forms of damping are very often present. Nonideal damping is characterized usually by nonlinearities that can be severe, especially where poorly lubricated parts move with metal-to-metal contact. In other cases the nonlinearity is owing to hydrodynamic flow effects, where internal inertia forces predominate, such as in the fluid coupling discussed briefly in Chapter 1 under the topic of linearization and in hydraulic shock absorbers employing orifice-type energy dissipation.

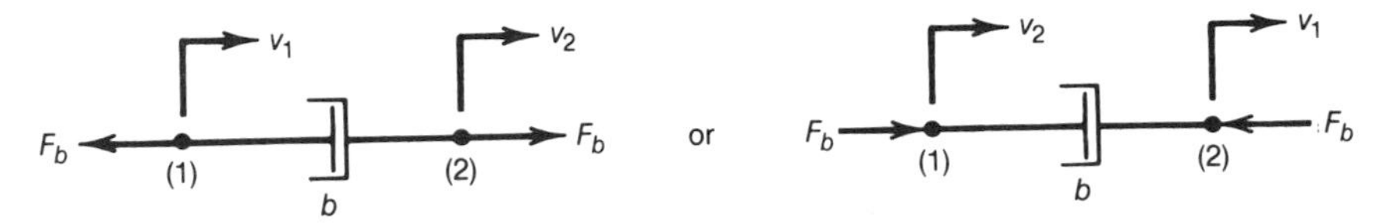

FIGURE 2.8 Free-body diagram of a ideal translational damper.

For the case when the damping is nonlinear, there is no damping constant, b, and the elemental equation is expressed as a nonlinear function of velocity.

$$F_{\mathrm{NLD}} = f_{\mathrm{NL}}(v_{12}) \tag{2.17}$$

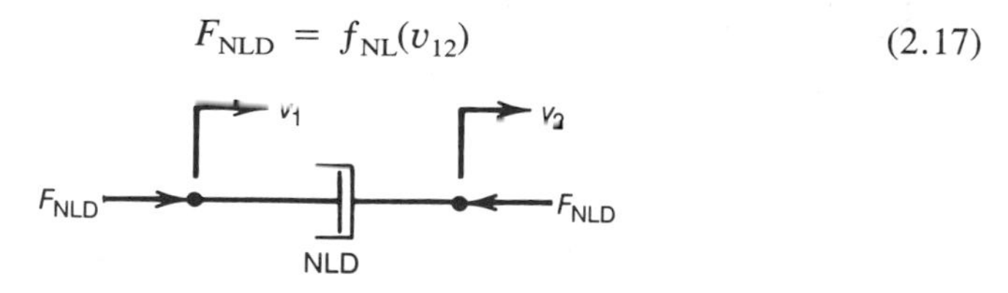

Linearization for variations around an operating point is often feasible, especially when no discontinuities exist in the F versus v characteristic. Otherwise computer simulation of the damper characteristic is required.

2.2.4 Elementary Systems—Combinations of Translational Elements

The equations used to describe a combination of interacting elements constitute a mathematical model for the system. Other types of models will be discussed in later chapters.

Newton's second law was employed to model mathematically the motion of an ideal mass in Section 2.2.1. Now Newton's third law will be employed to sum forces at interconnection points between the elements, assuring *continuity* of force in the system. Choice of common variables for common motions (position, velocity) at connection points assures *compatibility*[2] (i.e., no gaps between connected elements).

EXAMPLE 2.3

Consider the spring-damper system shown in Figure 2.9. This combination of elements is useful for absorbing the impulsive interaction with an impinging system, i.e., a kind of shock absorber. As developed here it is intended that an input force, F_i, be the forcing function, and the resulting motion, $x_1 - x_2$ (or $v_{1g} - v_{2g}$), is then to be considered the resulting output. The relationship between input and output is to be modelled mathematically. Qualitatively speaking, the system responds to the force, F_i, storing energy in the spring and dissipating energy in the damper until the force is reduced to zero, whereupon the spring gives up its stored energy, and the damper continues to dissipate energy until the system returns to its original state. The net result, after the force has been removed, is that energy that has been delivered to the system by the action of the force, F_i, has been dissipated by the damper and the system has returned to its original relaxed state.

[2]The principles of continuity and compatability are discussed in detail in *Introduction to System Dynamics*, by J. L. Shearer, A. T. Murphy, and H. H. Richardson, Addison-Wesley Publishing Co., Reading, Mass., 1967.

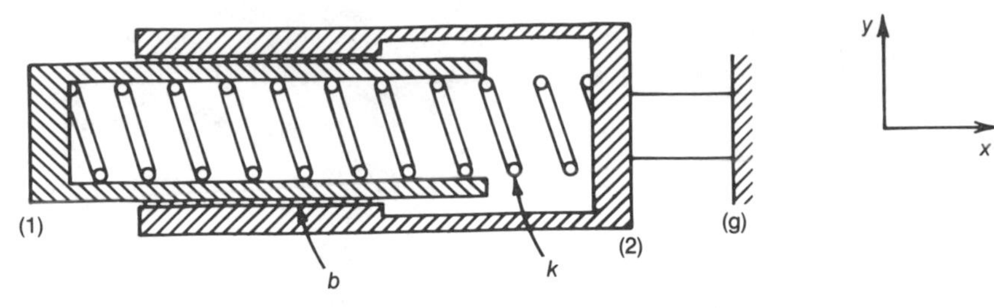

(a)

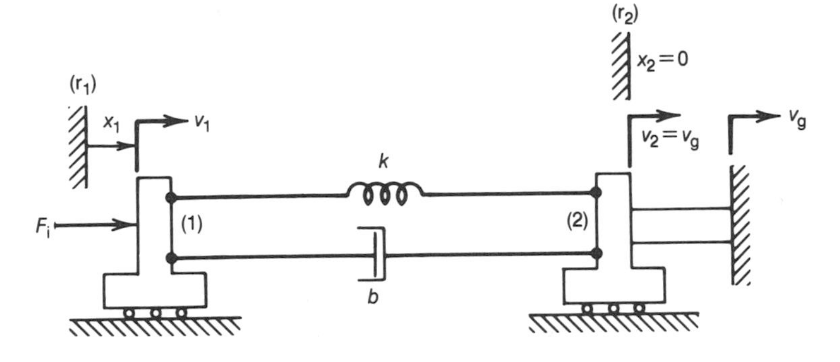

(b)

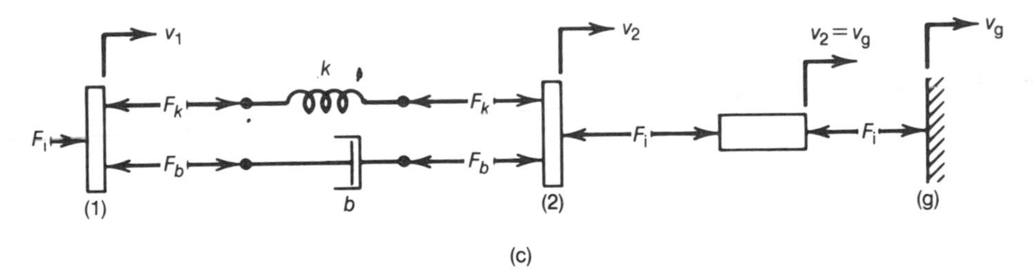

(c)

FIGURE 2.9 Diagrams for spring-damper system.

The object here is to develop a mathematical model relating the output motion to the input force. The use of this mathematical model in solving for the output motion as a function of time will be left to a later chapter.

Solution

As an introductory aid in visualizing the action of each member of the system and defining variables, Figure 2.9 shows three different kinds of diagram for this system: (a) A cross-sectioned mechanical drawing showing the system in its initial relaxed state with $F_i = 0$. (b) A stylized diagram showing the system in an active, displaced state when the force, F_i, is acting. (c) A free-body diagram of the system ''broken open'' to show the free-body diagram for each member of the system.

Applying Newton's third law at (1) yields

$$F_i = F_k + F_b \tag{2.18}$$

For the elemental equations

$$F_k = k(x_1 - x_2) \tag{2.19}$$

$$F_b = b(v_{1g} - v_{2g}) \tag{2.20}$$

Note that $v_{1g} = v_1$ and v_{2g} and v_2 only if $v_g = 0$.

Definitions:

$$v_{1g} \equiv \frac{dx_1}{dt} \tag{2.21}$$

$$v_{2g} \equiv \frac{dx_2}{dt} \tag{2.22}$$

The system is now described completely by a *necessary and sufficient set* of five equations containing the five unknown variables x_1, F_k, F_b, v_{1g}, v_{2g}. *Note: The number of independent describing equations must equal the number of unknown variables before proceeding to eliminate the unwanted unknown variables*. Combining Equations (2.18) through (2.22) to eliminate F_k, F_b, v_{1g}, and v_{2g} yields:

$$b\left(\frac{dx_1}{dt} - \frac{dx_2}{dt}\right) + k(x_1 - x_2) = F_i \tag{2.23}$$

Note that x_2 and dx_2/dt have been left in Equation (2.23) for the sake of generality, to cover the situation where the right-hand side of the system might be in motion as it could be in some systems. Since the right-hand side here is rigidly connected to the frame of reference g, these variables are zero, leaving

$$b\frac{dx_1}{dt} + kx_1 = F_i \tag{2.24}$$

This first-order differential equation is the desired mathematical model, describing in a very concise way the events described earlier in verbal form. It may be noted, in the context of state variables to be discussed in Chapter 4, that a first-order system such as this requires only one state variable, in this case x_1, to describe its state from instant to instant as time passes. ■

Given its initial state and the nature of F_i as a function of time, it is possible to solve for the response of this system as a function of time. The procedure for doing this will be discussed in later chapters.

The steps involved in producing mathematical models of simple mechanical systems are also illustrated in the following additional examples.

EXAMPLE 2.4 Mass-damper System with Input Force Applied to Mass

A mass m, supported only by a bearing having a pressurized film of viscous fluid, undergoes translation (i.e., motion along a straight line in the x direction) as the result of having a time-varying input force, F_i, applied to it as shown in Figure 2.10. The object is to develop a mathematical model that relates the velocity v_{1g} of the mass to the input force F_i. Expressed verbally, the system responds to the input force as follows: Initially the input force accelerates the mass so that its velocity increases, accompanied by an increase in its kinetic energy; however, as the velocity increases, the damper force increases, opposing the action of the input force and dissipatiing energy at an increasing rate. Thus the action of the damper is to reduce the acceleration of the mass resulting from the input force. If the input force is then removed, the damper force continues to oppose the motion of the mass until it comes to rest, having lost all of its kinetic energy due to dissipation in the damper.

Solution

Newton's second law applied to the mass, m, yields

$$F_i - F_b = m\frac{dv_{1g}}{dt} \tag{2.25}$$

The elemental equation for the damper is

$$F_b = b(v_{1g} - v_{2g}) \tag{2.26}$$

Equations (2.25) and (2.26) constitute a necessary and sufficient set of two

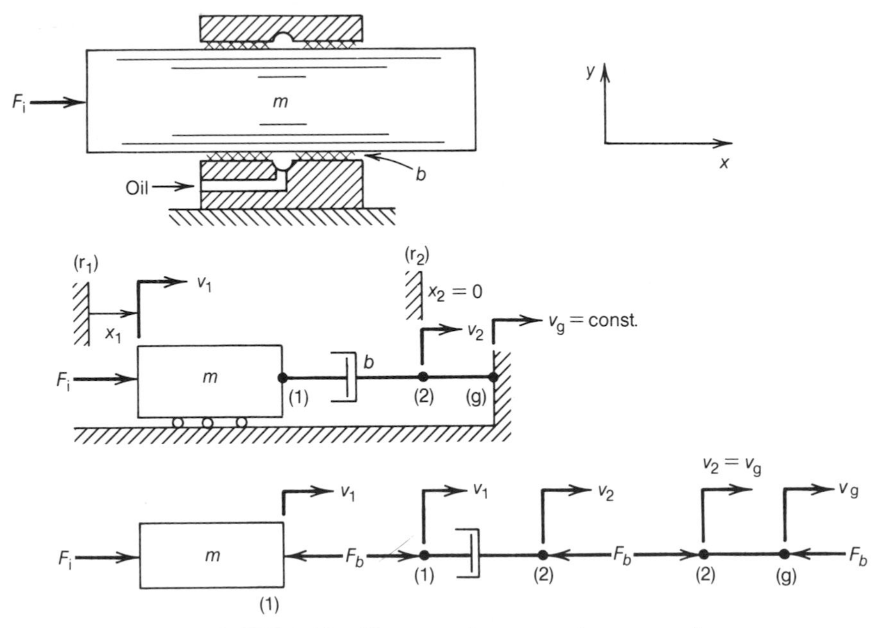

FIGURE 2.10 Diagrams for mass-damper system.

equations containing the two unknowns: F_b, and v_{1g} (v_{2g} is zero here). Combining (2.25) and (2.26) to eliminate F_b gives

$$m \frac{dv_{1g}}{dt} + b(v_{1g} - v_{2g}) = F_i \tag{2.27}$$

Since the bearing block is rigidly connected to ground, $v_{2g} = 0$, leaving

$$m \frac{dv_{1g}}{dt} + bv_{1g} = F_i \tag{2.28}$$

Again this is a simple first-order system, requiring only one state variable, v_{1g}, to describe its state as a function of time. ■

EXAMPLE 2.5 Mass-Spring-Damper System in a Gravity Field with an Input Force Applied to the Mass

This system, typical of spring-suspended mass systems, is shown in Figure 2.11. The object is to develop the mathematical model relating displacement of the mass, x_1, to the input force F_i. Expressed verbally, the system responds to the input in the following fashion: First the velocity of the mass increases, accompanied by an increase in its stored kinetic energy, while the rate of energy dissipated in the damper increases. Meanwhile the motion results in a displacement x_1, which is the time integral of its velocity, so that the spring is compressed, accompanied by an increase in its stored potential energy. If the damping coefficient is small enough, the spring will cause a rebounding action, transferring some of its potential energy back into kinetic energy of the mass on a cyclic basis; in this case the decaying oscillation associated with a lightly damped second-order system will occur, even after the input force has been removed.

Solution

Newton's second law applied to m, i.e., the elemental equation for m, yields

$$F_i + mg - F_k - F_b = m \frac{dv_{1g}}{dt} \tag{2.29}$$

The elemental equations for k and b are

$$F_b = bv_{1g} \tag{2.30}$$

$$F_k = k(x_1 + \Delta_1 - x_g) \tag{2.31}$$

Equations (2.29), (2.30), and (2.31) constitute the necessary and sufficient set of three equations containing unknowns: F_k, F_b, and x_1 for this system. (Here v_{1g} is a form of x_{1g} namely dx_1/dt.)

Combining (2.29) through (2.31) to eliminate F_k, F_b, mg, Δ_1, and v_{1g} gives

$$m\left(\frac{d^2x_1}{dt^2}\right) + b\left(\frac{dx_1}{dt}\right) + k(x_1 - x_g) = F_i \tag{2.32}$$

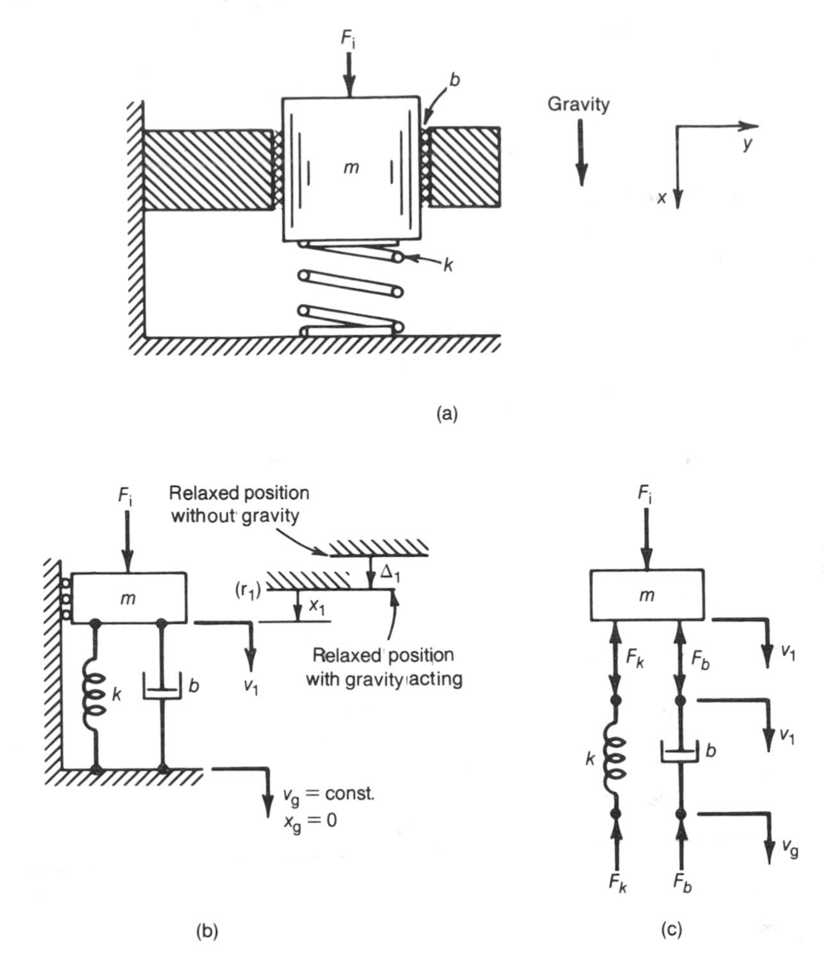

FIGURE 2.11 Diagrams for mass-spring-damper system in a gravity field.

Note that the terms mg and $k\ \Delta_1$ eliminate each other, so that they do not show up in Equation (2.32).

Also since x_g is zero because of the bearing being rigidly connected to the frame of reference g, the resulting mathematical model for this system is

$$m\left(\frac{d^2x_1}{dt^2}\right) + b\left(\frac{dx_1}{dt}\right) + kx_1 = F_i \tag{2.33}$$

where x_1 is the displacement of the mass from the relaxed state in the gravity field, i.e., the motion from the Δ_1-displaced reference. Mathematical solution of Equation (2.33) for specific forcing functions will be carried out in Chapter 5.

This second-order system contains two independent energy-storing elements, and it requires a set of two state variables (e.g., x_1 and dx_1/dt, or v_{1g} and F_k, or some other pair) to describe its state as a function of time. ■

The second-order differential equation of (2.33) expresses in a very succint way the action described earlier in a paragraph of many words. Moreover it is a more precise description, capable of providing a detailed picture of the system response to various inputs.

EXAMPLE 2.6

The six-element system shown in Figure 2.12 is a simplified representation of vibrating spring-mass assembly (k_1, m_1, b_1) with an attached vibration absorber, subjected to a displacement input x_1 as shown. The object is to develop a mathematical model capable of relating the motions x_2 and x_3 to the input displacement x_1.

Also shown in Figure 2.12 are diagrams showing the system in an active displaced state and "broken open" for free-body representation.

Solution

The elemental equation for the spring k_1 in derivative form is

$$\frac{dF_{k1}}{dt} = k_1(v_{1g} - v_{2g}) \tag{2.34}$$

Upon integrating Equation (2.34) with respect to time with x_1 and x_2, both zero in the relaxed state,

$$F_{k1} = k_1(x_1 - x_2) \tag{2.35}$$

For the mass m_1,

$$F_{k1} - F_{k2} - F_{b1} = m_1 \frac{d^2x_2}{dt^2} \tag{2.36}$$

For the damper b_1,

$$F_{b1} = b_1 v_{1g} \tag{2.37}$$

For the spring k_2,

$$F_{k2} = k_2(x_2 - x_3) \tag{2.38}$$

For the mass m_2,

$$F_{k2} - F_{\text{NLD}} = m_2 \frac{d^2x_3}{dt^2} \tag{2.39}$$

And for the nonlinear damper, NLD,

$$F_{\text{NLD}} = f_{\text{NL}}(v_{3g}) = f_{\text{NL}}\left(\frac{dx_3}{dt}\right) \tag{2.40}$$

Equations (2.35) through (2.38) may now be combined, giving

$$m_1 \frac{d^2x_2}{dt^2} + b_1 \frac{dx_2}{dt} + (k_1 + k_2)x_2 = k_1x_1 + k_2x_3 \tag{2.41}$$

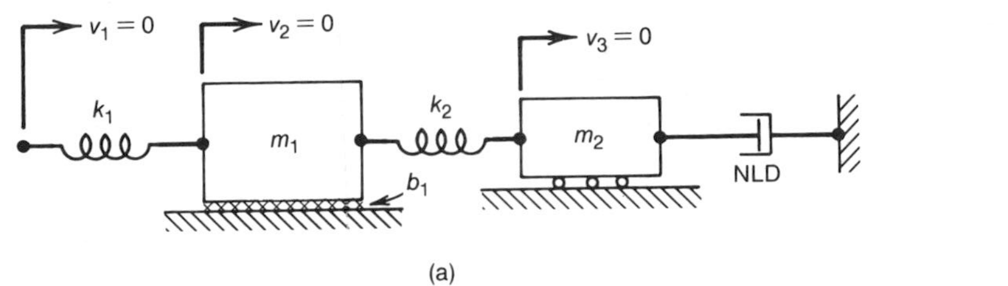

(a)

Input
x_1
v_1
x_2
v_2
x_3
v_3
$v_g = \text{const.}$
k_1
m_1
m_2
NLD
b_1

(b)

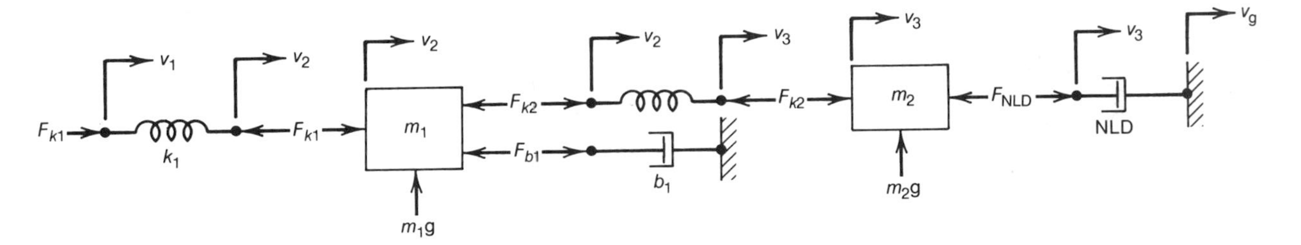

(c)

FIGURE 2.12 Six-element system responding to a displacement input.

And Equations (2.38) through (2.40) are combined to give

$$m_2 \frac{d^2x_3}{dt^2} + f_{\mathrm{NL}}\left(\frac{dx_3}{dt}\right) + k_2x_3 = k_2x_2 \tag{2.42}$$

It is seen that two second-order differential equations are needed to model this fourth-order (four independent energy storage elements) system, one of which is nonlinear. The nonlinear damping term in Equation (2.42) complicates the algebraic combination of Equations (2.41) and (2.42) into a single fourth-order differential equation model. In some cases the nonlinear damper characteristic may be linearized, making it possible to combine (2.41) and (2.42) into a single fourth-order differential equation for x_2 or x_3. Since this system has four independent energy storage elements, a set of four state variables is required to describe the state of this system (e.g., x_2, v_{2g}, x_3, v_{3g}, or F_{k1}, v_{2g}, F_{k2}, v_{3g}). The exchange of energy between the input source and the two springs and two masses, together with the energy dissipated by the dampers would require a very long complicated verbal description. Thus the mathematical model is a very compact, yet concise description of the system. Further discussion of the manipulation and solution of this mathematical model is deferred to later chapters. ■

2.3 Rotational Mechanical Systems

Corresponding to each of the translational elements, mass, spring, and damping, are rotational inertia, rotational spring, and rotational damping. These rotational elements are used in the modeling of systems in which each element rotates about a single nonaccelerating axis.

2.3.1 Rotational Inertias

An ideal inertia, depicted schematically in free-body diagram form in Figure 2.13, moves in relation to a nonaccelerating rotational frame of reference, which is usually taken to be the earth (ground). However, the frame of a steadily rotating space vehicle, for instance, could be used as a reference.

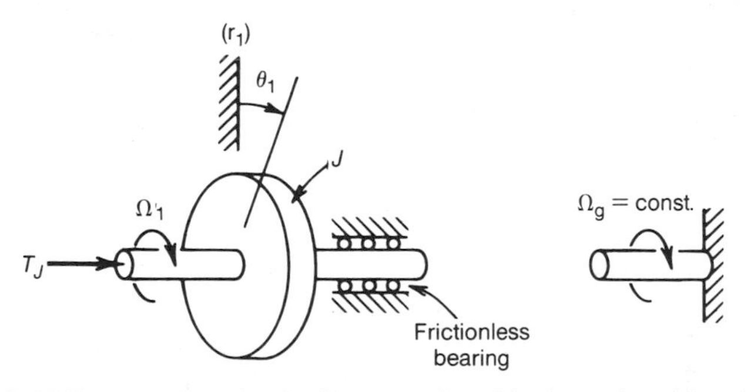

FIGURE 2.13 Free-body diagram of an ideal rotational inertia.

The elemental equation for an ideal inertia, J, based on Newton's second law for rotational motion is

$$T_J = J \frac{d\Omega_{1g}}{dt} \tag{2.43}$$

where Ω_{1g} is the angular velocity of the inertia relative to the ground reference (g) and T_J is the sum of all the external torques (twisting moments) applied to the inertia. Since $\Omega_{1g} = d\theta_1/dt$, the variation of θ_1 is related to T_J by

$$J \frac{d^2\theta_1}{dt^2} = T_J \tag{2.44}$$

The kinetic energy stored in an ideal rotational inertia is

$$\mathscr{E}_k = \left(\frac{J}{2}\right)\Omega_{1g}^2$$

Hence it is designated as an A-type element, storing energy as a function of the square of its A-variable Ω_{1g}.

The response of an inertia to an applied torque T_J is analogous to the response of a mass to an applied force F_m. It takes time for angular velocity, kinetic energy, and angular displacement to accumulate after the application of a finite torque, and it would not be realistic to try to impose a sudden change in angular velocity on a rotational inertia.

2.3.2 Rotational Springs

A rotating shaft may be modeled as an ideal spring if the torque required to accelerate its rotational inertia is negligible compared to the torque that it transmits. Sometimes, however, the torque transmitted by the shaft is small compared to that required to accelerate its own inertia so that it should be modeled as an inertia; and sometimes a combination of springs and inertias may be required to model a real shaft, as illustrated in Section 2.1 for a translational spring.

An ideal rotational spring stores potential energy as it is twisted (i.e., wound up) by the action of equal but opposite torques as shown in Figure 2.14. Here the rotational spring is shown (a) in a relaxed state with no torques acting, and (b) while transmitting torque T_K with both ends displaced rotationally from their local references r_1 and r_2 with T_K shown acting at both ends in free-body diagram form.

The elemental equation for a rotational spring is similar to Hooke's law for a translational spring given by

$$T_K = K(\theta_1 - \theta_2) \tag{2.45}$$

where θ_1 and θ_2 are the angular displacements of the ends from their local references r_1 and r_2. In derivative form this equation becomes

$$\frac{dT_K}{dt} = K(\Omega_{1g} - \Omega_{2g}) \tag{2.46}$$

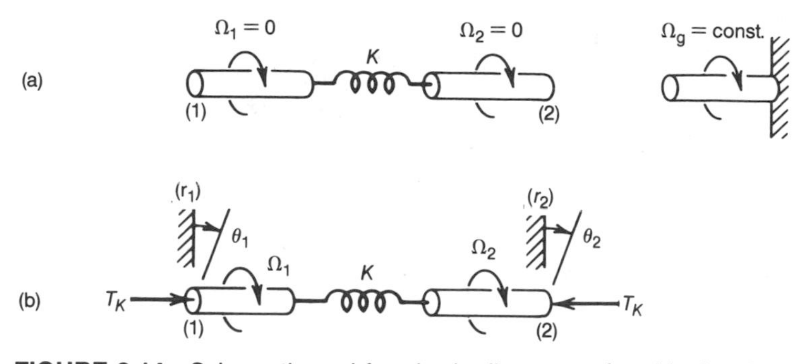

FIGURE 2.14 Schematic and free-body diagrams of an ideal spring.

where Ω_{1g} and Ω_{2g} are the angular velocities of the ends relative to a non-accelerating reference g. In each case the sign convention employed for motion is clockwise positive when viewed from the left, and for torque is clockwise positive when acting from the left (with the head of the torque vector toward the left end of the spring). The potential energy stored in a rotational spring is given by

$$\mathscr{E}_P = \frac{1}{2K} T_K^2 \tag{2.47}$$

Hence it is designated as a T-type element, storing energy as a function of the square of its T-variable T_K.

The comments about the response of a translational spring to a step change in velocity difference between its ends apply equally well to the response of a rotational spring to a step change in angular velocity difference between its ends. Thus it would be unreasonable to try to impose a step change of torque in a rotational spring because that would represent an attempt to change suddenly the energy stored in it in a real world that does not contain sources of infinite power.

When a rotational spring is nonlinear it does not have truly constant stiffness K, and it is advisable to employ a nonlinear symbol designation and a nonlinear function to describe it as follows.

$$T_{\text{NLS}} = f_{\text{NL}}(\theta_1 - \theta_2)$$

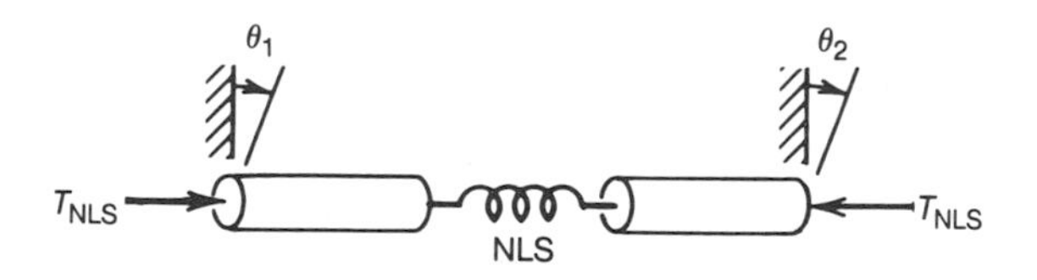

Linearization to achieve a local incremental stiffness K_{inc} is often feasible as in the case of a nonlinear translational spring discussed earlier.

2.3.3 Rotational Dampers

Just as friction between moving parts of a translational system gives rise to translational damping, friction between rotating parts in a rotational system is the source of rotational damping. When the interfaces are well lubricated, the friction is owing to the shearing of a thin film of viscous fluid, yielding a constant damping coefficient B as shown in Figure 2.15 employing (a) a cross-section diagram with no torque being transmitted so that $\Omega_{1g} = \Omega_{2g} = \Omega$, and (b) a free-body diagram with torque being transmitted.

The elemental equation of an ideal damper is given by

$$T_B = B(\Omega_{1g} - \Omega_{2g}) \tag{2.48}$$

where T_B is the torque transmitted by the damper.

A rotational damper is designated as a D-type element because it dissipates energy.

When the lubrication is imperfect, so that direct contact occurs between the two parts of the damper, dry friction becomes evident and the damping effect cannot be described by means of a simple damping constant. Fluid couplings (Sect 1.3) are also nonlinear. Instead a nonlinear function is required, and the elemental equation for a nonlinear damper NLD is expressed in nonlinear form

$$T_{\text{NLD}} = f_{\text{NL}}(\Omega_{1g}, \Omega_{2g}) \tag{2.49}$$

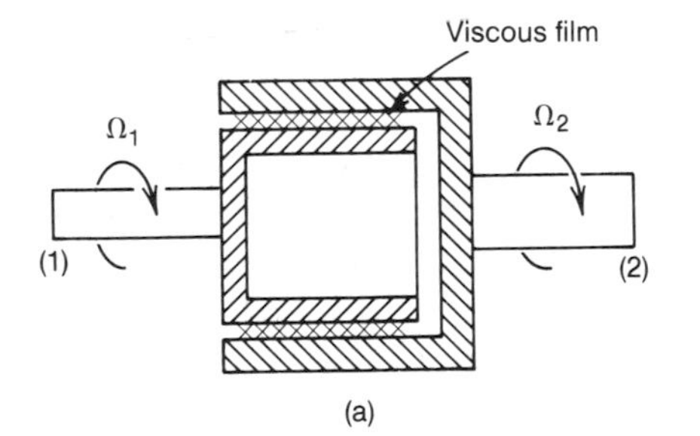

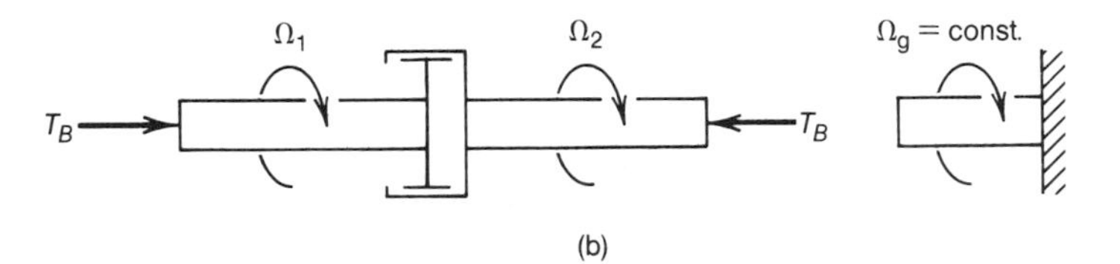

FIGURE 2.15 Cross-sectioned and free-body diagrams of an ideal rotational damper.

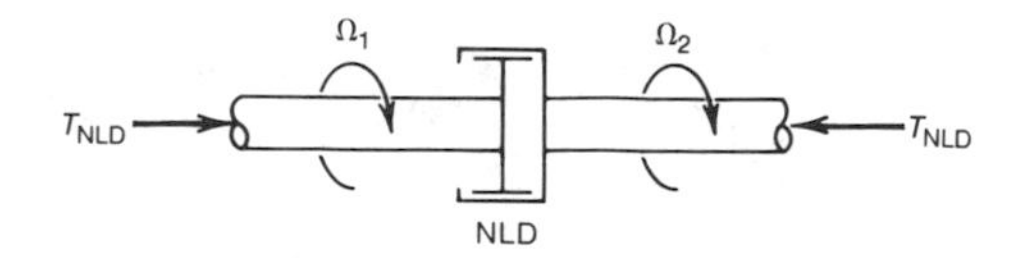

In some cases, the interaction between the two parts of the damper involves hydrodynamic fluid motion, as in a fluid coupling mentioned briefly in Chapter 1, where a square-law nonlinear function is involved.

EXAMPLE 2.7

The power transmission system from diesel engine to propellor for a ship is shown in simplified form in Figure 2.16. The role of the fluid coupling is to transmit the main flow of power from the engine to the propellor shaft without allowing excessive vibration, which would otherwise be caused by the pulsations of engine torque owing to the cyclic firing of its cylinders. The object here is to develop a mathematical model for this system in order to relate the shaft torque T_K to the inputs T_e and T_w.

Solution

The complete set of free-body diagrams for this system is shown in Figure 2.17. The system analysis will be developed by beginning at the left-hand end and writing the describing equation for each element and any necessary connecting point equations until the system has been completely described.

For the engine (moving parts and flywheel lumped together into an ideal inertia—negligible friction)

$$\frac{d\Omega_{1g}}{dt} = \left(\frac{1}{J_e}\right)(T_e - T_c) \tag{2.50}$$

For the fluid coupling (negligible inertia)

$$T_c = C_c(\Omega_{1g}^2 - \Omega_{2g}^2) \tag{2.51}$$

At the junction between the fluid coupling and drive shaft

$$T_c = T_K \tag{2.52}$$

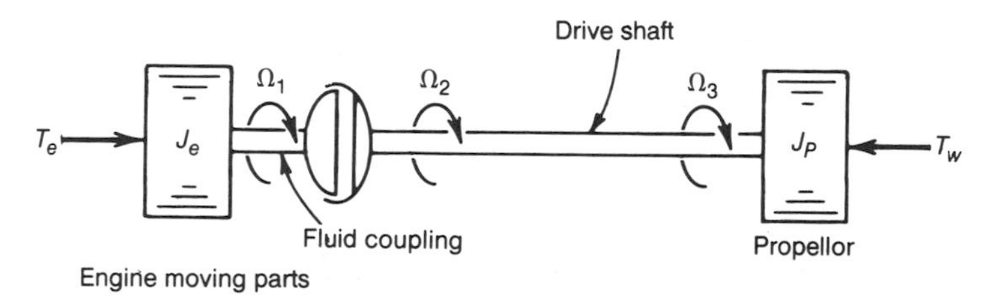

FIGURE 2.16 Schematic diagram of a simplified model of a ship propulsion system.

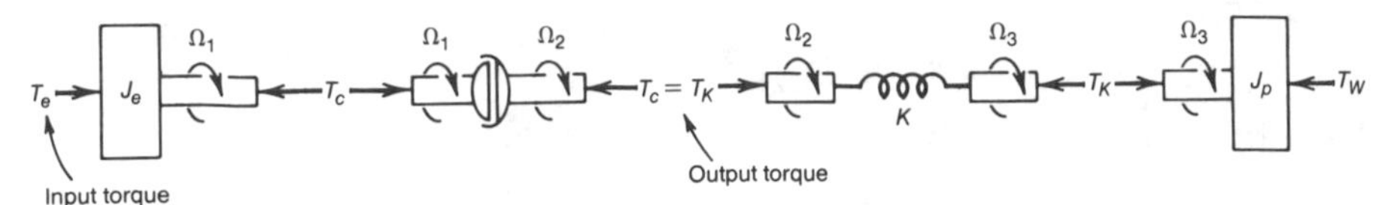

FIGURE 2.17 Complete set of free-body diagrams for a ship propulsion system.

For the drive shaft (ideal spring—negligible friction and inertia)

$$\frac{dT_K}{dt} = K(\Omega_{2g} - \Omega_{3g}) \tag{2.53}$$

For the propellor (ideal inertia—negligible friction)

$$\frac{d\Omega_{3g}}{dt} = \left(\frac{1}{J_p}\right)(T_K - T_w); \quad \text{state-variable equation 1} \tag{2.54}$$

Equations (2.50) through (2.54) constitute a necessary and sufficient set of five equations for this system containing five unknowns: Ω_{1g}, Ω_{2g}, T_c, T_K, and Ω_{3g}.

Note that additional dampers to ground at points (1), (2), and (3) would be required if bearing friction at these points were not negligible.

Rearranging Equation (2.51) into the form: $\Omega_{2g} = f_2(T_c, \Omega_{1g})$ gives

$$\Omega_{2g}^2 = \Omega_{1g}^2 - \frac{T_c}{C_c} \tag{2.51a}$$

or

$$\Omega_{2g} = \text{SSR}\left(\Omega_{1g}^2 - \frac{T_c}{C_c}\right) \tag{2.51b}$$

where SSR denotes 'signed square root', e.g.

$$\text{SSR}(X) = \frac{X}{\sqrt{|X|}}$$

Combining Equations (2.50) and (2.52)

$$\frac{d\Omega_{1g}}{dt} = \left(\frac{1}{J_e}\right)(T_e - T_K); \quad \text{state-variable equation 2} \tag{2.55}$$

and combining Equations (2.51b), (2.52), and (2.53)

$$\frac{dT_K}{dt} = K\left[\text{SSR}\left(\Omega_{1g}^2 - \frac{T_K}{C_c}\right) - \Omega_{3g}\right]; \quad \text{state-variable equation 3} \tag{2.56}$$

Equations (2.54), (2.55), and (2.56) are a set of three state-variable equations for the system described in Example 2.7 and, as such, constitute a necessary and sufficient set of equations for this system, containing the three unknown variables Ω_{3g}, T_K, and Ω_{1g}. Because of the nonlinearity in the fluid coupling, it is not possible to combine these equations algebraically into a single input-output differential equation. However, in their present form they are a complete set of state-variable equations ready to integrate numerically on a digital computer (see Chapter 6).

If Equation (2.51a) is linearized to give

$$\hat{T}_c = 2C_c\overline{\Omega}_{1g}\hat{\Omega}_{1g} - 2C_c\overline{\Omega}_{2g}\hat{\Omega}_{2g} \tag{2.51lin}$$

this may be solved for small perturbations of $\hat{\Omega}_{2g}$ in terms of $\hat{T}_c$ and $\hat{\Omega}_{1g}$ to give

$$\hat{\Omega}_{2g} = \left(\frac{\overline{\Omega}_{1g}}{\overline{\Omega}_{2g}}\right)\hat{\Omega}_{1g} - \left(\frac{1}{2C_c\overline{\Omega}_{2g}}\right)\hat{T}_c \tag{2.51sp}$$

Now Equations (2.50), (2.52), (2.53), and (2.54), which are already linear, may be expressed in terms of small perturbations of the unknown variables

$$\frac{d\hat{\Omega}_{1g}}{dt} = \left(\frac{1}{J_e}\right)(\hat{T}_e - \hat{T}_c) \tag{2.50sp}$$

$$\hat{T}_c = \hat{T}_K \tag{2.52sp}$$

$$\frac{d\hat{T}_K}{dt} = K(\hat{\Omega}_{2g} - \hat{\Omega}_{3g}) \tag{2.53sp}$$

$$\frac{d\hat{\Omega}_{3g}}{dt} = \left(\frac{1}{J_p}\right)(\hat{T}_K - \hat{T}_w) \tag{2.54sp}$$

Equations (2.50sp), (2.51sp), . . . (2.54sp) constitute a complete set of five linear 'small perturbation' equations for this system involving the five unknown variables $\hat{\Omega}_{1g}$, $\hat{T}_c$, $\hat{T}_K$, $\hat{\Omega}_{2g}$, and $\hat{\Omega}_{3g}$, which may be combined as follows to obtain the input-output differential equation relating $\hat{T}_K$ to $\hat{T}_p$ and $\hat{T}_w$.

Differentiating (2.53sp) with respect to time and substituting from (2.52sp) and (2.51sp) and (2.50sp)

$$\frac{d^2\hat{T}_K}{dt^2} = K\left[\left(\frac{\overline{\Omega}_{1g}}{\overline{\Omega}_{2g}}\right)\frac{d\hat{\Omega}_{1g}}{dt} - \left(\frac{1}{2C_c\overline{\Omega}_{2g}}\right)\frac{d\hat{T}_c}{dt} - \frac{d\hat{\Omega}_{3g}}{dt}\right] \tag{2.55sp}$$

$$= K\left[\left(\frac{\overline{\Omega}_{1g}}{\overline{\Omega}_{2g}}\right)\left(\frac{1}{J_e}\right)(\hat{T}_e - \hat{T}_K) - \left(\frac{1}{2C_c\overline{\Omega}_{2g}}\right)\frac{d\hat{T}_K}{dt} - \frac{d\hat{\Omega}_{3g}}{dt}\right] \tag{2.56sp}$$

Transposing (2.56sp)

$$\frac{d^2\hat{T}_K}{dt^2} + \left(\frac{K}{2C_c\overline{\Omega}_{2g}}\right)\frac{d\hat{T}_K}{dt} + \left(\frac{\overline{\Omega}_{1g}}{\overline{\Omega}_{2g}}\right)\left(\frac{K}{J_e}\right)\hat{T}_K$$

$$= \left(\frac{K}{J_e}\right)\left(\frac{\overline{\Omega}_{1g}}{\overline{\Omega}_{2g}}\right)\hat{T}_e - K\frac{d\hat{\Omega}_{3g}}{dt} \tag{2.57sp}$$

Combining (2.54sp) and (2.57sp) yields

$$\frac{d^2\hat{T}_K}{dt^2} + \left(\frac{K}{2C_c\overline{\Omega}_{2g}}\right)\frac{d\hat{T}_K}{dt} + \left[\left(\frac{\overline{\Omega}_{1g}}{\overline{\Omega}_{2g}}\right)\frac{K}{J_e} + \frac{K}{J_p}\right]\hat{T}_K$$

$$= \left(\frac{\overline{\Omega}_{1g}}{\overline{\Omega}_{2g}}\right)\left(\frac{K}{J_e}\right)\hat{T}_e + \left(\frac{K}{J_p}\right)\hat{T}_w \tag{2.58sp}$$

■

2.4 Synopsis

This chapter has demonstrated the principles involved in developing simplified lumped parameter mathematical models of mechanical systems of two basic types: (a) translational systems, and (b) rotational systems. In each case the system model has been developed through the use of Newton's laws dealing with summation of forces at a massless point (torques at an inertialess point), and acceleration of a lumped mass (lumped inertia), together with elemental equations for springs, and/or dampers. When carried out properly this results in a set of n equations containing n unknown variables. Subsequent mathematical manipulation of these equations was carried out to eliminate unwanted variables, producing the desired model involving the variables of greatest interest.

Usually this desired model consisted of a single input-output differential equation relating a desired output to one or more given inputs. In some cases a reduced set of first order equations, called state-variable equations was developed as part of the process of eliminating unwanted variables. This was done because in some instances in the future this is all the reduction needed to proceed with a computer simulation or analysis of the system. The definition of state variables is left to Chapter 4, which covers the topic of this aspect of system modeling in considerable detail.

It was then shown how linearization of nonlinearities helps to expedite the mathematical manipulation of the system equations to produce a single input-output differential equation.

The energy convertors required to couple translational with rotational systems will be discussed in Chapter 10, in which the general topic of energy convertors in "mixed" systems is covered in some detail.

During the development of this chapter, the basic system elements have been classified as to their energy-storing or energy-dissipating traits. Furthermore, in the case of energy-storing elements, precautions are emphasized relative to the impossibility of storing a finite amount of energy in an energy-storing element with finite sources of force (or torque) and velocity in zero time. Knowledge of these limitations on energy storage establishes limits on the kinds of sudden changes of input that are physically realizable when only finite sources of force (or torque) and velocity are available. Knowledge of these limitations on energy storage is also essential to the determination of the initial conditions of a system (in other words the values of the system's variables) after a sudden change of a system input occurs.

The determination of initial conditions after sudden changes of input and the role of initial conditions in the classical solution of the system differential equations will be discussed in Chapter 5.

Problems

2.1 A rather heavy compression spring weighing 1.0 lb has a stiffness k_s of 2000 lb/in. To the casual observer it looks like a spring but 'feels' like a mass. This problem deals with the choice of a suitable lumped-parameter model for such a spring.

According to vibration theory, this spring, containing both mass and stiffness itself, will respond with different kinds of oscillations, depending on how it is forced (i.e., its boundary conditions)[3]. This means that the choice of an approximate lumped-parameter model for this element will depend partly on the elements with which it interacts and partly on the range of frequencies, or rate of variation, of inputs applied to the system containing it. Obviously, in a simple system containing this heavy spring and an attached mass, the spring may be modeled to a good degree of approximation as a pure spring if the attached mass is at least an order of magnitude greater than the self-mass of the spring and if the portion of the force applied to the spring required to accelerate its self-mass is small compared to the force required to deflect the spring. Likewise, the spring might be approximately modeled as a pure mass if it interacts with another spring having a stiffness at least an order of magnitude smaller than its own stiffness and if its self-deflection is small compared to the deflection of the other spring due to acceleration force of its mass.

You are asked to propose approximate lumped-parameter models for such a spring in the following situations:

(a) The heavy spring supports a mass m weighing 5.0 lb with a force source F_s acting on the mass, as shown in Figure P2.1a. The maximum frequency of a possible sinusoidal variation of F_s is about half the lowest natural frequency of the heavy spring itself, operating in free-free or clamped-clamped mode.[3]

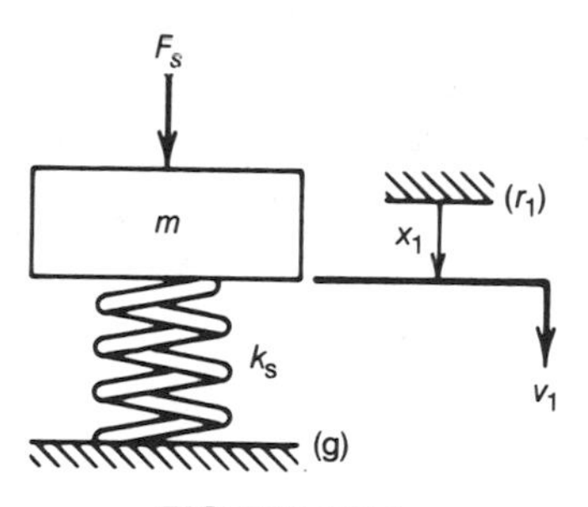

FIGURE P2.1a

(b) The spring acts between a mass weighing 3.0 lb and a force source, as shown in Figure P2.1b, with a maximum possible frequency of sinusoidal

[3]See T. Baumeister, *Mark's Standard Handbook for Mechanical Engineers*, 8th ed. McGraw-Hill Publishing Co., New York, 1978, p. 5–74.

variation that is about one half the lowest natural frequency of the heavy spring itself, operating in free-clamped mode.[3]

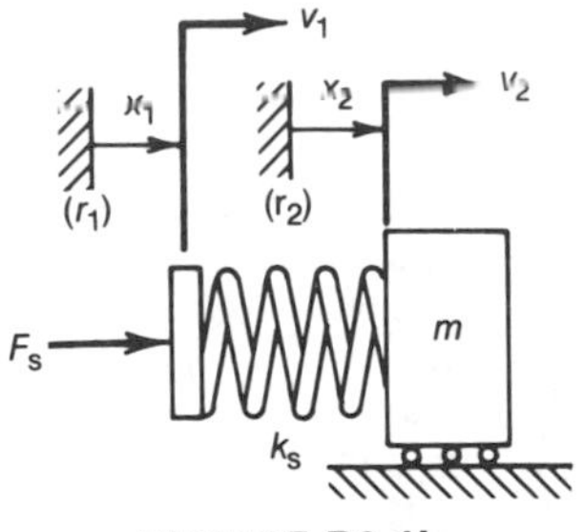

FIGURE P2.1b

2.2 An automobile weighing 3000 lb has been put in motion on a level highway and is then allowed to coast to rest. Its speed is measured at successive increments of time as recorded in the following table:

Time, sec	*Speed, ft/sec (approx.)*
0	15.2
10	10.1
20	6.6
30	4.5
40	2.9
50	2.1
60	1.25

(a) Using a mass-damper model for this system, draw the system diagram and set up the differential equation for the velocity v_{1g} of the vehicle.

(b) Evaluate m and then estimate b using a graph of the data given in the above table. (*Hint:* Employ the slope dv_{1g}/dt and v_{1g} itself at any time t.)

2.3 An automobile weighing 2200 lb is released on a 5.8 deg slope (i.e., 1/10 rad) and its speed at successive increments of time is recorded in the table below. Estimate the effective damping constant b for this system, expressed in Newton-seconds per meter (N-sec/m). (See *hint* in Problem 2.2. Also check first for coulomb friction.)

Time, sec	*Speed, m/sec*
0	0
10	2.05
20	3.30
30	4.15
40	4.85
50	5.20
60	5.55

2.4 An electric motor has been disconnected from its electrical driving circuit and set up to be driven mechanically by a variable-speed electric hand-drill mounted in a simple dynamometer arrangement. The torque versus speed data given below were obtained. Then starting at a high speed, the motor was allowed to coast so that the speed versus time data given below could be taken.

Driven		**Coasting**	
Shaft Speed, rpm	*Torque, N-m*	*Time, sec*	*Shaft Speed, rpm*
100	0.85	0	600
200	1.35	10	395
300	2.10	20	270
400	2.70	30	180
500	3.70	40	110
600	4.50	50	70
—	—	60	40

(a) Evaluate the coulomb friction torque T_c and the linear rotational damping coefficient B for the electric motor.
(b) Draw a system model for the electric motor, and set up the differential equation for the shaft speed Ω_{1g} during the coasting interval.
(c) Estimate the rotational inertia J for the rotating parts of the electric motor.

2.5 **(a)** Draw a complete free-body diagram for all the elements of the system shown in Figure P2.5.
(b) When the input force F_i is zero, the displacement x_1 is zero, the force in each spring is F_s, and the system is motionless. Find the spring forces F_{k1} and F_{k2} and the displacement x_1 when the input force is 5.0 lb and the system is again motionless.
(c) Derive the differential equation relating the displacement x_1 to the input force F_i.
(d) Find the initial values of v_{1g} and x_1 at $t = 0^+$, i.e., $v_{1g}(0^+)$ and $x_1(0^+)$, after the input force F_i is suddently reduced to zero from 5.0 lb.

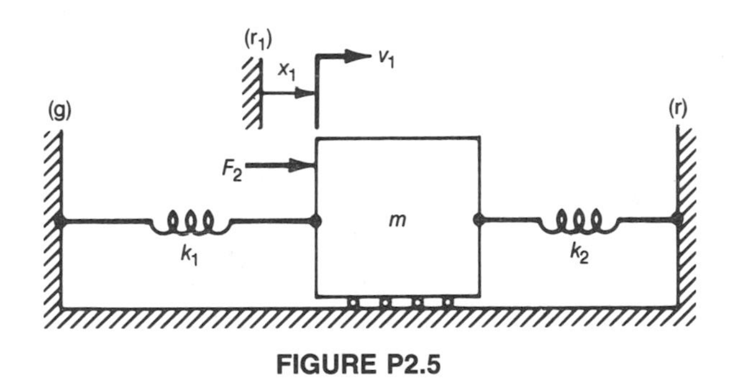

FIGURE P2.5

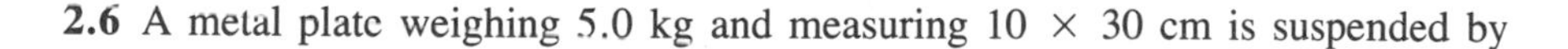
2.6 A metal plate weighing 5.0 kg and measuring 10 × 30 cm is suspended by

means of a 30-cm long 5.0-mm diameter steel rod as shown in Figure P2.6.

(a) Draw a complete free-body diagram for all the elements of this system and write the differential equation for the angular displacement θ_1 of the plate, using a simple inertia-spring model for this system.

(b) Using your knowledge about vibrations, find the natural frequency of oscillation for this system model in radians per second.

(c) Comment on the suitability of employing this simple model for the system. Take into consideration that the input torque T_i will never change with sinusoidal variations that have frequencies exceeding about 1/10 the lowest clamped-clamped natural frequency of the suspension rod itself.

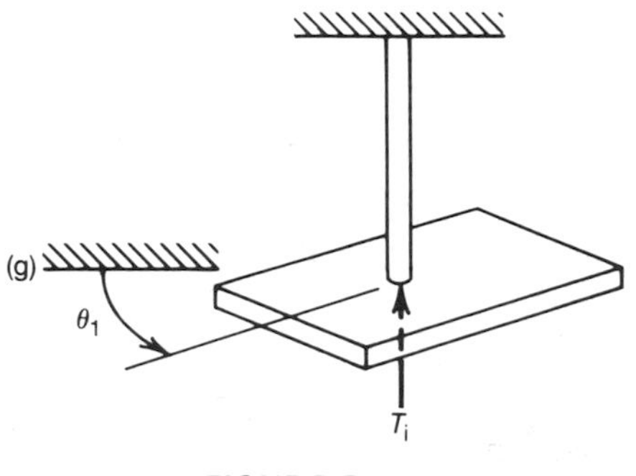

FIGURE P2.6

2.7 A simplified schematic diagram of one fourth of a vehicle suspension system is shown in Figure P2.7.

(a) Derive the differential equation relating the motion x_3 to the road profile motion x_1.

(b) Derive the differential equation relating the spring force F_{ks} to the road profile motion x_1.

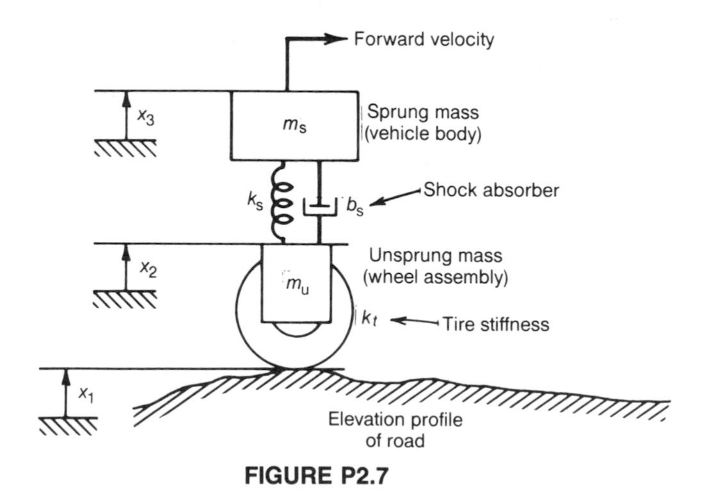

FIGURE P2.7

2.8 The rotational system shown schematically in Figure P2.8 is an idealized model of a machine-tool drive system. Although friction in the bearings is considered negligible in this case, the nonlinear friction effect NLD between the inertia J_2 and ground is significant. It is described by

$$T_{\mathrm{NLD}} = \Omega_{2g}\left[\left(\frac{T_o}{|\Omega_{2g}|}\right) + C|\Omega_{2g}|\right].$$

(a) Draw the complete free-body diagram for this system, showing each element separately and clearly delineating all variables.

(b) Write the necessary and sufficient set of describing equations for this complete system.

(c) Write corresponding equations for small perturbations of all variables for the case when Ω_{2g} is always positive and varies about a mean value $\overline{\Omega}_{2g}$. Find corresponding mean values of T_s and Ω_{1g} in terms of $\overline{\Omega}_{2g}$.

(d) Combine the small perturbation equations to eliminate unwanted variables and develop the system differential equation, relating the output $\hat{\Omega}_{2g}$ to the input $\hat{T}_s(t)$.

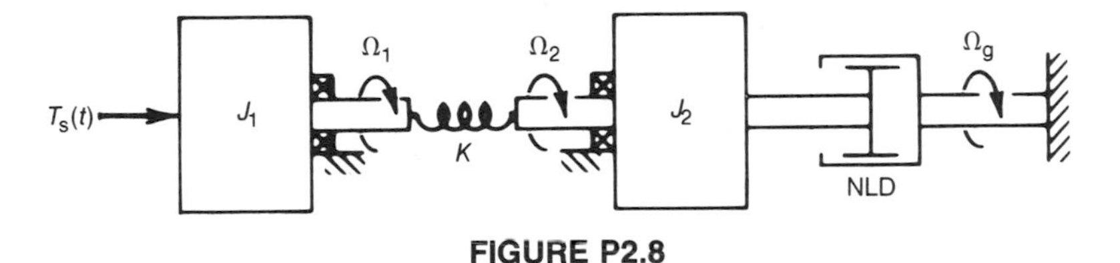

FIGURE P2.8

2.9 The system shown schematically in Figure P2.9 is an idealized model of a cable lift system where the springs k_1 and k_2 represent the compliances of connecting flexible cables.

(a) Draw a complete free-body diagram for the system showing each element separately, including the massless pulley, and clearly delineating all variables.

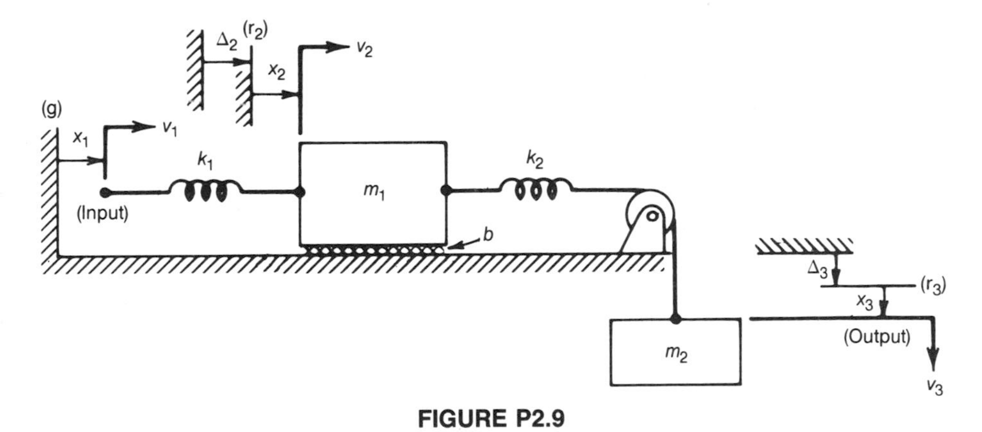

FIGURE P2.9

(b) Develop expressions for Δ_2 and Δ_3, the displaced references owing to the action of gravity on the mass m_2 when $x_1 = 0$.

(c) Write the necessary and sufficient set of describing equations for this system.

(d) Combine these equations to remove unwanted variables and develop the system differential equation relating x_3 to x_1.

2.10 Two masses are connected via springs to a rotating lever as shown schematically in Figure P2.10.

(a) Draw a complete free-body diagram for this system showing each element, including the massless lever, and delineating clearly all variables.

(b) Write the necessary and sufficient set of equations describing this system, assuming that the angle θ never changes by more than a few degrees.

(c) Combine equations to remove unwanted variables and obtain the system differential equation relating x_3 to $F_s(t)$.

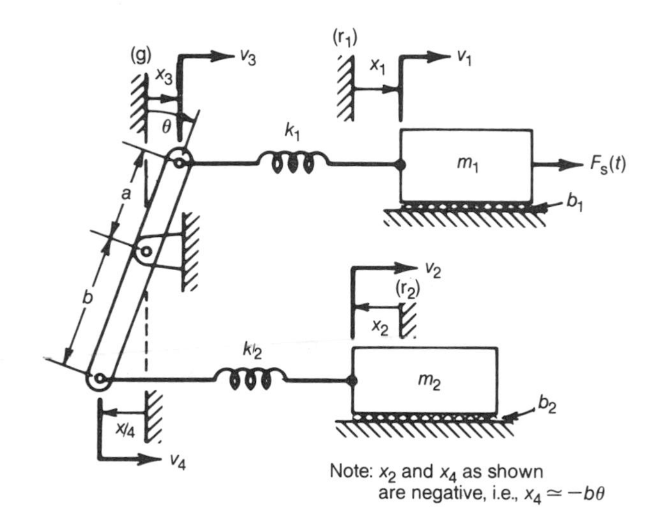

FIGURE P2.10

3

Simulation Block Diagrams

3.1 Introduction

Although the dynamic interactions between system elements are properly described by mathematical equations of the type developed in Chapter 2 (elemental and Newton's law, state-variable, or input-output differential equations), the use of functional block diagrams is often very helpful in visualizing how these interactions occur. Each block of such a diagram represents a single mathematical operation employed in the mathematical describing equations. These block diagrams were employed first when analog computers were developed to simulate the performance of dynamic systems. Later they became very helpful in "sizing-up" and making a preliminary analysis of a system without using a computer. Currently such diagrams are also very helpful in preparing programs for digital computers to produce digital simulations of system models.

In this chapter each of the functional blocks will be shown and defined for its corresponding mathematical function; then simple connections or combinations will be employed to represent typical describing equations. Finally, examples have been chosen to illustrate how these diagrams are used to represent complete mechanical systems, showing all the significant interactions involved in the response of the system model to input disturbances. Thus, in addition to specific output variables, all the other system variables are shown as well. Applications to other systems, such as electrical, thermal, fluid, and mixed systems, will be left to illustrations, examples, and home problems in later chapters.

The use of simulation block diagrams helps to reveal the key role of sequential time integrations, coupled with associated feedback effects, in shaping and/or describing the dynamic response of a system. A thoughtful review of the flow of information (i.e., signals) through such a diagram helps one to understand the key role played by time integrations in producing the time lags observed when a system responds to a given input disturbance. It might even be said that time integration is the essence of dynamic response.

3.2 Functional Blocks and Their Mathematical Descriptions

3.2.1 Coefficient Blocks

When a system variable (or one of its derivatives—which is also a system variable) is multiplied by a coefficient, this function is represented by a coefficient block as shown in Figure 3.1. The figure includes the mathematical operation represented by the block and its input and output variables depicted as signals.

The arrow-directed input and output signal lines are intended only to represent signal flow; they are not necessarily intended to represent physical connections (although they sometimes do so). Note that the multiplier function ($\times$) is needed when the coefficient C is time varying, either as a known, or unknown function of time. Similarly a divider function ($\div$) is occasionally needed when x is to be divided by a time-varying coefficient C.

3.2.2 Summation Blocks

When a system variable is equal the sum of two or more other system variables, the relationship is depicted by means of a small circle with the inputs and outputs arranged as shown in Figure 3.2. The figure illustrates the three most commonly encountered forms for this block diagram.

3.2.3 Integration and Differentiation Blocks

When a system variable is the time-integral of another system variable, the integration block is used to describe this functional relatinship as shown in Figure 3.3a. Figure 3.3a also shows the initial value of the output at time $t = 0^-$ being summed with the integrator output, along with the corresponding mathematical equation. The inverse operation, the function of differentiation, is represented by means of a differentiator block as shown in Figure 3.3b. Since true differentiation is not physically achievable for step inputs, simulation diagrams usually do not employ

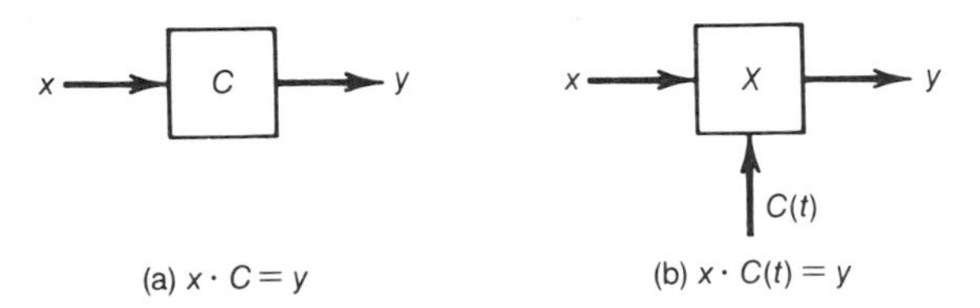

FIGURE 3.1 Coefficient block. (a) Constant coefficient, and (b) time-varying coefficient, i.e., a multiplier.

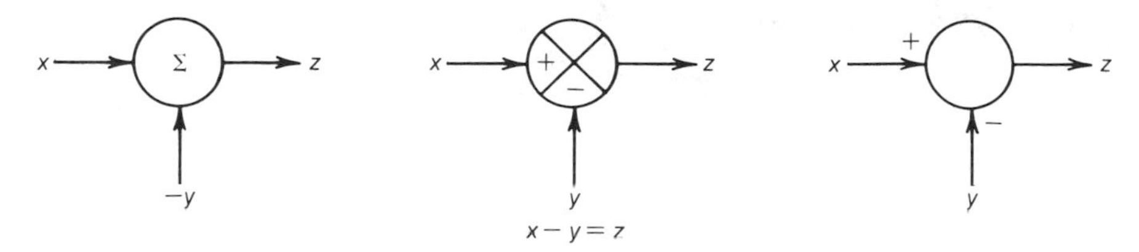

FIGURE 3.2 Three commonly employed versions of the summation block diagram.

differentiator blocks. It seems ironic that derivatives with respect to time are commonly used in differential equations to describe dynamic systems, whereas in point of fact the dynamic behavior is more truly described physically in terms of time integrations[1].

3.2.4 Drawing Complete Diagrams from Describing Equations

Usually a typical simulation block diagram is developed by starting with the input signal and employing successive integrators combined with other functional blocks to arrive at the output. This approach will always work for describing physically realizable models of dynamic systems. If it is impossible to model a set of system equations without resorting to the use of one or more differentiators, it can be assumed that these equations represent a system physically unrealizable for step inputs; moreover, the system input-output equation will have one or more right-hand side (input) terms involving derivatives that are of higher order than the highest derivative term of the left-hand (output) side. The following examples are employed to illustrate the development of simulation block diagrams for physically realizable system models.

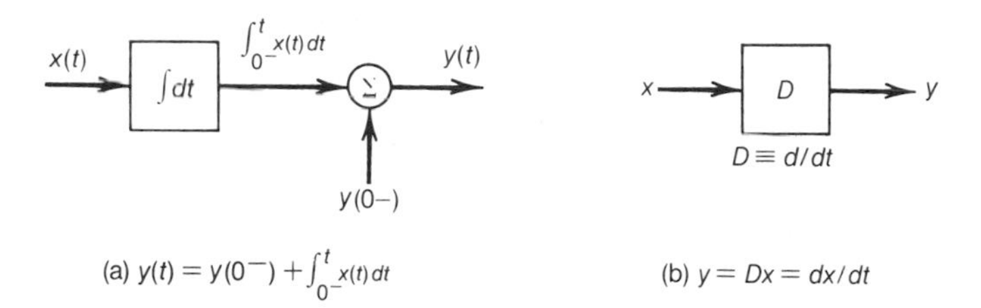

FIGURE 3.3 Block diagram symbols for integration and differentiation with respect to time.

[1]It may be noted that achieving true time differentiation by physical means would be equivalent to predicting the future; this is implicitly involved in a system input-output differential equation that has a highest time derivative of the input variable of higher order than the highest time derivative of the output variable. Such an equation also represents a situation that is impossible to achieve physically when step changes occur in the input variable.

EXAMPLE 3.1.

Develop the simulation block diagram for the mass-spring-damper system shown modeled schematically in Figure 3.4.

Solution

Applying Newton's law to the mass m yields

$$F_i - F_k - F_b = m \frac{dv_{1g}}{dt} \tag{3.1}$$

Equation (3.1) may be combined with the elemental equations for the spring and damper forces and rearranged to provide the expression for dv_{1g}/dt.

$$\frac{dv_{1g}}{dt} = \frac{1}{m}(F_i - kx_1 - bv_{1g}) \tag{3.2}$$

Starting with the input force F_i, a summing block is used to incorporate the other force terms kx_1 and bv_{1g} into the net force F_m, acting to accelerate the mass m which is then multiplied by the coefficient $(1/m)$ in a coefficient block to produce dv_{1g}/dt as shown in Figure 3.5.

Now since v_{1g} is the time integral of dv_{1g}/dt, expressed by the equation

$$v_{1g}(t) = \int_{0^-}^{t} \left(\frac{dv_{1g}}{dt}\right) dt + v_{1g}(0^-) \tag{3.3}$$

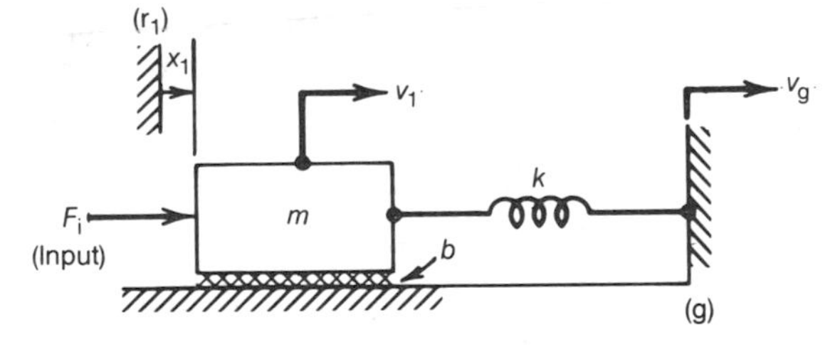

FIGURE 3.4 Simple mass-spring-damper system.

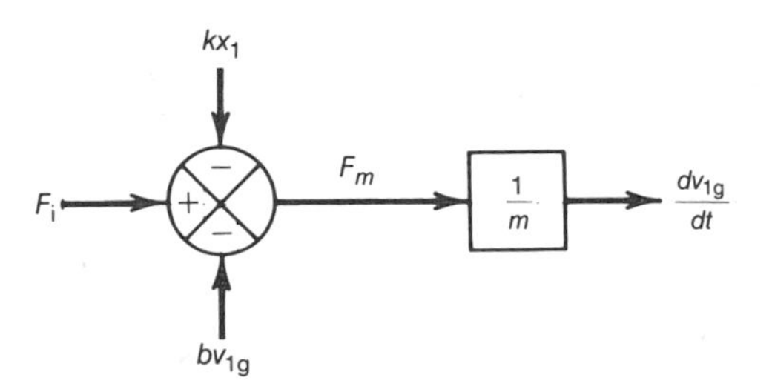

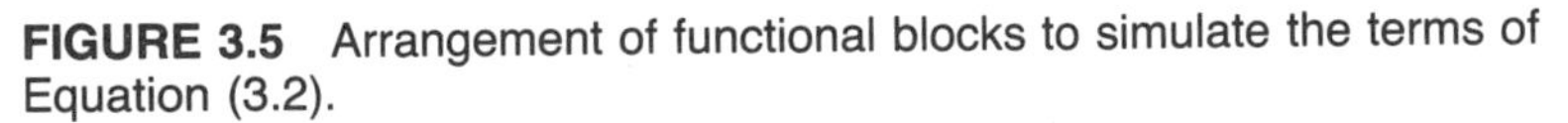
FIGURE 3.5 Arrangement of functional blocks to simulate the terms of Equation (3.2).

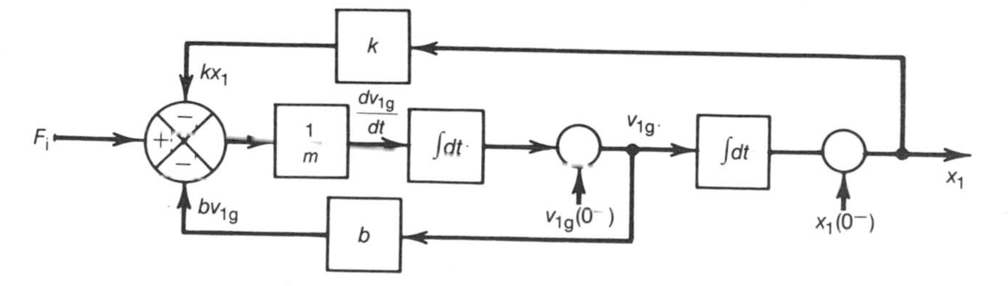

FIGURE 3.6 Simulation block diagram for mass-spring-damper system of Example 3.1.

an integration block is employed to produce v_{1g}. Then since

$$x_1(t) = \int_{0^-}^{t} (v_{1g})dt + x_1(0^-) \tag{3.4}$$

another integration block is employed to produce x_1, as shown in Figure 3.6. Then it is a simple matter to use coefficient blocks to multiply v_{1g} by b and x_1 by k to complete the simulation diagram for this system. ■

The dynamic response of this system is readily traced from the input F_i through the summer, the coefficient $1/m$, the first integrator, and then the second integrator. The two feedback paths through the coefficients b and k represent nature's way of keeping things under control.[2] The effects of such feedback effects in dynamic systems will be discussed at length in future chapters.

EXAMPLE 3.2

Develop the simulation block diagram for the two-mass system model shown in Figure 3.7.

Solution

For the first mass m_1,

$$F_i - F_{k1} - F_{b1} = m_1 \frac{dv_{1g}}{dt} \tag{3.5}$$

or

$$\frac{dv_{1g}}{dt} = \left(\frac{1}{m_1}\right) [F_i - k_1(x_1 - x_2) - b_1(v_{1g} - v_{2g})] \tag{3.6}$$

Similarly for the second mass m_2,

$$\frac{dv_{2g}}{dt} = \left(\frac{1}{m_2}\right) (F_{k1} + F_{b1} - F_{k2} - F_{b2}) \tag{3.7}$$

[2]Integrators without feedback exhibit curious traits like having outputs that keep on growing with time as long as even very small inputs remain, and holding their output values when their inputs are zero.

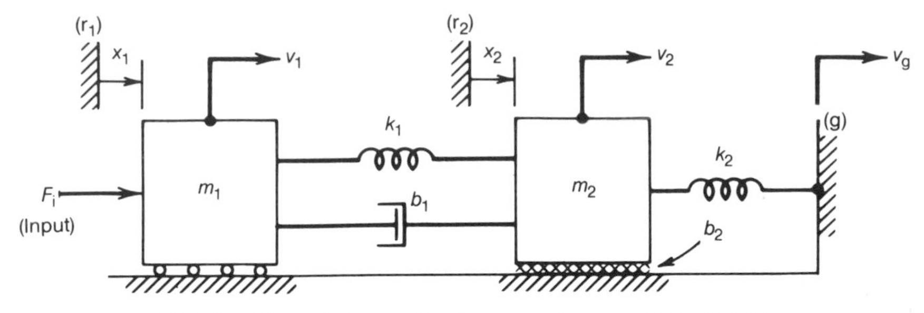

FIGURE 3.7 Schematic diagram of two-mass system.

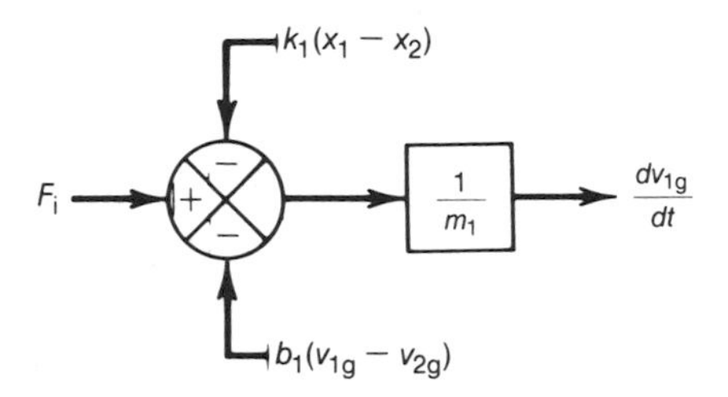

FIGURE 3.8 Block diagram segment to simulate Equation (3.6).

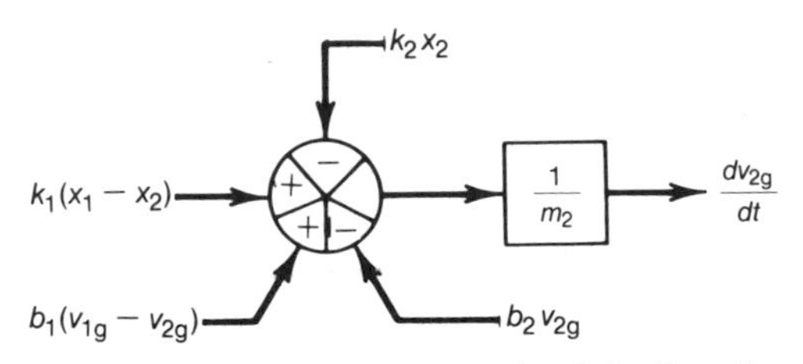

FIGURE 3.9 Block diagram segment to simulate Equation (3.8).

or

$$\frac{dv_{2g}}{dt} = \left(\frac{1}{m_2}\right) [k_1(x_1 - x_2) + b_1(v_{1g} - v_{2g}) - k_2x_2 - b_2v_{2g}] \quad (3.8)$$

Starting with the input F_i involves the need to simulate the summation described in Equation (3.6), multiplied by the constant $1/m_1$, as shown by the block diagram segment shown in Figure 3.8.

Similarly Equation (3.8) is simulated by means of the block diagram segment shown in Figure 3.9.

Next the two segments can be combined after inserting the successive integrators needed to obtain v_{1g} and x_1 from dv_{1g}/dt and to obtain v_{2g} and x_2 from dv_{2g}/dt, as shown in Figure 3.10.

Finally the diagram is completed with the use of coefficient blocks for k_1, b_1, k_2, and b_2 and a third summer to close the feedback paths, as shown in the complete system block diagram given in Figure 3.11. ■

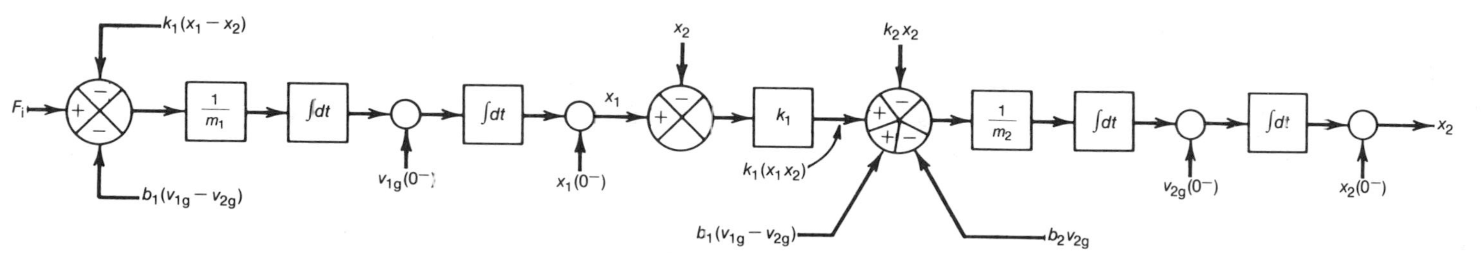

FIGURE 3.10 Combined segments after inserting integrators.

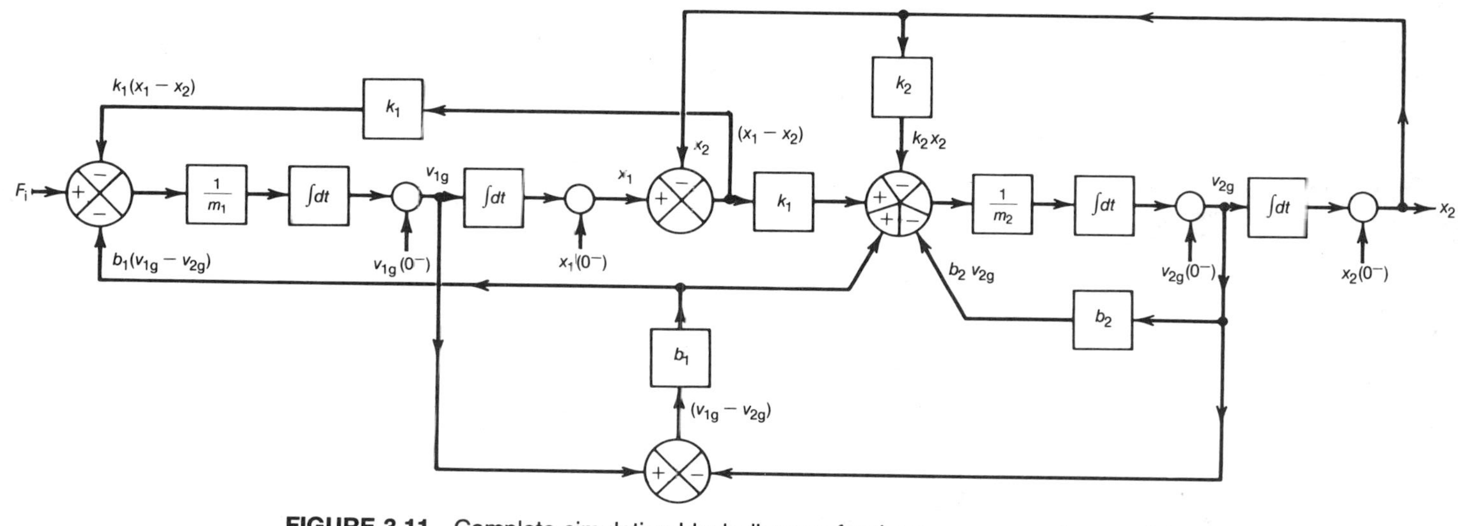

FIGURE 3.11 Complete simulation block diagram for the two-mass system model.

The dynamic response of this system is now readily traced from the input F_i through the successive summations, coefficients, and integrators to the remotest output x_2. Note that each integrator produces an additional lag in the response of each of the successive variables v_{1g}, x_1, v_{2g}, and x_2. All of the remaining connecting pathways involve feedbacks around integrators, with the exception of the feedforward path from v_{1g} through the third summer and b_1.

EXAMPLE 3.3

Develop the simulation block diagram for a system described by an input-output equation such as that given in Equation (3.9).

$$c_2 \frac{d^2y}{dt^2} + c_1 \frac{dy}{dt} + c_0 y = d_1 \frac{du}{dt} + d_0 u \tag{3.9}$$

First, it should be noted the derivative term on the right-hand side of Equation (3.9) creates a problem. If the output y is employed directly in the development of a simulation block diagram, a differentiator will be needed to create the term $d_1\,(du/dt)$ (readers are encouraged to demonstrate this to themselves). Because of our aversion to the use of differentiators, this problem is easily by-passed by employing the auxiliary differential equation[3] having the auxiliary variable x as follows:

$$c_2 \frac{d^2x}{dt^2} + c_1 \frac{dx}{dt} + c_0 x = u \tag{3.10}$$

The new variable x can be considered as one of the alternative state variables for this system, and of course an additional output equation will be needed to recover the output variable y.

The simulation block diagram for Equation (3.10) is developed as shown in Figure 3.12 after rearranging (3.10) as follows:

$$\frac{d^2x}{dt^2} = \left(\frac{1}{c_2}\right)\left(u - c_1 \frac{dx}{dt} - c_0 x\right) \tag{3.11}$$

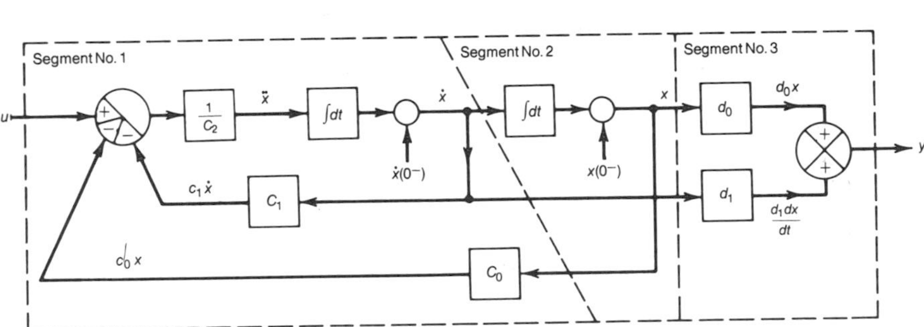

FIGURE 3.12 Simulation block diagram for system of example 3.3. Based on the auxiliary variable x.

[3]See further discussion of the use of an auxiliary variable in this connection in Section 4.4.

Segment No. 1 of Figure 3.12 simulates (3.10), and Segment No. 2 provides the integration for converting dx/dt to x.

In order to recover the output variable y, an output equation is developed as follows:

The right-hand side of Equation (3.9) may be thought of as the operator $(d_1\,(d/dt) + d_0)$ times the input variable u. If all of the terms of Equation (3.10) are multiplied by this same operator,

$$c_2 \frac{d^2\left(d_1 \frac{d}{dt} + d_0\right) x}{dt^2} + c_1 \frac{d\left(d_1 \frac{d}{dt} + d_0\right) x}{dt} + c_0 \left(d_1 \frac{d}{dt} + d_0\right) x = \left(d_1 \frac{d}{dt} + d_0\right) u \qquad (3.12)$$

it then becomes clear, comparing Equations (3.9) and (3.12), that y is given by

$$y = d_1 \frac{dx}{dt} + d_0 x \qquad (3.13)$$

Equation (3.13) is then employed to develop Segment No. 3 of Figure 3.12. Although the physical significance of the auxiliary variable x is not evident here, it often happens that it is one of the real system variables that appeared in the process of developing the input-output equation of the type that was used in this example. ■

3.2.5 A Step-by-Step Approach for a System Described by n State Variables

When the mathematical model consists of a set of n state-variable equations and appropriate output equations (see Chapter 4), the following step-by-step procedure is useful:

- *Step 1.* Draw n integrator blocks, one for each state variable, with inputs $\dot{q}_1, \dot{q}_2, \ldots \dot{q}_n$, and outputs $q_1, q_2, \ldots q_n$, respectively.
- *Step 2.* Draw a summer before each integrator (each summer to represent an equal sign in a state-variable equation).
- *Step 3.* Draw interconnecting lines from u's and q's through appropriate coefficient blocks to represent terms on the right-hand sides of the state-variable equations.
- *Step 3a.* Eliminate redundant summers (the ones with only one input).
- *Step 4.* Draw a summer for each output variable.
- *Step 5.* Draw interconnecting lines from proper u's and/or q's through appropriate coefficient blocks to the output summers to represent terms on the right-hand sides of the output equations.
- *Step 5a.* Eliminate redundant summers.

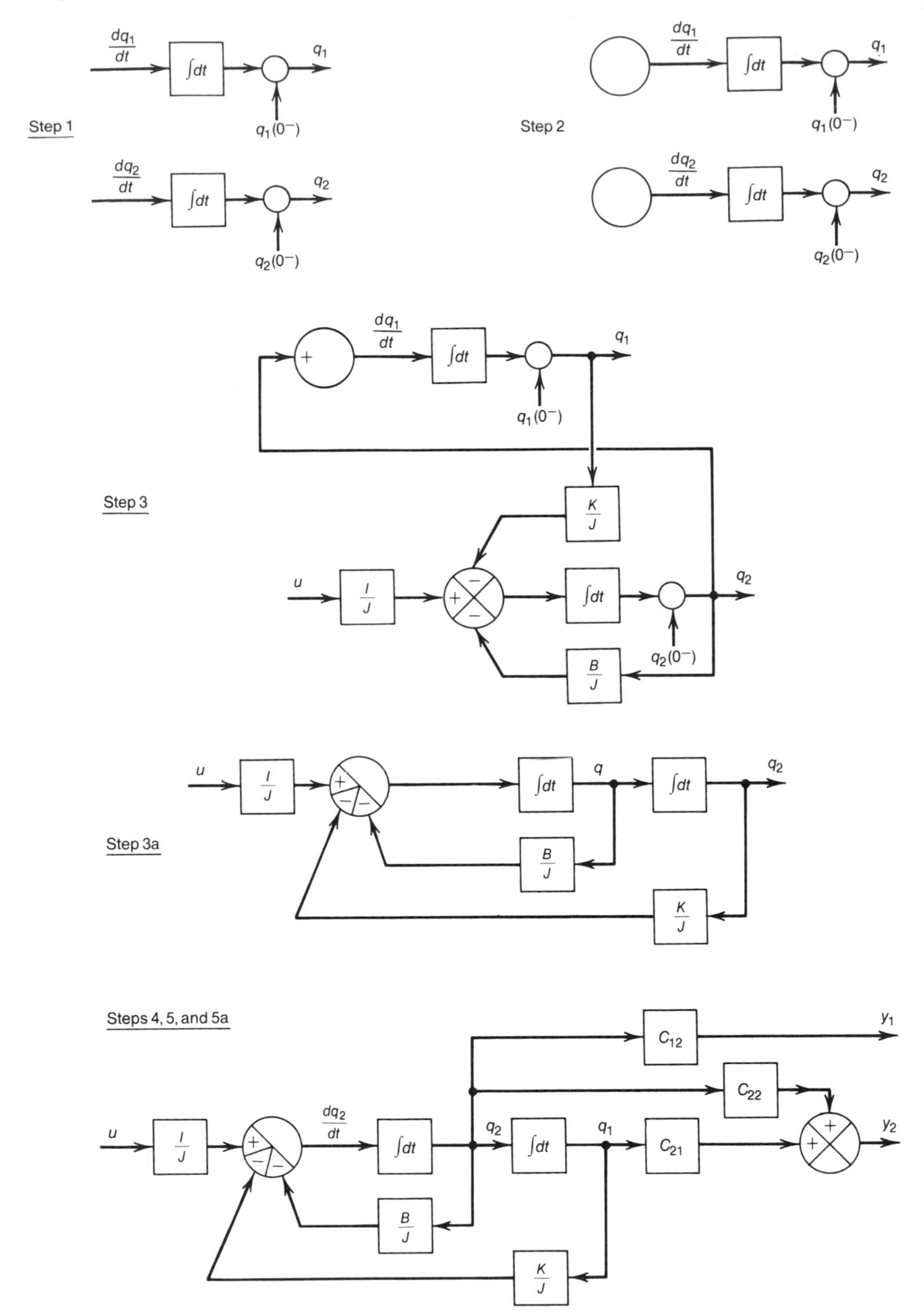

FIGURE 3.13 Illustration of step-by-step procedure for n state-variable equations.

EXAMPLE 3.4

A given rotational system has the following state-variable and output equations.

$$dq_1/dt = q_2$$
$$dq_2/dt = (-K/J)q_1 + (-B/J)q_2 + (1/J)u$$
$$y_1 = c_{12}q_2 \qquad y_2 = c_{22}q_2 + c_{21}q_1$$

The initial conditions are

$$q_1(0^-) = 0 \qquad q_2(0^-) = 0$$

Draw the simulation block diagram for this system.

Solution

The steps of this solution are given in Figure 3.13.

In this case, individual parameters are not shown in separate blocks, but the diagrams are more compact and have a minimum number of blocks. ■

3.3 Synopsis

The various kinds of blocks for simulating the functions of summation, multiplying by a constant, and integration were introduced by means of symbolic diagrams shown together with their describing equations.

Examples of mechanical systems were then provided to demonstrate the interconnection of these basic functional blocks to produce complete simulation diagrams. In each case the simulation block diagram formed a comprehensive picture or map of that system, revealing all the significant internal dynamic interactions via internal natural closed loops which determine the overall dynamic response of the system as it interacts with its environment.

One of the examples involved the case of a system which had already been modeled in the form of an input-output differential equation having a derivative term on the input side of the differential equation, and it was shown how the use of an auxiliary state variable together with its derivatives could lead to a simulation block diagram without having to resort to the use of a differentiation block. (Differentiation having been set aside previously as an unfeasible function for a simulation block for reasons of physical unrealizability when dealing with discontinuous input functions.)

Further use of these simulation block diagrams has been left to later chapters and home study problems, when many different types of systems will have been introduced so that they are available for use as examples and problem segments.

Lest the reader may have the impression at this point that these diagrams are an academic frill, it should be stressed here that the use of these simulation block diagrams is often very helpful in finding where the faster and slower parts of a system are located, in finding ways to improve system performance, and in debugging simulation programs.

In some cases they are also useful in visualizing how to combine the system equations to eliminate unwanted variables; and they can be helpful in checking the accuracy of terms in a set of state-variable equations, including the discovery of terms that "accidentally disappear" during the complex algebraic manipulations often needed to achieve desired input-output relationships.

Problems

3.1 **(a)** Draw the simulation block diagram for the spring-damper system shown in Figure 2.9, using separate blocks for the system parameters b and k, with x_2 and F_i as system inputs.

(b) Rearrange the diagram so that b and k are shown together as a quotient in the feedback block.

3.2 Draw a simulation block diagram for the mass-damper system shown in Figure 2.10, with F_i as the input. Compare this diagram with those drawn for Problem 3.1.

3.3 **(a)** Draw the simulation block diagram for the mass-spring-damper system shown in Figure 2.11, showing the parameters m, b, and k in separate blocks and showing F_i as the input.

(b) Rearrange the diagram so that the outer feedback block has unity gain and by adding a coefficient block after the input F_i.

(c) Compare your diagrams for this system with your diagram for Problem 3.3.

3.4 **(a)** Draw a simulation block diagram for the nonlinear mechanical system shown in Figure 2.12, using a block labelled "$f_{NL}(v_{3g})$" to obtain the damper force F_{NLD} from the velocity v_{3g}.

(b) Prepare a similar diagram for the set of linearized equations developed in Chapter 2 for this system.

3.5 **(a)** Draw a simulation block diagram for the nonlinear ship propulsion system modeled in Figure 2.16, employing a block labelled "$f_{NL}(T_K, \Omega_{1g})$" to obtain the shaft speed Ω_{2g} from the drive shaft torque T_K and the engine speed Ω_{1g}.

(b) Prepare a similar diagram for the set of linearized equations developed in Chapter 2 for this system.

(c) Show how you would implement "$f_{NL}(T_K, \Omega_{1g})$," used in part (a), with a coefficient block, a summer, a multiplier block, and a signed square-root block. (An SSR block provides an output that has the same sign as its input and a magnitude that is the square root of the magnitude of the input.)

4

Mathematical Models

4.1 Introduction

In almost all areas of engineering, and certainly in all those areas where new processes or devices are being developed, considerable efforts are directed towards acquiring information on various aspects of system performance. This process is generally referred to as a system analysis. Traditionally, the system analysis was carried out by investigating the performance of an existing physical object subjected to selected test input signals. Although there is no doubt that such an experimental approach provides extremely valuable and most reliable information about the system characteristics, experimenting with an actual full-scale system is not always feasible and is very often practically impossible, especially in the early stages of the system analysis. There are several reasons: First, an actual system must be available for testing or a new test object must be constructed, which may involve high cost in terms of time and money. Second, the extent to which the parameters of an existing engineering system can be varied in order to observe their effects on system performance is usually very limited. Third, the experimental results are always "object specific," since they represent only a particular system under investigation, and may be difficult to generalize.

In order to overcome these problems, researchers develop simplified representations of actual systems called system models. The system models may be physical in nature (down-scaled actual systems), or they may be developed in a form of abstract descriptions of the relationships existing among the system variables. In the latter case a dynamic system can be described using verbal text, plots and graphs, tables of relevant numerical data, or mathematical equations. The language of mathematics is preferred in modeling engineering systems because of its superior precision and generality of expression. Mathematical methods of system modeling lead to better ability to generalize the results and to apply them to solving control and design problems. A representation of the relationships existing among the system variables in a form of mathematical equations is a mathematical model of the system.

Two types of mathematical models are introduced in this chapter, input-output models and state models. Both types of models carry essentially the same information about the system dynamics, but the sets of model differential equations are different from each other in several respects. These differences carry serious implications on practical usefulness and applicability of the types of models in various engineering problems.

4.2 Input-Output Models

Consider the single-input single-output dynamic system shown in Figure 4.1. In general, the relationship between the input and output signals of the system can be represented by a nth-order differential equation of the following form

$$a_n\left(\frac{d^n y}{dt^n}\right) + a_{n-1}\left(\frac{d^{n-1}y}{dt^{n-1}}\right) + \cdots + a_1\left(\frac{dy}{dt}\right) + a_0 y = f\left(\frac{d^m u}{dt^m}, \frac{d^{m-1}u}{dt^{m-1}}, \ldots, \frac{du}{dt}, u, t\right) \tag{4.1}$$

where $m \leq n$ for existing and realizable engineering systems because of inherent inertia of those systems. Also, having $m > n$ is physically impossible because it would imply the ability to "predict the future" of the system input. A set of n initial conditions, $y(0^+)$, $\dot{y}(0^+)$, . . . , $y^{(n-1)}(0^+)$, must be known in order to solve the equation for a given input $u(t)$, $t \geq 0$.

The model parameters, a_0, a_1, . . . , a_n, may be functions of y, u, and/or t, in which case the system is nonlinear. If the system is stationary, the model parameters are constant and the model differential equation can then be written as

$$a_n \frac{d^n y}{dt^n} + a_{n-1}\frac{d^{n-1}y}{dt^{n-1}} + \cdots + a_1\frac{dy}{dt} + a_0 y = f\left(\frac{d^m u}{dt^m}, \frac{d^{m-1}u}{dt^{m-1}}, \ldots, \frac{du}{dt}, u\right) \tag{4.2}$$

where f may be a nonlinear function.

For a stationary linear model, the function f on the right hand side of Equation (4.2) is a sum of terms linear with respect to u and its derivatives.

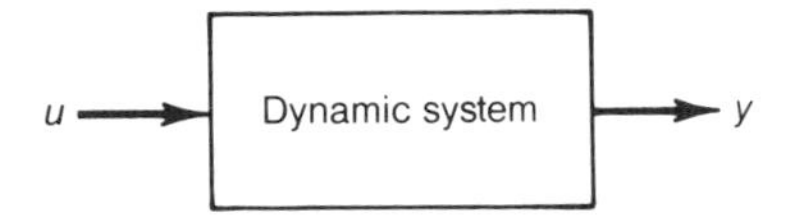

FIGURE 4.1 Single-input single-output dynamic system.

$$a_n \frac{d^n y}{dt^n} + a_{n-1} \frac{d^{n-1} y}{dt^{n-1}} + \cdots + a_1 \frac{dy}{dt} + a_0 y =$$

$$b_m \frac{d^m u}{dt^m} + b_{m-1} \frac{d^{m-1} u}{dt^{m-1}} + \cdots + b_1 \frac{du}{dt} + b_0 u \quad (4.3)$$

where $b_0, b_1, \ldots, b_m$ are constants. As before, the order of the highest derivative of the input variable cannot be greater than the order of the highest derivative of the output variable, $m \leq n$.

Equations (4.1), (4.2), and (4.3) represent general forms of input-output models for single-input single-output dynamic systems.

EXAMPLE 4.1

Derive input-output equations for the mechanical system shown in Figure 4.2 using force $F_1(t)$ as the input variable and displacements $x_1(t)$ and $x_2(t)$ as the output variables. The symbols r_1 and r_2 represent spring relaxed positions in the gravity field.

Solution

The equation of motion for mass m_1 is

$$m_1\ddot{x}_1 + b_1\dot{x}_1 + (k_1 + k_2)x_1 - k_2x_2 = F_1(t)$$

The equation of motion for mass m_2 is

$$m_2\ddot{x}_2 + k_2x_2 - k_2x_1 = 0$$

Combining the equations for the two masses and eliminating x_1 yields the input-output equation for the system.

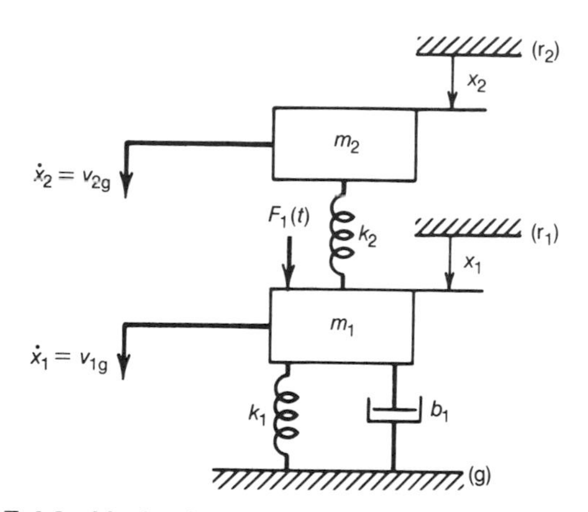

FIGURE 4.2 Mechanical system considered in Example 4.1.

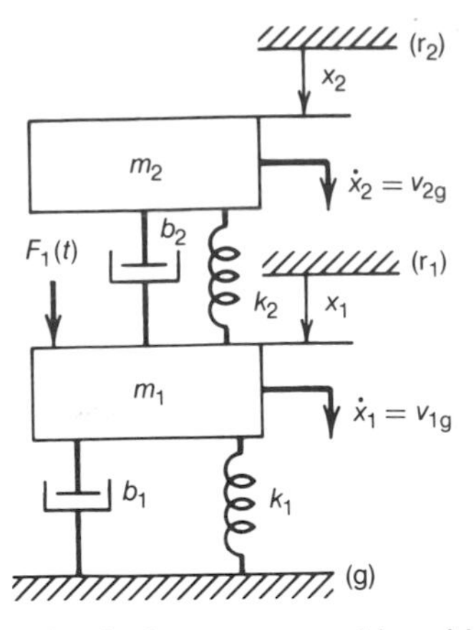

FIGURE 4.3 Mechanical system considered in Example 4.2.

$$\frac{d^4x_2}{dt^4} + \left(\frac{b_1}{m_1}\right)\frac{d^3x_2}{dt^3} + \left(\frac{k_2}{m_2} + \frac{k_1}{m_1} + \frac{k_2}{m_1}\right)\frac{d^2x_2}{dt^2} + \left(\frac{b_1k_2}{m_1m_2}\right)\frac{dx_2}{dt} + \left(\frac{k_1k_2}{m_1m_2}\right)x_2 = \left(\frac{k_2}{m_1m_2}\right)F_1(t)$$

The preceding fourth-order differential equation can be solved provided the initial conditions, $x_2(0)$, $(dx_2/dt)|_{t=0}$, $(d^2x_2/dt^2)|_{t=0}$, $(d^3x_2/dt^3)|_{t=0}$, and the input variable, $F_1(t)$ for $t \geq 0$, are known.

Similarly, the input-output equation relating x_1 to $F_1(t)$ is

$$\frac{d^4x_1}{dt^4} + \left(\frac{b_1}{m_1}\right)\frac{d^3x_1}{dt^3} + \left(\frac{k_2}{m_2} + \frac{k_1}{m_1} + \frac{k_2}{m_1}\right)\frac{d^2x_1}{dt^2} + \left(\frac{b_1k_2}{m_1m_2}\right)\frac{dx_1}{dt} + \left(\frac{k_1k_2}{m_1m_2}\right)x_1 = \left(\frac{1}{m_1}\right)\frac{d^2F_1}{dt^2} + \left(\frac{k_2}{m_1m_2}\right)F_1(t)$$

■

The process of deriving the input-output equation in Example 4.1 will become considerably more complicated if an additional damper b_2 is included between the two masses, as shown in Figure 4.3.

EXAMPLE 4.2

Derive an input-output equation for the mechanical system shown in Figure 4.3, using x_2 as the output variable.

Solution

The equations of motion for masses m_1 and m_2 now take the form

$$m_1\frac{d^2x_1}{dt^2} + (b_1 + b_2)\frac{dx_1}{dt} + (k_1 + k_2)x_1 - b_2\frac{dx_2}{dt} - k_2x_2 = F_1(t)$$

$$m_2 \frac{d^2x_2}{dt^2} + b_2 \frac{dx_2}{dt} + k_2x_2 - b_2 \frac{dx_1}{dt} - k_2x_1 = 0$$

The unwanted variable x_1 cannot be eliminated from these equations using simple substitutions as in Example 4.1 because the derivatives of both x_1 and x_2 are present in each equation. In such case an operator D can be introduced, defined as

$$D^k x(t) = \frac{d^k x(t)}{dt^k}$$

Using the D operator, the differential equations of motion can be rearranged into the following form

$$m_1D^2x_1 + (b_1 + b_2)Dx_1 + (k_1 + k_2)x_1 - b_2Dx_2 - k_2x_2 = F_1(t)$$

$$m_2D^2x_2 + b_2Dx_2 + k_2x_2 - b_2Dx_1 - k_2x_1 = 0$$

From the last equation, x_1 can be expressed as

$$x_1 = \left(\frac{m_2D^2 + b_2D + k_2}{b_2D + k_2}\right) x_2$$

Substituting into the operator equation for mass m_1 yields

$$m_1m_2D^4x_2 + (m_2b_1 + m_2b_2 + m_1b_2)D^3x_2 + (m_1k_2 + m_2k_1 + m_2k_2 + b_1b_2)D^2x_2$$
$$+ (b_1k_2 + b_2k_1)Dx_2 + k_1k_2x_2 = b_2DF_1 + k_2F_1$$

Using the inverse of the definition of the D operator to transform this equation back to the time domain gives the input-output equation for the system.

$$(m_1m_2) \frac{d^4x_2}{dt^4} + (m_2b_1 + m_2b_2 + m_1b_2) \frac{d^3x_2}{dt^3}$$
$$+ (m_1k_2 + m_2k_1 + m_2k_2 + b_1b_2) \frac{d^2x_2}{dt^2}$$
$$+ (b_1k_2 + b_2k_1) \left(\frac{dx_2}{dt}\right) + k_1k_2x_2 = b_2 \frac{dF_1}{dt} + k_2F_1$$

■

Example 4.2 demonstrates that by inserting a damper between the two masses the process of deriving the input-output equation becomes considerably more complicated. In the next example a two-input, two-output system will be considered. Two separate input-output equations, one for each output variable, will have to be derived.

EXAMPLE 4.3

Consider again the mechanical system shown in Figure 4.2. The system is now subjected to two external forces, $F_1(t)$ and $F_2(t)$. The displacements of both

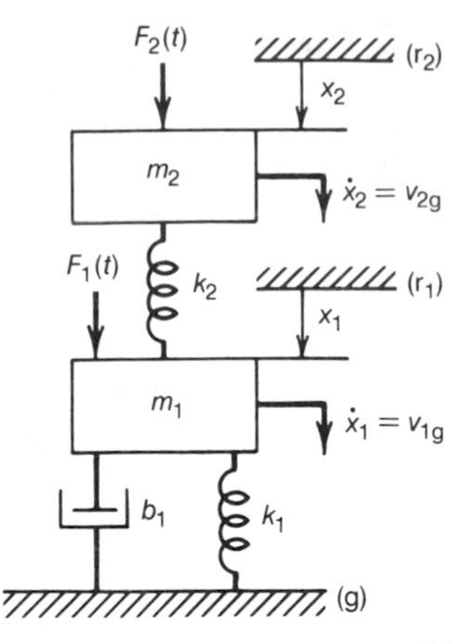

FIGURE 4.4 Two-input two-output system considered in Example 4.3.

masses are of interest and therefore $x_1(t)$ and $x_2(t)$ will be the two output variables of this system. The system is shown in Figure 4.4. The pair of differential equations of motion for the two masses are

$$m_1 \frac{d^2x_1}{dt^2} + b_1 \frac{dx_1}{dt} + (k_1 + k_2)x_1 - k_2x_2 = F_1$$

$$m_2 \frac{d^2x_2}{dt^2} + k_2x_2 - k_2x_1 = F_2$$

By combining the above two equations the separate input-output equations for $x_1(t)$ and $x_2(t)$ are obtained

$$m_1m_2 \frac{d^4x_1}{dt^4} + m_2b_1 \frac{d^3x_1}{dt^3} + (m_1k_2 + m_2k_1 + m_2k_2) \frac{d^2x_1}{dt^2} + b_1k_2 \frac{dx_1}{dt} + k_1k_2x_1 = m_2 \frac{d^2F_1}{dt^2} + k_2F_1 + k_2F_2$$

and

$$m_1m_2 \frac{d^4x_2}{dt^4} + m_2b_1 \frac{d^3x_2}{dt^3} + (m_1k_2 + m_2k_1 + m_2k_2) \frac{d^2x_2}{dt^2} + b_1k_2 \frac{dx_2}{dt} + k_1k_2x_2 = k_2F_1 + m_1 \frac{d^2F_2}{dt^2} + b_1 \frac{dF_2}{dt} + (k_1 + k_2)F_2$$

Note that the two input-output equations are independent of each other and can be solved separately. *On the other hand, the coefficients of each of the terms on the left-hand sides are the same, regardless of which system variable is chosen as the output.*

■

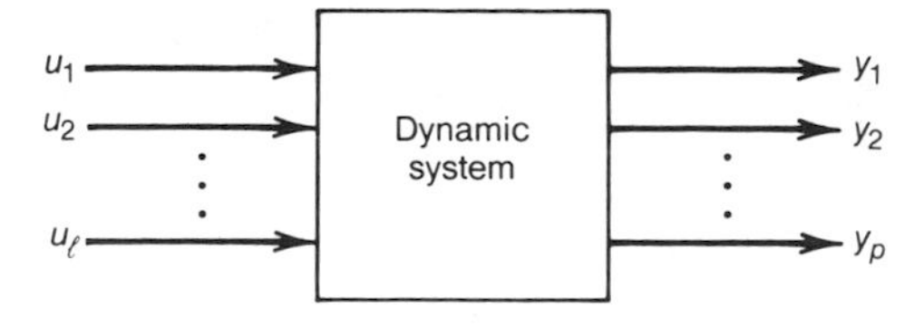

FIGURE 4.5 Multi-input multi-output system.

In general, a linear system with l inputs and p outputs shown schematically in Figure 4.5 is described by p independent input-output equations given below.

$$
\begin{aligned}
a_{1n}y_1^{(n)} + \cdots + a_{11}\dot{y}_1 + a_{10}y_1 &= f_1(u_1^{(m)}, \ldots, \dot{u}_1, u_1, u_2^{(m)}, \ldots, \dot{u}_2, \\
&\qquad u_2, \ldots, u_l^{(m)}, \ldots, \dot{u}_l, u_l, t) \\
a_{2n}y_2^{(n)} + \cdots + a_{21}\dot{y}_2 + a_{20}y_2 &= f_2(u_1^{(m)}, \ldots, \dot{u}_1, u_1, u_2^{(m)}, \ldots, \dot{u}_2, \\
&\qquad u_2, \ldots, u_l^{(m)}, \ldots, \dot{u}_l, u_l, t) \\
&\ \vdots \\
a_{pn}y_p^{(n)} + \cdots + a_{p1}\dot{y}_p + a_{p0}y_p &= f_p(u_1^{(m)}, \ldots, \dot{u}_1, u_1, u_2^{(m)}, \ldots, \dot{u}_2, \\
&\qquad u_2, \ldots, u_l^{(m)}, \ldots, \dot{u}_l, u_l, t)
\end{aligned}
\tag{4.4}
$$

where $m \leq n$ and a superscript enclosed in parentheses denotes the order of a derivative. If a system is assumed to be stationary, time t does not appear explicitly on the right hand side of Equation (4.4).

The functions $f_1, f_2, \ldots, f_p$ arc linear combinations of terms involving the system inputs and their derivatives. The input-output model for a linear, stationary, multi-input multi-output system can be presented in a more compact form

$$
\begin{aligned}
\sum_{i=0}^{n} a_{1i}y_1^{(i)} &= \sum_{j=1}^{l}\sum_{k=0}^{m} b_{1jk}u_j^{(m)} \\
\sum_{i=0}^{n} a_{2i}y_2^{(i)} &= \sum_{j=1}^{l}\sum_{k=0}^{m} b_{2jk}u_j^{(m)} \\
&\ \vdots \\
\sum_{i=0}^{n} a_{pi}y_p^{(i)} &= \sum_{j=1}^{l}\sum_{k=0}^{m} b_{pjk}u_j^{(m)}
\end{aligned}
\tag{4.5}
$$

Note that some of the a and b coefficients may be equal to zero; also that $a_{1i} = a_{2i} = \cdots a_{pi}$, $i = 1, 2, \ldots, n$.

The input-output equations, even for relatively simple multi-input multi-output models become extremely complicated. Moreover, as will be shown in Chapter 5, analytical methods for solving input-output equations are practical only for low-order, single-input single-output models. In fact, most numerical methods for solving high-order differential equations require that those equations be replaced by an equivalent set of first-order equations. Also, many quite powerful concepts and methodologies based on input-output models, such as the transfer function, are applicable to linear, stationary models only. The conceptual simplicity of using the input-output representation of a dynamic system is lost in the complexity of the mathematical forms with models that are nonlinear, have many inputs and/or outputs, or simply are of higher order than three. Moreover, it is not even possible to obtain an input-output differential equation for most nonlinear systems since the presence of nonlinearities inhibits combination of model equations to eliminate unwanted variables.

4.3 State Models

A concept of a state of a dynamic system was promoted in the 1950s. Its significance has grown since then, and today the state approach to modeling is the most powerful and dominant technique used in analysis of engineering systems.

Several basic terms associated with state models will first be defined.

- *State* of a dynamic system is defined by the smallest set of variables such that the knowledge of these variables at time $t = t_0$ together with the knowledge of the input for $t > t_0$ completely determines the behavior of the system for time $t \geq t_0$.
- *State variables*, $q_1, q_2, \ldots, q_n$, are the elements of the smallest set of variables required to completely describe the state of the system. One important implication of this definition is that state variables are independent of each other. If it were possible to express any of the state variables in terms of others, those variables would not be necessary to uniquely describe the system dynamics, and such a set of variables would not constitute the smallest set of variables as required by the definition of the state of a dynamic system.
- *State vector* of a dynamic system is the column vector $\mathbf{q}$ whose components are the state variables, $q_1, q_2, \ldots, q_n$.
- *State space* is an n-dimensional space containing the n system state variables. The state of a dynamic system at any instant of time t is represented by a single point in the state space.
- *State trajectory* is the path over time of the point representing the state of the system in a state space. Figure 4.6 shows a state trajectory for a rock thrown vertically with the initial velocity v_0 at the location x_0, reaching the maximum height, x_f, with velocity v_f equal to zero and falling down to the initial position x_0 with the final velocity equal to $-v_0$ if losses of energy due to air resistance are negligible. The position of the rock and its velocity are the state variables and the plane determined by the coordinate system (x, v) is the two-dimensional state space.

Mathematically, state models take the form of sets of first-order differential equations as follows

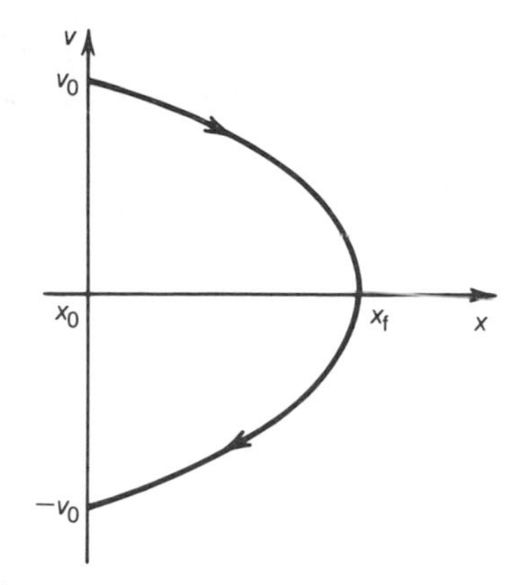

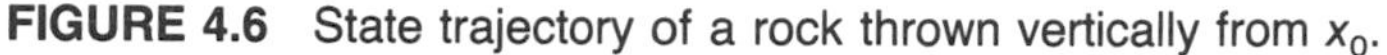

FIGURE 4.6 State trajectory of a rock thrown vertically from x_0.

$$
\begin{aligned}
\dot{q}_1 &= f_1(q_1, q_2, \ldots, q_n, u_1, u_2, \ldots, u_l, t) \\
\dot{q}_2 &= f_2(q_1, q_2, \ldots, q_n, u_1, u_2, \ldots, u_l, t) \\
&\;\;\vdots \\
\dot{q}_n &= f_n(q_1, q_2, \ldots, q_n, u_1, u_2, \ldots, u_l, t)
\end{aligned} \tag{4.6}
$$

where $q_1, q_2, \ldots, q_n$ are the system state variables and $u_1, u_2, \ldots, u_l$ are the input variables. If the model is nonlinear, at least one of the functions f_i, $i = 1, 2, \ldots, n$, is nonlinear.

Although the state vector completely represents the system dynamics, the selected state variables are not necessarily the same as the system outputs. However, each output variable or, in general any system variable that is of interest, can be expressed mathematically in terms of the system state variables and the input variables. A block diagram of a state representation of a multi-input multi-output dynamic system is shown in Figure 4.7.

In general, the system output equations can be written in the following form.

$$
\begin{aligned}
y_1 &= g_1(q_1, q_2, \ldots, q_n, u_1, u_2, \ldots, u_l, t) \\
y_2 &= g_2(q_1, q_2, \ldots, q_n, u_1, u_2, \ldots, u_l, t) \\
&\;\;\vdots \\
y_p &= g_p(q_1, q_2, \ldots, q_n, u_1, u_2, \ldots, u_l, t)
\end{aligned} \tag{4.7}
$$

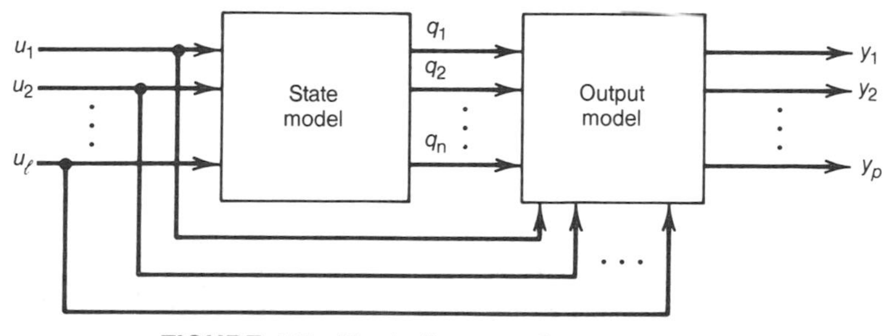

FIGURE 4.7 Block diagram of a state model.

If a system model is linear, all functions on the right-hand side of the state equations (4.6), f_i for $i = 1, 2, \ldots, n$, and all functions on the right-hand side of the output equations (4.7), g_j for $j = 1, 2, \ldots, p$ are linear. Also, in stationary model equations system parameters do not vary with time so that linear, stationary-state model equations take the form

$$\begin{aligned}
\dot{q}_1 &= a_{11}q_1 + a_{12}q_2 + \cdots + a_{1n}q_n + b_{11}u_1 + b_{12}u_2 + \cdots + b_{1l}u_l \\
\dot{q}_2 &= a_{21}q_1 + a_{22}q_2 + \cdots + a_{2n}q_n + b_{21}u_1 + b_{22}u_2 + \cdots + b_{2l}u_l \\
&\vdots \\
\dot{q}_n &= a_{n1}q_1 + a_{n2}q_2 + \cdots + a_{nn}q_n + b_{n1}u_1 + b_{n2}u_2 + \cdots + b_{nl}u_l
\end{aligned} \tag{4.8}$$

The linear output equations are

$$\begin{aligned}
y_1 &= c_{11}q_1 + c_{12}q_2 + \cdots + c_{1n}q_n + d_{11}u_1 + d_{12}u_2 + \cdots + d_{1l}u_l \\
y_2 &= c_{21}q_2 + c_{22}q_2 + \cdots + c_{2n}q_n + d_{21}u_1 + d_{22}u_2 + \cdots + d_{2l}u_l \\
&\vdots \\
y_p &= c_{p1}q_1 + c_{p2}q_2 + \cdots + c_{pn}q_n + d_{p1}u_1 + d_{p2}u_2 + \cdots + d_{pl}u_l
\end{aligned} \tag{4.9}$$

Equations (4.8) and (4.9) can be written using matrix-vector notation

$$\begin{bmatrix} \dot{q}_1 \\ \dot{q}_2 \\ \vdots \\ \dot{q}_n \end{bmatrix} = \begin{bmatrix} a_{11} & a_{12} & \cdots & a_{1n} \\ a_{21} & a_{22} & \cdots & a_{2n} \\ \vdots & & & \\ a_{n1} & a_{n2} & \cdots & a_{nn} \end{bmatrix} \begin{bmatrix} q_1 \\ q_2 \\ \vdots \\ q_n \end{bmatrix} + \begin{bmatrix} b_{11} & b_{12} & \cdots & b_{1l} \\ b_{21} & b_{22} & \cdots & b_{2l} \\ \vdots & & & \\ b_{n1} & b_{n2} & \cdots & b_{nl} \end{bmatrix} \begin{bmatrix} u_1 \\ u_2 \\ \vdots \\ u_l \end{bmatrix} \tag{4.10}$$

and

$$\begin{bmatrix} y_1 \\ y_2 \\ \vdots \\ y_p \end{bmatrix} = \begin{bmatrix} c_{11} & c_{12} & \cdots & c_{1n} \\ c_{21} & c_{22} & \cdots & c_{2n} \\ \vdots & & & \\ c_{p1} & c_{p2} & \cdots & c_{pn} \end{bmatrix} \begin{bmatrix} q_1 \\ q_2 \\ \vdots \\ q_n \end{bmatrix} + \begin{bmatrix} d_{11} & d_{12} & \cdots & d_{1l} \\ d_{21} & d_{22} & \cdots & d_{2l} \\ \vdots & & & \\ d_{p1} & d_{p2} & \cdots & d_{pl} \end{bmatrix} \begin{bmatrix} u_1 \\ u_2 \\ \vdots \\ u_l \end{bmatrix} \tag{4.11}$$

Rewriting these equations in a more compact form

$$\dot{\mathbf{q}} = \mathbb{A}\mathbf{q} + \mathbb{B}\mathbf{u} \tag{4.10$'$}$$

$$\mathbf{y} = \mathbb{C}\mathbf{q} + \mathbb{D}\mathbf{u} \tag{4.11$'$}$$

where $\mathbb{A}$ is a $n \times n$ state matrix, $\mathbb{B}$ is a $n \times l$ input matrix, $\mathbb{C}$ is a $p \times n$ output matrix, $\mathbb{D}$ is a $p \times l$ direct transmission matrix, $\mathbf{q}$ is a state vector, $\mathbf{u}$ is an input vector, and $\mathbf{y}$ is an output vector.

As mentioned before, the selection of state variables constitutes a nontrivial problem since for each system there are usually many different sets of variables that uniquely represent the system dynamics. The following sets of variables are most commonly used as state variables:

(1) sets of T-type and A-type variables associated with T and A energy-storing elements of the system
(2) sets including one variable and its successive derivatives
(3) sets including two or more variables and their derivatives
(4) sets including an auxiliary variable and its successive derivatives

In addition, there are still other, relatively less common types of state variables, such as the variables associated with the roots of the system characteristic equation obtained by means of manipulation of the system matrix or sets of variables obtained as nonredundant algebraic combinations of other state variables.[1]

In general, a state of a dynamic system evolves from the process of storing energy in those system components which are capable of it. In fact, the number of state variables is always equal to the number of independent energy-storing elements in the system, regardless of the type of the state-variables employed.

In Example 4.4 the first three most common types of the state models listed above will be derived. The use of an auxiliary variable and its derivatives as the state variables will be demonstrated in Section 4.4.

EXAMPLE 4.4

Derive state models of the types (1), (2), and (3) for the mechanical system shown in Figure 4.2.

Type (1): T-type and A-type variables

Solution

There are four independent energy-storing elements in the system shown in Figure 4.2: masses m_1 and m_2 and springs k_1 and k_2. Masses in mechanical systems are A-type elements, which can store kinetic energy, whereas springs are T-type elements, capable of storing potential energy. The respective A-type and T-type variables are the velocities of the two masses, v_{1g} and v_{2g}, and the forces exerted by the springs, F_{k1} and F_{k2}. Hence, the four variables selected to represent the state of the system are $q_1 = F_{k1}$, $q_2 = v_{1g}$, $q_3 = F_{k2}$, and $q_4 = v_{2g}$. To derive the state equations, first consider the forces exerted by the springs F_{k1} and F_{k2}. The equations defining these forces are

$$F_{k1} = k_1 x_1$$

$$F_{k2} = k_2(x_2 - x_1)$$

Differentiating both sides of the above equations with respect to time gives the first two state-variable equations

$$\dot{F}_{k1} = (k_1)v_{1g}$$

$$\dot{F}_{k2} = (-k_2)v_{1g} + (k_2)v_{2g}$$

[1]Y. Takahashi, M. J. Rabins, and D. M. Auslander, *Control and Dynamic Systems*, Addison-Wesley Publishing Company, Reading, Mass., 1972.

The equations of motion for masses m_1 and m_2 derived in Example 4.1 were

$$m_1 \underbrace{\frac{d^2x_1}{dt^2}}_{dv_{1g}\, dt} + b_1 \underbrace{\frac{dx_1}{dt}}_{v_{1g}} + \underbrace{k_1x_1}_{F_{k1}} - \underbrace{k_2(x_2 - x_1)}_{F_{k2}} = F_1(t)$$

and

$$m_2 \underbrace{\frac{d^2x_2}{dt^2}}_{dv_{2g}\, dt} + \underbrace{k_2(x_2 - x_1)}_{F_{k2}} = 0$$

which give the other two state variable equations

$$\dot{v}_{1g} = -\frac{1}{m_1} F_{k1} - \frac{b_1}{m_1} v_{1g} + \frac{1}{m_1} F_{k2} + \frac{1}{m_1} F_1(t)$$

$$\dot{v}_{2g} = -\frac{1}{m_2} F_{k2}$$

Rewriting the state variable equations in a vector-matrix form yields

$$\begin{bmatrix} \dot{F}_{k1} \\ \dot{v}_{1g} \\ \dot{F}_{k2} \\ \dot{v}_{2g} \end{bmatrix} = \begin{bmatrix} 0 & k_1 & 0 & 0 \\ -1/m_1 & -b_1/m_1 & 1/m_1 & 0 \\ 0 & -k_2 & 0 & k_2 \\ 0 & 0 & -1/m_2 & 0 \end{bmatrix} \begin{bmatrix} F_{k1} \\ v_{1g} \\ F_{k2} \\ v_{2g} \end{bmatrix} + \begin{bmatrix} 0 \\ 1/m_1 \\ 0 \\ 0 \end{bmatrix} F_1(t)$$

The output equations for the displacements x_1 and x_2 are

$$x_1 = \frac{1}{k_1} F_{k1}$$

$$x_2 = \frac{1}{k_1} F_{k1} + \frac{1}{k_2} F_{k2}$$

In a vector-matrix form the output model equations are

$$\begin{bmatrix} x_1 \\ x_2 \end{bmatrix} = \begin{bmatrix} 1/k_1 & 0 & 0 & 0 \\ 1/k_1 & 0 & 1/k_2 & 0 \end{bmatrix} \begin{bmatrix} F_{k1} \\ v_{1g} \\ F_{k2} \\ v_{2g} \end{bmatrix} + \begin{bmatrix} 0 \\ 0 \end{bmatrix} F_1(t)$$

Type (2): One variable and its successive derivatives This type of state variables is particularly convenient when an input-output equation is available. Very often in such cases the output variable and successive derivatives are selected as the state variables. In Example 4.1 the input-output equations for the system shown in Figure 4.2 were derived relating displacements x_1 and x_2 to the input force $F_1(t)$. Both equations are of fourth order, so four state variables are needed to uniquely represent the system dynamics. Let x_2 and its first three derivatives be selected as the state variables, i.e., $q_1 = x_2$, $q_2 = dx_2/dt$, $q_3 = d^2x_2/dt^2$, $q_4 = d^3x_2/dt^3$.

The input-output equation relating x_2 to the input force, $F_1(t)$ was

$$\frac{d^4x_2}{dt^4} + \left(\frac{b_1}{m_1}\right)\frac{d^3x_2}{dt^3} + \left(\frac{k_2}{m_2} + \frac{k_1}{m_1} + \frac{k_2}{m_1}\right)\frac{d^2x_2}{dt^2} + \left(\frac{b_1k_2}{m_1m_2}\right)\frac{dx_2}{dt} + \left(\frac{k_1k_2}{m_1m_2}\right)x_2 = \left(\frac{k_2}{m_1m_2}\right)F_1(t)$$

Using the preceding equation and the definitions of the selected state variables, the following state equations are formed

$$\dot{q}_1 = q_2$$

$$\dot{q}_2 = q_3$$

$$\dot{q}_3 = q_4$$

$$\dot{q}_4 = -\left(\frac{k_1k_2}{m_1m_2}\right)q_1 - \left(\frac{b_1k_2}{m_1m_2}\right)q_2 - \left(\frac{k_1}{m_1} + \frac{k_2}{m_2} + \frac{k_2}{m_1}\right)q_3 - \left(\frac{b_1}{m_1}\right)q_4 + \left(\frac{k_2}{m_1m_2}\right)F_1(t)$$

Recognizing that $x_1 = x_2 - (x_2 - x_1) = x_2 - F_{k2}/k_2 = x_2 + (m_2/k_2)(d^2x_2/dt^2)$, then if x_1 and x_2 are selected as the output variables, y_1, and y_2 respectively, the output equations become

$$y_1 = q_1 + \left(\frac{m_2}{k_2}\right)q_3 = x_1$$

$$y_2 = q_1 = x_2$$

Type (3): Two variables and their derivatives This type of state variables is used most often in modeling mechanical systems. For the system considered in this example, x_1 and x_2 and their derivatives, $\dot{x}_1$, $\dot{x}_2$, are selected as the four state variables. Noting that $\dot{x}_1 = v_{1g}$ and $\dot{x}_2 = v_{2g}$ and using the equations of motion for masses m_1, m_2, derived in Example 4.1, the following state equations are obtained

$$\dot{q}_1 = q_3$$

$$\dot{q}_2 = q_4$$

$$\dot{q}_3 = -\left(\frac{k_1}{m_1} + \frac{k_2}{m_1}\right) q_1 + \left(\frac{k_2}{m_1}\right) q_2 - \left(\frac{b_1}{m_1}\right) q_3 + \left(\frac{1}{m_1}\right) F_1(t)$$

$$\dot{q}_4 = \left(\frac{k_2}{m_2}\right) q_1 - \left(\frac{k_2}{m_2}\right) q_2$$

where $q_1 = x_1$, $q_2 = x_2$, $q_3 = v_{1g}$, and $q_4 = v_{2g}$.

The output equations are simply

$$y_1 = q_1$$
$$y_2 = q_2$$

■

4.4 Transition Between Input-Output and State Models

At the beginning of this chapter both input-output and state models were said to be equivalent in the sense that each form completely represents the dynamics of the same system. It is, therefore, natural to expect that there is a corresponding state model involving successive derivatives of one state variable for each input-output model and vice versa. The transition between the two forms of models is indeed possible, although it is not always straightforward.

Consider first a simple case of a single-input single-output system model with no derivatives of input variable present in the input-output equation

$$a_n \frac{d^n y}{dt^n} + \cdots + a_1 \frac{dy}{dt} + a_0 y = b_0 u \tag{4.12}$$

If the following state variables are selected [as in Type (2) Sect. 4.3]

$$q_1 = y, \quad q_2 = \frac{dy}{dt}, \quad \ldots, \quad q_n = \frac{d^{n-1}y}{dt^{n-1}} \tag{4.13}$$

the equivalent set of state model equations is

$$\begin{aligned} \dot{q}_1 &= q_2 \\ \dot{q}_2 &= q_3 \\ &\vdots \\ \dot{q}_{n-1} &= q_n \\ \dot{q}_n &= -\left(\frac{a_0}{a_n}\right) q_1 - \left(\frac{a_1}{a_n}\right) q_2 - \cdots - \left(\frac{a_{n-1}}{a_n}\right) q_n + \left(\frac{b_0}{a_n}\right) u \end{aligned} \tag{4.14}$$

The transition procedure between an input-output model and an equivalent state model becomes more complicated if derivatives of input variables are present on the right-hand side of the input-output equation. The input-output equation of a single-input single-output model in this case is

$$a_n \frac{d^n y}{dt^n} + \cdots + a_1 \frac{dy}{dt} + a_0 y = b_m \frac{d^m u}{dt^m} + \cdots + b_1 \frac{du}{dt} + b_0 u \tag{4.15}$$

where $m \leq n$ for physically realizable systems. An equivalent state model will in this case consist of a set of state variable equations based on a state variable and $(n - 1)$ of its derivatives together with the output equations, as illustrated in Figure 4.8.

In order to obtain an equivalent state model, first the higher derivatives of the input variable in Equation (4.15) are ignored to yield an auxiliary input-output differential equation

$$a_n \frac{d^n x}{dt^n} + \cdots + a_1 \frac{dx}{dt} + a_0 x = u \tag{4.16}$$

where x is an auxiliary state variable. The set of equivalent state equations for the auxiliary input-output Equation (4.16) can be obtained as before, using the auxiliary variable and its successive derivatives as the state variables [type (4) of state variables discussed in Section 4.3]

$$q_1 = x, \quad q_2 = \dot{x}, \quad \ldots, \quad q_n = \frac{d^{n-1}x}{dt^{n-1}} \tag{4.17}$$

The state equations become

$$\begin{aligned}
\dot{q}_1 &= q_2 \\
\dot{q}_2 &= q_3 \\
&\vdots \\
\dot{q}_{n-1} &= q_n \\
\dot{q}_n &= -\frac{a_0}{a_n} q_1 - \frac{a_1}{a_n} q_2 - \cdots - \frac{a_{n-1}}{a_n} q_n + \frac{1}{a_n} u
\end{aligned} \tag{4.18}$$

Applying the differentiation operator introduced in Section 4.2 to Equations (4.15) and (4.16) gives

$$y(a_n D^n + \cdots + a_1 D + a_0) = u(b_m D^m + \cdots + b_1 D + b_0) \tag{4.19}$$

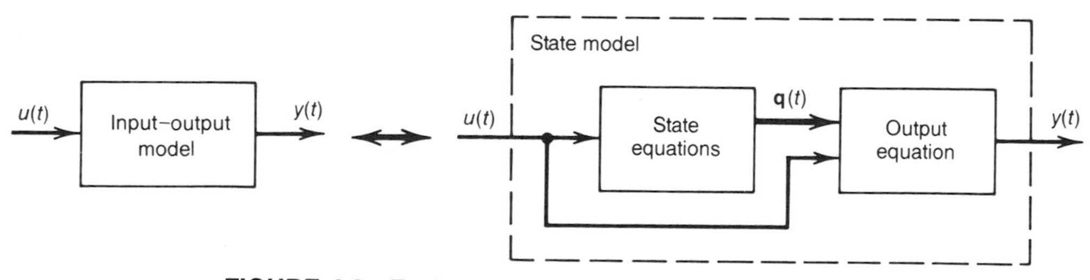

FIGURE 4.8 Equivalent input-output and state models.

and

$$x(a_nD^n + \cdots + a_1D + a_0) = u \tag{4.20}$$

Substitution of the expression on the left-hand side of Equation (4.20) for u in Equation (4.19) yields

$$y = x(b_mD^m + \cdots + b_1D + b_0) \tag{4.21}$$

Now use the definition of the D operator to transform Equation (4.21) back to the time domain.

$$y = b_m \frac{d^m x}{dt^m} + \cdots + b_1 \frac{dx}{dt} + b_0 x \tag{4.22}$$

The auxiliary variable x and its derivatives can be replaced by the state variables defined by (4.17) to produce the following output equation.

$$y = b_0q_1 + b_1q_2 + \cdots + b_mq_{m+1} \quad \text{for } m < n \tag{4.23}$$

Equation (4.23) holds for $m < n$. If both sides of the input-output equation are of the same order, $m = n$, then q_{m+1} in Equation (4.23) is replaced by the expression for $\dot{q}_n$ given by the last of Equations (4.18) and the output equation becomes

$$\begin{aligned} y = &\left(b_0 - \frac{b_m a_0}{a_n}\right) q_1 + \left(b_1 - \frac{b_m a_1}{a_n}\right) q_2 + \cdots \\ &+ \left(b_{m-1} - \frac{b_m a_{n-1}}{a_n}\right) q_n + \left(\frac{b_m}{a_n}\right) u \quad \text{for } m = n \end{aligned} \tag{4.24}$$

The input-output Equation (4.15) involving derivatives of the input variable is therefore equivalent to the state model, consisting of the auxiliary state variable Equations (4.18) and output Equations (4.23) or (4.24) for $m < n$ and $m = n$, respectively.

The procedure for transforming input-output equations into an equivalent state model is illustrated in Example 4.5.

EXAMPLE 4.5

Consider again the mechanical system shown in Figure 4.3. The input-output equation for this system, derived in Example 4.2, was

$$\begin{aligned} m_1m_2 \frac{d^4x_2}{dt^4} &+ (m_2b_1 + m_2b_2 + m_1b_2) \frac{d^3x_2}{dt^3} \\ &+ (m_1k_2 + m_2k_1 + m_2k_2 + b_1b_2) \frac{d^2x_2}{dt^2} \\ &+ (b_1k_2 + b_2k_1) \frac{dx_2}{dt} \\ &+ k_1k_2x_2 = b_2 \frac{dF_1}{dt} + k_2F_1 \end{aligned}$$

Derive an equivalent state model for this system.

Solution

First, the derivative of F_1 is ignored and a simplified input-output equation is obtained using the auxiliary output variable x.

$$m_1 m_2 \frac{d^4x}{dt^4} + (m_2 b_1 + m_2 b_2 + m_1 b_2) \frac{d^3x}{dt^3} + (m_1 k_2 + m_2 k_1 + m_2 k_2 + b_1 b_2) \frac{d^2x}{dt^2} + (b_1 k_2 + b_2 k_1) \frac{dx}{dt} + k_1 k_2 x = F_1$$

The state variables are

$$q_1 = x, \quad q_2 = \frac{dx}{dt}, \quad q_3 = \frac{d^2x}{dt^2}, \quad q_4 = \frac{d^3x}{dt^3}$$

and the state variable equations are

$$\dot{q}_1 = q_2$$

$$\dot{q}_2 = q_3$$

$$\dot{q}_3 = q_4$$

$$\dot{q}_4 = -\left(\frac{k_1 k_2}{m_1 m_2}\right) q_1 - \left[\frac{(b_1 k_2 + b_2 k_1)}{m_1 m_2}\right] q_2 - \left[\frac{(m_1 k_2 + m_2 k_1 + m_2 k_2 + b_1 b_2)}{m_1 m_2}\right] q_3 - \left[\frac{(m_2 b_1 + m_2 b_2 + m_1 b_2)}{m_1 m_2}\right] q_4 + \left(\frac{1}{m_1 m_2}\right) F_1$$

The output equation is

$$y_2 = k_2 q_1 + b_2 q_2$$

which completes the system mathematical model in a state form. ■

4.5 Nonlinearities in Input-Output and State Models

Very often in modeling dynamic systems, nonlinear characteristics of some of the system elements cannot be linearized either because the linearization error is not acceptable or because a particular nonlinearity may be essential for the system performance and must not be replaced with a linear approximation. The superiority of state models over input-output models in such cases is particularly pronounced. Derivation of input-output differential equations for systems in which nonlinearities are to be modeled without linearization is usually very cumbersome or even impossible, to say nothing about solving those equations. The derivation of state models, on the other hand, is barely affected by the presence of nonlinearities in the system. Furthermore, most modern computer programs for solving sets of state variable equations are capable of handling both linear and nonlinear models.

The effect of nonlinearities on the process of derivation of the two forms of mathematical models is illustrated in Example 4.6.

EXAMPLE 4.6

The mechanical system shown in Figure 4.9 is similar to the system considered earlier in Examples 4.1 and 4.4, except that the linear viscous friction element b_1 is here replaced with a nonlinear damper NLD.

The force generated by the damper, F_{NLD} is

$$F_{NLD} = f_{NL}(v_{1g}) = a|v_{1g}|v_{1g}$$

The square law expression embodied in the function f_{NL} is a fairly common type of nonlinearity, although the specific form of the nonlinear function is of no significance here.

Derive an input-output equation using x_2 as the output variable and a set of state equations for this system.

Solution

The equations of motion for masses m_1 and m_2 are

$$m_1 \frac{d^2x_1}{dt^2} + a \left|\frac{dx_1}{dt}\right| \frac{dx_1}{dt} + k_1x_1 - k_2(x_2 - x_1) = F_1(t)$$

and

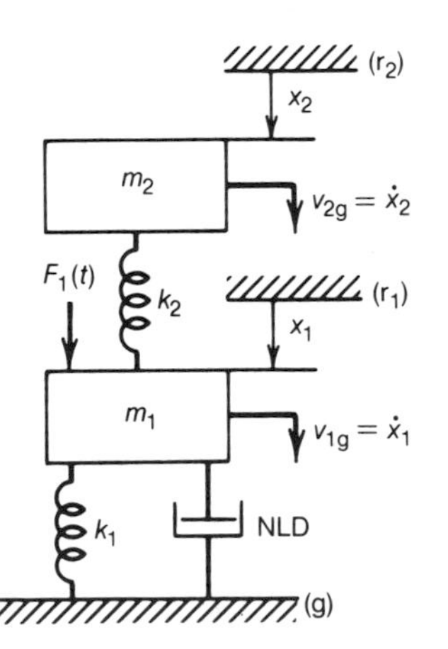

FIGURE 4.9 Mechanical system with nonlinear damper.

$$m_2 \frac{d^2x_2}{dt^2} + k_2(x_2 - x_1) = 0$$

From the equation for mass m_2

$$x_1 = \left(\frac{m_2}{k_2}\right) \frac{d^2x_2}{dt^2} + x_2$$

Taking derivatives of x_1 and substituting into the equation for mass m_1 gives the following input-output equation

$$\frac{d^4x_2}{dt^4} + \frac{k_2 a}{m_1 m_2} \left| \frac{m_2}{k_2} \frac{d^3x_2}{dt^3} + \frac{dx_2}{dt} \right| \left(\frac{m_2}{k_2} \frac{d^3x_2}{dt^3} + \frac{dx_2}{dt} \right) + \left(\frac{k_1 + k_2}{m_1}\right) \frac{d^2x_2}{dt^2} + \frac{k_1 k_2}{m_1 m_2} x_2 = F_1(t)$$

Only those readers not yet convinced about the superiority of state models over input-output models are encouraged to attempt to solve this equation.

To obtain an equivalent state model, select displacements x_1 and x_2 and their derivatives v_{1g} and v_{2g} as the state variables. The state variable equations are

$$\dot{q}_1 = q_2$$

$$\dot{q}_2 = -\left[\frac{(k_1 + k_2)}{m_1}\right] q_1 - \left(\frac{a}{m_1}\right) |q_2| q_2 + \left(\frac{k_2}{m_1}\right) q_3 + \left(\frac{1}{m_1}\right) F_1(t)$$

$$\dot{q}_3 = q_4$$

$$\dot{q}_4 = \left(\frac{k_2}{m_2}\right) q_1 - \left(\frac{k_2}{m_2}\right) q_3$$

where $q_1 = x_1$, $q_2 = v_{1g}$, $q_3 = x_2$, and $q_4 = v_{2g}$.

The preceding state variable equations are simpler and can be solved numerically using one of the methods described in Chapter 6. ■

The examples presented in this chapter illustrate the effectiveness of the state approach to system modeling. The state equations are much simpler, and the entire process of derivation is less vulnerable to so-called "stupid mistakes," which may often be made when the input-output model is developed. Although, as the next chapter will show, there are certain advantages associated with using input-output models in analysis of low-order linear systems, the state-variable approach is in general superior and will be used as much as possible in modeling systems throughout the rest of this book.

4.6 Synopsis

Two types of mathematical models of dynamic systems were presented, input-output models and state models. Although both models are essentially equivalent with regard to the information about the system behavior incorporated in the model equations, the techniques employed in their derivation and solution are principally different. In most cases the state models, consisting of sets of first order differential equations, are much easier to derive and to solve by computer simulation than the input-output equations. This is especially true when the mathematical models involve nonlinearities or when systems modeled have many inputs and outputs and are to be simulated on the computer.

The state models are based on the concept of state variables. The choice of state variables for a given system is not unique. Four different types of state variables were used in example problems presented in this chapter. It was shown that the number of state variables necessary to describe the system dynamics is always equal to the number of independent energy-storing elements regardless of the type of state variables used.

Problems

4.1. Derive a complete set of state model equations for the mechanical rotational system shown in Figure P4.1. Select the following state variables:
(a) T-type and A-type variables
(b) one variable and its derivative
Use torque $T_i(t)$ as the input variable and angular displacement $\Theta_{1g}(t)$ as the output variable in each model.

4.2. Derive an input-output model equation for the system shown in Figure P4.1, using torque $T_i(t)$ as the input variable and angular displacement $\Theta_1(t)$ as the output variable.

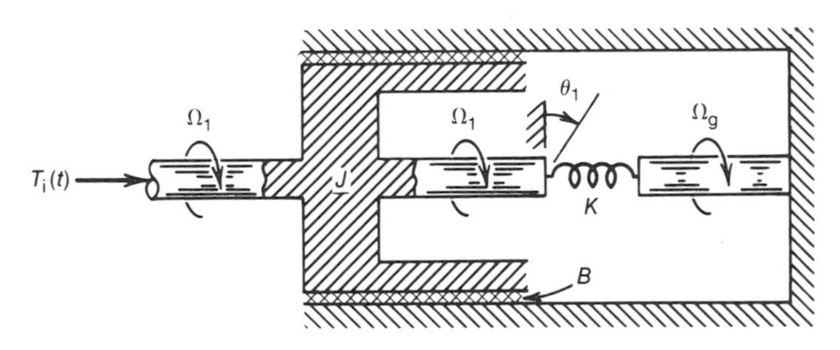

FIGURE P4.1 Mechanical system considered in Problem 4.1.

4.3. Derive complete sets of state-model equations for the system shown in Figure P4.3 using three different sets of state variables. The input variable is torque $T_m(t)$ and the output variables are displacements $x_1(t)$ and $x_2(t)$. Present the state models in a matrix form.

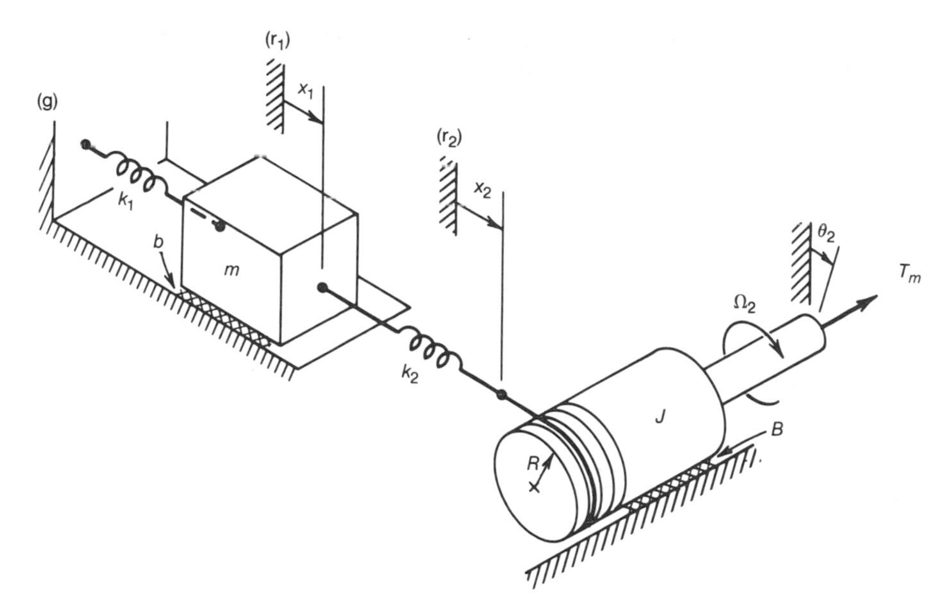

FIGURE P4.3 Mechanical system considered in Problem 4.3.

4.4. A nonlinear dynamic system has been modeled by the following state-variable equations:

$$\dot{q}_1 = -4q_1 + 10q_2 + 4u_1$$
$$\dot{q}_2 = 0.5q_1 - 2|q_2|q_2$$

where q_2 is also the output variable and u_1 is the input variable.
(a) Linearize the state-variable equations.
(b) Find the input-output equation for the linearized model.

4.5. The state-model matrices of a single-input single-output linear dynamic system are

$$\mathbb{A} = \begin{bmatrix} -3 & -19 \\ 1 & -2 \end{bmatrix} \quad \mathbb{B} = \begin{bmatrix} 0 \\ -1 \end{bmatrix}$$

$$\mathbb{C} = \begin{bmatrix} 0 \\ 2 \end{bmatrix} \qquad \mathbb{D} = 0$$

The column vector $\mathbf{q}$ of the system state variables contains q_1, q_2. Find the input-output model for this system.

4.6. A linear dynamic system is described by the following differential equations:

$$\ddot{y} + 4\dot{y} + 4y = 2\dot{x} + 2x$$
$$2\dot{x} + x - y = u(t)$$

(a) Derive the input-output model for this system using $y(t)$ as the output and $u(t)$ as the input variable.

(b) Derive a state model and present it in a matrix form.

4.7. Derive a set of state variable equations for the mechanical system shown in Figure P4.7.

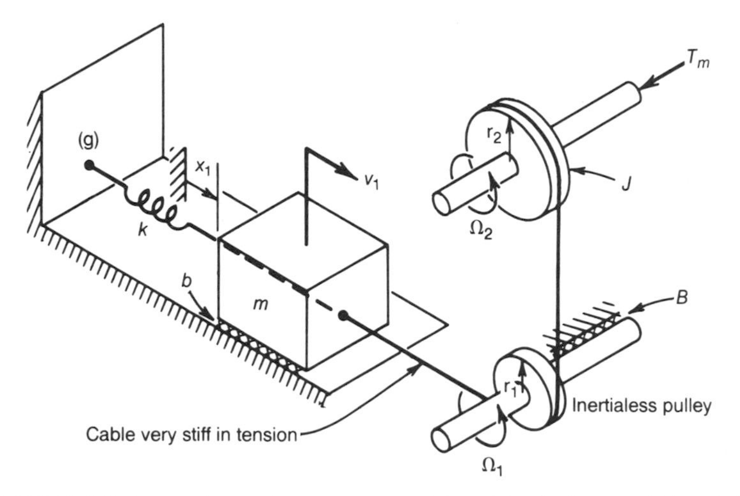

FIGURE P4.7 Mechanical systems considered in Problems 4.7 and 4.8.

4.8. Derive the input-output model equation for the system shown in Figure P4.7 using torque $T_m(t)$ as the input variable and angular velocity Ω_{2g} as the output variable.

4.9. A lumped model of a machine-tool drive system is shown in Figure P4.9. The system parameters are $J_m = 0.2$ N-m-sec^2/rad, $B = 30$ N-m-sec/rad, $m = 16$ kg, $k = 20$ N/m, and $R = 0.5$ m. The stiffness of the shaft is represented by a nonlinear torque, $T_{\mathrm{NLK}} = 2|\Theta_1 - \Theta_2|(\Theta_1 - \Theta_2)$. The force of the nonlinear friction device NLD is $F_{\mathrm{NLD}} = f_{\mathrm{NL}}(v_{1g}) = 2v_{1g}^3 + 4v_{1g}$. The system is driven by a torque consisting of a constant and an incremental component, $T_i(t) = 0.8 + 0.02 \sin (0.1t)$ N-m.

(a) Select state variables and derive nonlinear state-variable equations.

(b) Find the normal operating point values for all state variables.

(c) Linearize the state-model equations in the vicinity of the normal operating point. Present the linearized state model in a matrix form.

(d) Derive the input-output model equation for the linearized system using the incremental torque, $\hat{T}_i(t)$, as the input variable and incremental displacement, $\hat{x}_1(t)$, as the output variable.

(e) Draw a simulation block diagram for this system.

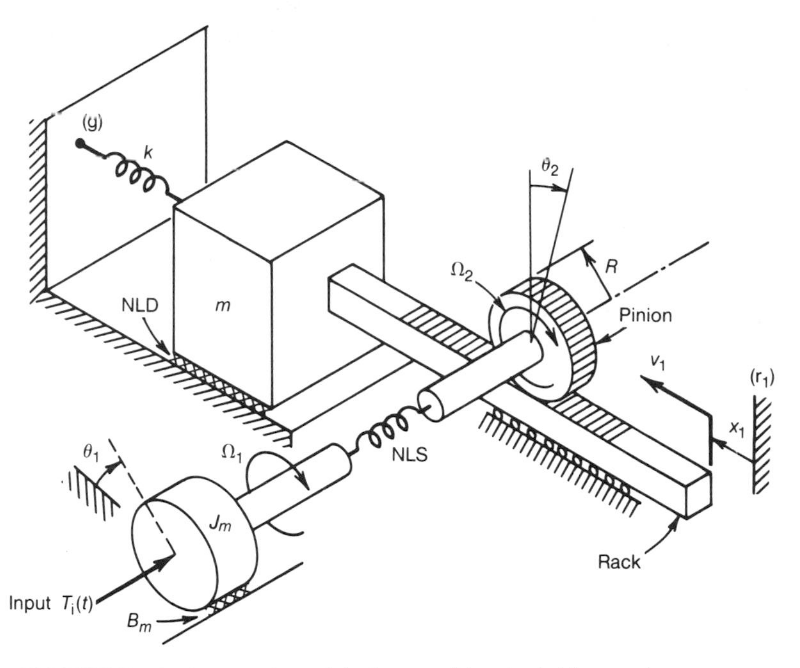

FIGURE P4.9 Lumped model of a machine-tool drive system.

4.10. Derive state-variable equations for the system shown in Figure 2.15. Use velocities v_{1g} and v_{2g} and the spring force F_{k1} as the state variables. Note that the two springs in this system are not independent as long as the lever is massless. Combine the state-variable equations to obtain an input-output equation relating velocity v_{2g} to force $F_s(t)$.

5

Analytical Solutions of System Input-Output Equations

5.1. Introduction

In Chapter 4 the state representation of system dynamics was introduced and the derivation of state equations was a relatively simple and straightforward process. Moreover, the state models take the form of sets of first-order differential equations that can be readily solved using one of many available computer programs. Having all those unquestionable advantages of state-variable models in mind, one might wonder whether devoting an entire chapter to the methods for solving the old-fashioned input-output model equations is justified. Despite all its limitations, the classical input-output approach still plays an important role in analysis of dynamic systems because many of the systems to be analyzed are neither very complex nor nonlinear. Such systems can be adequately described by low-order linear differential equations. Also, even in those cases where a low-order linear model is too crude to produce an accurate solution and a computer-based method is necessary, an analytical solution of an approximate linearized input-output equation can be used to verify the computer solution.

Section 5.2 gives a brief review of methods for solving linear differential equations. Initial conditions constitute an important but often overlooked aspect of modeling dynamic system. Section 5.3 presents a method for determining the initial conditions for input-output models. The next three sections deal with application of analytical methods to systems represented by first-, second-, and higher-order linear differential equations.

5.2. Analytical Solutions of Linear Differential Equations

An input-output equation for a linear, stationary, single-input single-output model has the following general form.

$$a_n \frac{d^n y}{dt^n} + a_{n-1} \frac{d^{n-1} y}{dt^{n-1}} + \cdots + a_1 \frac{dy}{dt} + a_0 y = u(t) \tag{5.1}$$

where the model parameters, $a_0, a_1, \ldots, a_n$, are constant. Equation (5.1) can be solved for a given input signal $u(t)$ for $t \geq 0$ if the initial conditions for y and its $n - 1$ derivatives are known.

$$y(0^+) = y_0, \quad \left.\frac{dy}{dt}\right|_{t=0^+} = \dot{y}_0, \quad \ldots, \quad \left.\frac{d^{n-1}y}{dt^{n-1}}\right|_{t=0^+} = y_0^{(n-1)} \tag{5.2}$$

A complete solution, $y(t)$, of Equation (5.1), consists of two parts.

$$y(t) = y_h(t) + y_p(t) \tag{5.3}$$

where $y_h(t)$ represents a homogenous (characteristic) solution and $y_p(t)$ is a particular (forced) solution. The homogenous solution gives the model response for the input signal equal to zero, $u(t) = 0$, and thus satisfies Equation (5.4).

$$a_n \frac{d^n y_h}{dt^n} + a_{n-1} \frac{d^{n-1} y_h}{dt^{n-1}} + \cdots + a_1 \frac{dy_h}{dt} + a_0 y_h = 0 \tag{5.4}$$

The particular solution, $y_p(t)$, satisfies the nonhomogenous equation

$$a_n \frac{d^n y_p}{dt^n} + a_{n-1} \frac{d^{n-1} y_p}{dt^{n-1}} + \cdots + a_1 \frac{dy_p}{dt} + a_0 y_p = u(t) \tag{5.5}$$

Note that by adding up Equations (5.4) and (5.5) the original model Equation (5.1) is obtained.

$$a_n \frac{d^n(y_h + y_p)}{dt^n} + a_{n-1} \frac{d^{n-1}(y_h + y_p)}{dt^{n-1}} + \cdots + a_1 \frac{d(y_h + y_p)}{dt} + a_0(y_h + y_p) = u(t) \tag{5.6}$$

Both the homogenous and the particular solution components have to be found. First, consider the homogenous solution. A characteristic equation is obtained from Equation (5.4) by putting $d^k y_h/dt^k = p^k$ for $k = 0, 1, \ldots, n$, which yields

$$a_n p^n + a_{n-1} p^{n-1} + \cdots + a_1 p + a_0 = 0 \tag{5.7}$$

The nth order algebraic Equation (5.7) possesses n roots, some of which may be identical. Assume that there are m distinct roots ($0 \leq m \leq n$) and therefore the number of multiple roots is $n - m$. To simplify further derivations all multiple roots of the characteristic Equation (5.7) are left to the end and are denoted by p_n where p_n is the value of the identical roots,

$$p_{m+1} = p_{m+2} = \cdots = p_n \tag{5.8}$$

The general form of the homogenous solution is

$$y_h(t) = K_1e^{p_1t} + K_2e^{p_2t} + \cdots + K_me^{p_mt} + K_{m+1}e^{p_nt} + K_{m+2}te^{p_nt} + \cdots + K_nt^{n-m-1}e^{p_nt} \tag{5.9}$$

or, in more compact form,

$$y_h(t) = \sum_{i=1}^{m} K_ie^{p_it} + \sum_{i=m+1}^{n} K_it^{i-m-1}e^{p_it} \tag{5.10}$$

The integration constants $K_1, K_2, \ldots, K_n$ will be determined after the particular solution is found.

Solving for the particular solution is more difficult to generalize because it always depends on the form of the forcing function $u(t)$. Generally speaking, the form of $y_p(t)$ has to be guessed, based on the form of $u(t)$ and/or its derivatives.[1] A method of undetermined coefficients provides a more systematic approach to solving for the particular solution. In many cases where the forcing function $u(t)$ reaches a steady-state value, say u_{ss}, for time approaching infinity, that is if

$$\lim_{t\to\infty} u(t) = u_{ss} \tag{5.11}$$

and if $a_0 \neq 0$, the particular solution can be simply calculated as

$$y_p(t) = u_{ss}/a_0 \tag{5.12}$$

Once the general forms of both parts of the solution are found, the constants that appear in the two expressions must be determined.

If there are unknown constants in the particular solution, that is, if Equation (5.12) cannot be applied, the general expression for the particular solution is substituted into Equation (5.5) and the constants are found by solving equations created by equating corresponding terms on both sides of the equation.

To find the integration constants $K_1, K_2, \ldots, K_n$ in the homogenous solution, the set of n initial conditions, given by Equation (5.2) is used to form the following n equations

$$\begin{aligned} y_h(0^+) + y_p(0^+) &= y(0^+) \\ \left(\frac{dy_h}{dt}\right)\bigg|_{t=0^+} + \left(\frac{dy_p}{dt}\right)\bigg|_{t=0^+} &= \left(\frac{dy}{dt}\right)\bigg|_{t=0^+} \\ &\vdots \end{aligned} \tag{5.13}$$

[1]When the input is of the form e^{pt} where p is one of the roots of the system characteristic equations, then the particular solution may contain an exponential term of the form Cte^{pt}, where C is determined by substituting Cte^{pt} in the system differential equation and solving for C. See Ref. 2(a), p. 37.

$$\left(\frac{d^{n-1}y_h}{dt^{n-1}}\right)\bigg|_{t=0^+} + \left(\frac{d^{n-1}y_p}{dt^{n-1}}\right)\bigg|_{t=0^+} = \left(\frac{d^{n-1}y}{dt^{n-1}}\right)\bigg|_{t=0^+}$$

where the terms on the right-hand sides represent the initial conditions.

The method for solving differential equations of the general form given by Equation (5.1) can be summarized by the following step by step procedure.

- Step 1. Obtain the characteristic Equation (5.7).
- Step 2. Find roots of the characteristic equation, $p_1, p_2, \ldots, p_n$.
- Step 3. Write the general expression for the homogenous solution, Equation (5.10).
- Step 4. Determine the general expression for the particular solution.[2]
- Step 5. Determine constants in the particular solution by equating corresponding terms on both sides of Equation (5.5). This step can be skipped if the particular solution was found using Equation (5.12).
- Step 6. Find integration constants $K_1, K_2, \ldots, K_n$ by solving the set of Equations (5.13) involving the initial conditions.

Several examples involving analytical solution of linear input-output equations are given in Sections 5.4, 5.5, and 5.6.

5.3 Determining Initial Conditions at $t = 0^+$

In order to use analytical methods to solve an input-output differential equation and to find the system response to an input beginning at time $t = 0$, it is necessary to have the initial conditions for the system just after the input starts, i.e., at $t = 0^+$. Usually the initial conditions are known only as they existed just before the input started to change, i.e., at $t = 0^-$. The required initial conditions at $t = 0^+$ can usually be readily determined, using the initial conditions at $t = 0^-$ and information about the system model and its inputs, through use of the elemental equations, the interconnecting equations, and energy-storage limitations when they are all available. However, there are times when only the input-output model differential equation is available.

The method presented in this section allows for finding the required initial conditions at $t = 0^+$ when the input signal is at all times bounded and starts at $t = 0$, using only the system input-output differential equation. This method is most frequently employed when the system is subjected to a step input at $t = 0$, but it is applicable to any other type of input, as long as the input signal is bounded. The method cannot, therefore, be applied to systems subjected to impulsive inputs but this should not be a serious drawback because, as will be shown later, the response to a unit impulse is the time derivative of the response to a unit step.

[2]The method of undetermined coefficients is described in detail in the following references: (a) D.K. Chang, *Analysis of Linear Systems*, Addison-Wesley Publishing Company, Reading, Mass., 1960, pp. 26–40. (b) J.L. Shearer, A.T. Murphy, and H.H. Richardson, *Introduction to System Dynamics*, Addison-Wesley Publishing Company, Reading, Mass., 1967, pp. 287–289.

Thus, for the case of an impulse input, it is only necessary to find the step response and differentiate it with respect to time.

Consider the second-order input-output differential equation

$$a_2 \frac{d^2y}{dt^2} + a_1 \frac{dy}{dt} + a_0 y = b_2 \frac{d^2u}{dt^2} + b_1 \frac{du}{dt} + b_0 u \tag{5.14}$$

This equation may be integrated twice in succession with respect to time between the limits of $t = 0^-$ and $t = 0^+$ to yield

$$a_2 y(t) \Big|_{y(0^-)}^{y(0^+)} + a_1 \int_{0^-}^{0^+} y(t)\, dt + a_0 \int_{0^-}^{0^+} \int_{0^-}^{t} y(t_D)\, dt_D\, dt =$$

$$b_2 u(t) \Big|_{u(0^-)}^{u(0^+)} + b_1 \int_{0^-}^{0^+} u(t)\, dt + b_0 \int_{0^-}^{0^+} \int_{0^-}^{t} u(t_D)\, dt_D\, dt \tag{5.15}$$

where t_D is a dummy time variable for the first integration from 0^- to t.

Since $u(t)$ is assumed bounded for all t, the last two terms on the right-hand side of Equation (5.15) are zero (the time integral of a finite integrand over an infinitesimal time interval is zero). This means that none of the terms on the left-hand side can be infinite, so that if y is finite at $t = 0^-$, it will also be finite over the infinitesimal interval from $t = 0^-$ to $t = 0^+$, and the other terms on the left-hand side of Equation (5.15) will also be zero. Equation (5.15) thus becomes

$$a_2[y(0^+) - y(0^-)] = b_2[u(0^+) - u(0^-)] \tag{5.16}$$

which yields

$$y(0^+) = y(0^-) + (b_2/a_2)[u(0^+) - u(0^-)] \tag{5.17}$$

Note that if $b_2 = 0$, then $y(0^+) = y(0^-)$.

Next, to find $\dot{y}(0^+)$, it is necessary to integrate Equation (5.14) only once instead of twice which yields

$$a_2 \dot{y}(t) \Big|_{\dot{y}(0^-)}^{\dot{y}(0^+)} + a_1 y(t) \Big|_{y(0^-)}^{y(0^+)} + a_0 \int_{0^-}^{0^+} y(t)\, dt =$$

$$b_2 \dot{u}(t) \Big|_{\dot{u}(0^-)}^{\dot{u}(0^+)} + b_1 u(t) \Big|_{u(0^-)}^{u(0^+)} + b_0 \int_{0^-}^{0^+} u(t)\, dt \tag{5.18}$$

By reasoning similar to that used to obtain Equation (5.16) from Equation (5.15), the third terms on each side of Equation (5.18) are both zero, which gives

$$a_2[\dot{y}(0^+) - \dot{y}(0^-)] + a_1[y(0^+) - y(0^-)] = b_2[\dot{u}(0^+) - \dot{u}(0^-)] + b_1[u(0^+) - u(0^-)] \tag{5.19}$$

Rearranging, the initital condition for $\dot{y}$ at time $t = 0^+$ is

$$\dot{y}(0^+) = \dot{y}(0^-) + \left(\frac{b_2}{a_2}\right)[\dot{u}(0^+) - \dot{u}(0^-)] + \left(\frac{(a_2b_1 - a_1b_2)}{a_2^2}\right)[u(0^+) - u(0^-)] \quad (5.20)$$

Note that if both $b_2 = 0$ and $b_1 = 0$, then $\dot{y}(0^+) = \dot{y}(0^-)$.

5.4 First-Order Models

A linear, stationary, single-input, single-output system model is described by a first-order differential equation

$$a_1\dot{y} + a_0y = u(t) \quad (5.21)$$

with the initial condition, $y(0^+) = y_0$. The characteristic equation is

$$a_1p + a_0 = 0 \quad (5.22)$$

A general form for the model homogeneous solution is

$$y_h(t) = K_1e^{p_1t} \quad (5.23)$$

where a single root of the characteristic equation, p_1, is

$$p_1 = \frac{-a_0}{a_1} \quad (5.24)$$

A particular solution of Equation (5.21) depends on the type of the input signal $u(t)$. Three types of input signal will be considered: zero, step function, and impulse function. The corresponding system outputs will be: free response,[3] step response, and impulse response, respectively.

First, an input will be assumed to be equal to zero, $u(t) = 0$ for $t \geq 0$. The homogenous solution constitutes, in this case, a complete solution given by the following expression

$$y(t) = K_1e^{-(a_0/a_1)t} \quad (5.25)$$

The integration constant, K_1, can be found by putting $t = 0^+$ in Equation (5.25) and using the initial condition to obtain

$$K_1 = y_0 \quad (5.26)$$

where $y_0 = y(0^+) = y(0^-)$. Hence the free response of the first order model is

$$y(t) = y_0e^{-(a_0/a_1)t} \quad (5.27)$$

[3]The term "free response", adapted from vibration theory, is somewhat of a misnomer. The response is free only in the sense that the system input $u(t)$ is zero for the time interval of interest, here for $t > 0$. In all cases a variation of the output occurs only in response to some past input forcing function or system change such as switch or circuit breaker activation, shaft failure, pipe rupture, etc. In this case the system is responding to a non-zero initial output condition, which must be the remainder of a response to a previous input. This non-zero initial condition at $t = 0^-$ is sometimes referred to as a non-equilibrium initial condition.

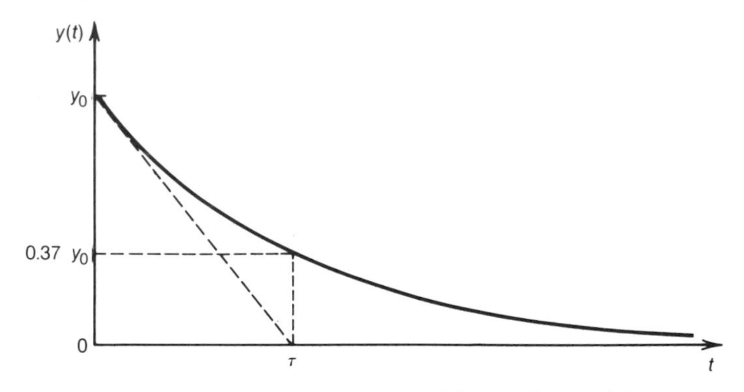

FIGURE 5.1 Free response of first order model.

The free response curve is shown in Figure 5.1. The curve starts at time $t = 0^+$ from the initial value, y_0, and then decays exponentially to zero as time approaches infinity. The rate of the exponential decay is determined by the model time constant, τ, defined as

$$\tau = a_1/a_0 \tag{5.28}$$

The time constant is equal to the time during which the first order model free response decreases by 63.1 percent of its initial value. A normalized free response of the first order model, $y(t)/y_0$, versus normalized time, t/τ, is plotted in Figure 5.2, and its numerical values are given in Table 5.1.

TABLE 5.1. Normalized Free Response of First Order Model.

t/τ	0	0.5	1.0	1.5	2.0	2.5	3.0	4.0
$y(t)/y_0$	0	0.6065	0.3679	0.2231	0.1353	0.0821	0.0498	0.0183

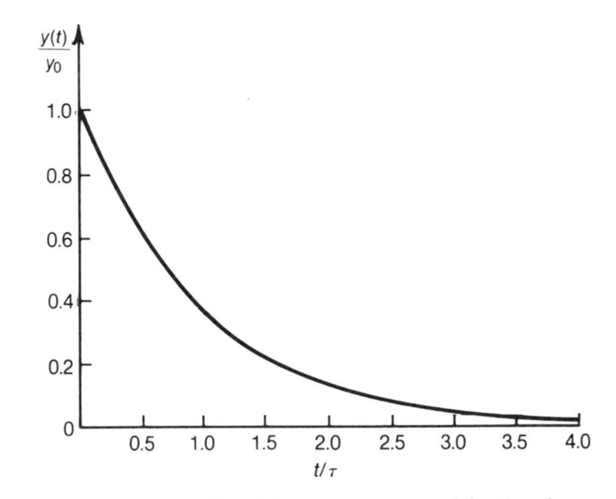

FIGURE 5.2 Normalized free response of first order model.

The second type of input signal considered here is a step perturbation of $u(t)$ expressed by the step function $U_s(t)$

$$\hat{u}(t) = AU_s(t) \tag{5.29}$$

where A is the magnitude of the step function and $\hat{u}(t) = u(t) - \bar{u}$. The system normal operating part (N.O.P.) of u is $\bar{u}$ and the (N.O.P.) of y is $\bar{y}$. The unit step function $U_s(t)$ is defined by Equation (5.30)

$$U_s(t) = \begin{cases} 0 & \text{for } t < 0 \\ 1 & \text{for } t > 0 \end{cases} \tag{5.30}$$

The complete input $u(t)$ is then given by

$$u(t) = \bar{u} + \hat{u}(t) = \bar{u} + AU_s(t) \tag{5.31}$$

And the complete output $y(t)$ is given by

$$y(t) = \bar{y} + \hat{y}(t) \tag{5.32}$$

where

$$\bar{y} = \bar{u}/a_0 \tag{5.33}$$

The homogenous solution in this case is, of course, of the same form as before (Equation 5.25). Here, the particular solution must satisfy Equation (5.34).

$$a_1\dot{\hat{y}}_p + a_0\hat{y}_p = A \qquad \text{for } t > 0 \tag{5.34}$$

Using Equation (5.12), the particular solution for a step input perturbation of magnitude A from $\bar{u}$ is

$$\hat{y}_p(t) = A/a_0 \tag{5.35}$$

Hence, a complete solution for $\hat{y}(t)$ is

$$\hat{y}(t) = \hat{y}_h(t) + \hat{y}_p(t) = K_1e^{-t/\tau} + A/a_0 \tag{5.36}$$

Substituting $\hat{y}(t) = (y_0 - \bar{y})$ for $t = 0^+$, the constant K_1 is found to be

$$K_1 = (y_0 - \bar{y}) - A/a_0 \tag{5.37}$$

where K_1 now includes a term due to $u(t)$. The complete response of the first order model is finally obtained as

$$y(t) = \underbrace{\bar{u}/a_0}_{\text{N.O.P.}} + \underbrace{(y_0 - \bar{u}/a_0)e^{-t/\tau}}_{\text{free response part}} + \underbrace{A/a_0(1 - e^{-t/\tau})}_{\text{step response part}} \tag{5.38}$$

where $y_0 = y(0^+)$ is the same as $y(0^-)$[no du/dt term in Equation (5.21)]. From Equation (5.36) the steady-state value of the step perturbation response, $\hat{y}_{ss}$ is

$$\hat{y}_{ss} = \lim_{t\to\infty} \hat{y}(t) = A/a_0 \tag{5.39}$$

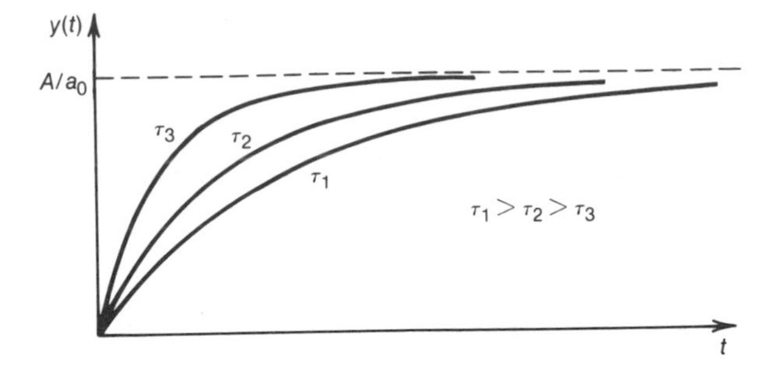

FIGURE 5.3 Step response curves of first-order model with zero initial condition $y_0 = \bar{y} = 0$.

The perturbation response can now be expressed in terms of its initial and steady-state values

$$\hat{y}(t) = \hat{y}_{ss} - (\hat{y}_{ss} - y_0)e^{-t/\tau} \tag{5.40}$$

Several step response curves for different values of time constant and zero initial condition, $y_0 - \bar{y} = 0$, are plotted in Figure 5.3.

The third type of input to be considered here is an impulse function $u(t)$ expressed by

$$u(t) = PU_i(t) \tag{5.41}$$

where P is the strength of the impulse function (in other words, its area). The unit impulse function $U_i(t)$ occurring at time $t = 0$ is defined by Equations (5.42) and (5.43)

$$U_i(t) = \begin{cases} \infty & \text{for } t = 0 \\ 0 & \text{for } t \neq 0 \end{cases} \tag{5.42}$$

$$\int_{-\infty}^{+\infty} U_i(t)\, dt = \int_{0-}^{0+} U_i(t)\, dt = 1.0 \tag{5.43}$$

In general, a unit impulse, also called Dirac's delta function, which occurs at time $t = t_1$ is denoted by $U_i(t - t_1)$ and is defined by Equations (5.44) and (5.45).

$$U_i(t - t_1) = \begin{cases} \infty & \text{for } t = t_1 \\ 0 & \text{for } t \neq t_1 \end{cases} \tag{5.44}$$

$$\int_{-\infty}^{+\infty} U_i(t - t_1)\, dt = \int_{t_1^-}^{t_1^+} U_i(t - t_1)\, dt = 1.0 \tag{5.45}$$

An ideal unit impulse function as just defined cannot be physically generated. It can be thought of as a limit of a unit pulse function, $U_p(t)$, for the pulse duration T/n approaching zero as n approaches infinity, as illustrated in Figure 5.4.

The unit impulse function is widely used in theoretical system analysis because of its useful mathematical properties, which make it a very desirable type of input. One such property, called a filtering property, is mathematically expressed by the sifting integral, Equation (5.46).

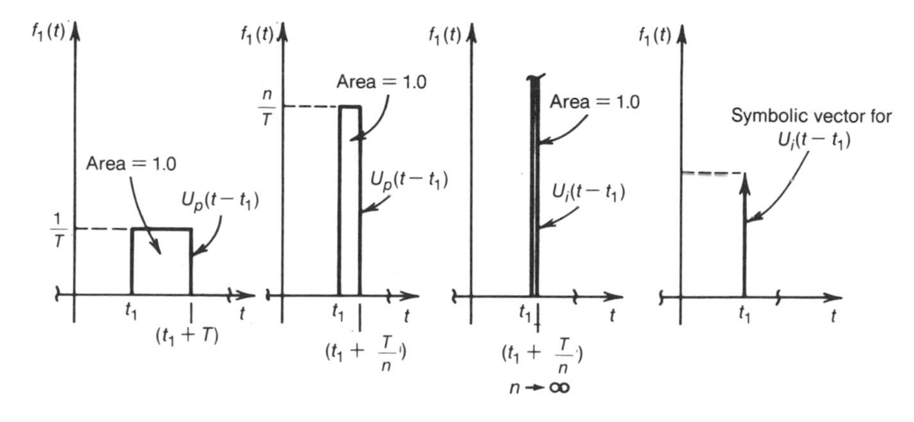

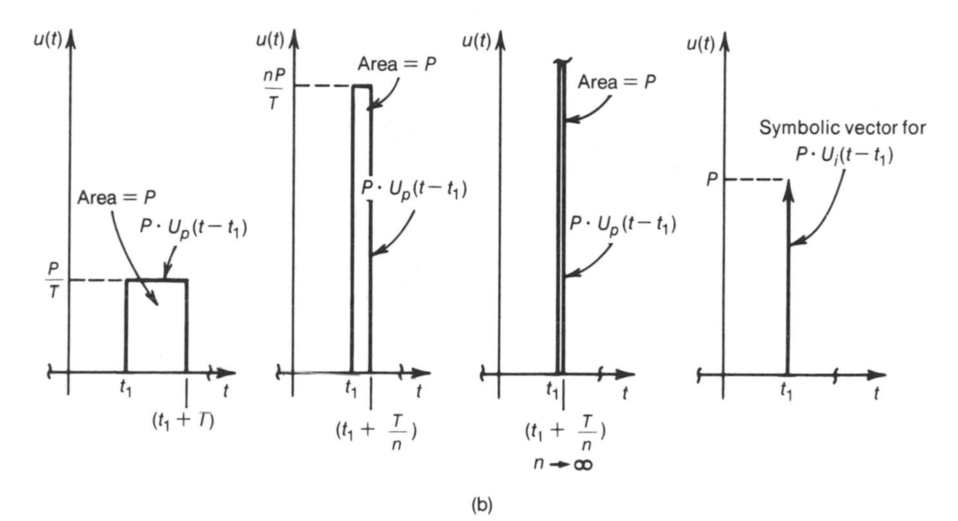

FIGURE 5.4 Transition from a Pulse to an Impulse, (a) Unit Pulse to a Unit Impulse and (b) Pulse of Strength P to an Impulse of Strength P, together with Symbolic Vector Representations of Impulses.

$$\int_{-\infty}^{+\infty} U_i(t - t_1)f(t)\,dt = f(t_1) \tag{5.46}$$

Equation (5.46) implies that

$$\int_{-\infty}^{+\infty} U_i(t)f(t)\,dt = f(0) \tag{5.47}$$

Although a rigorous mathematical proof is difficult, a unit impulse function is often considered as the derivative of a unit step function[4]

[4]D. K. Chang, *Analysis of Linear Systems*, Addison-Wesley Publishing Co., Reading, Massachusetts, 1959, pp. 220–224.

$$U_i(t) = \frac{dU_s(t)}{dt} \tag{5.48}$$

Equation (5.48) implies that the dimension of the amplitude of an impulse function is [1/time] if the amplitude of a step function is dimensionless. However, since the amplitude of $U_i(t)$ is infinite at $t = 0$, an impulse is measured instead by its strength, P, equal to its integral from $t = 0^-$ to $t = 0^+$ [in other words, the area under the "spike-like" curve representing $PU_i(t)$]. Hence the dimension of (P/A) is time and P is numerically equal to A. An impulse input can thus be expressed by

$$u(t) = PU_i(t) = \frac{P}{A}\frac{d[AU_s(t)]}{dt} \tag{5.49}$$

Therefore, if the input to a system is an impulse $PU_i(t)$ instead of a step $AU_s(t)$, the system response to the impulse $PU_i(t)$ will be (P/A) times the time-derivative of the response to the step $AU_s(t)$. It should be emphasized that only the part of $y(t)$ that is generated by the step input should be differentiated to obtain an impulse response (in other words, the free response part of the perturbation response due to non-zero initial condition should not be included). Differentiating the part of the step response given by Equation (5.38) due to the input $AU_s(t)$ with respect to time, the impulse response $\hat{y}_i(t)$ of the first order model is obtained as follows

$$\hat{y}_i(t) = (P/A)\frac{d[(A/a_0)(1 - e^{-t/\tau})]}{dt} + (y_0 - \bar{y})e^{-t/\tau} \tag{5.50}$$

and hence

$$\hat{y}_i(t) = \frac{P}{A}\frac{A}{a_1}e^{-t/\tau} + \left(y_0 - \frac{\bar{u}}{a_0}\right)e^{-t/\tau} \tag{5.51}$$

Using the definition of the time constant given by Equation (5.28), the impulse response can be expressed by

$$\hat{y}_i(t) = \frac{P}{a_1}e^{-t/\tau} + \left(y_0 - \frac{\bar{u}}{a_0}\right)e^{-t/\tau} \tag{5.52}$$

Figure 5.5 shows an impulse response curve together with the response to a pulse of width T and strength P for a first order model for the case when $y_0 = \bar{u}/a_0$.

Note that employing the derivative of the step response eliminates the problem of finding $y(0^+)$ when the input is an impulse.

Just as the response to an impulse input of strength P is (P/A) times the derivative of the response to a step input of magnitude A, it can be shown that the response to a ramp of slope S starting at $t = 0$ is (S/A) times the time integral of the response to a step input.

By employing the principle of superposition, which applies to all linear systems, the response to several inputs and/or nonzero initial conditions can be obtained as the sum of the responses to the individual inputs and/or nonzero initial conditions found separately.

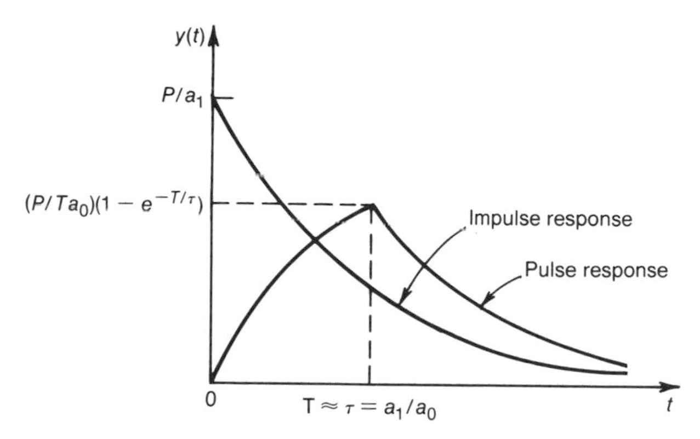

FIGURE 5.5 Impulse response of first-order model together with response to a pulse of strength P and duration T.

EXAMPLE 5.1.

In the mechanical system shown in Figure 5.6 mass m is initially subjected to a constant force $\overline{F}$ and is moving with initial velocity $\overline{v}$. At time t_0 the force changes suddenly from $\overline{F}$ to $\overline{F} + \Delta F$. Find velocity and acceleration of the mass for time $t > t_0$.

Solution

Using force $F(t)$ applied to the mass as the input and the velocity of the mass, v_{1g}, as the output variable, the model input-output equation can be written as

$$m\dot{v}_{1g} + bv_{1g} = F(t)$$

Because the velocity of the mass cannot change suddenly, the initial condition is

$$v_{1g}(t_0^+) = v_{1g}(t_0^-) = \overline{v}_{1g}$$

A mathematical form describing the input force is

$$F(t) = \overline{F} + \Delta F U_s(t - t_0)$$

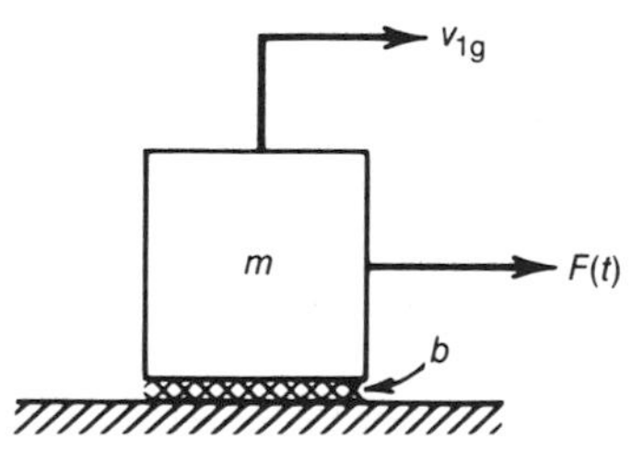

FIGURE 5.6 Mechanical system considered in Example 5.1.

The input-output equation can be rewritten in the following form

$$\dot{v}_{1g} + \left(\frac{b}{m}\right) v_{1g} = \left(\frac{1}{m}\right) F(t)$$

Using Equation (5.28) the time constant can be identified as

$$\tau = m/b$$

The steady-state value of the mass velocity can be found by taking limits of both sides of the input-output equation for time approaching infinity

$$\lim_{t\to\infty} \left(\dot{v}_{1g} + \frac{b}{m} v_{1g}\right) = \lim_{t\to\infty} \frac{1}{m} [\bar{F} + \Delta F U_s(t - t_0)]$$

Assuming that v_{1g} will reach the steady-state value of v_{ss} for time approaching infinity, the above equation yields

$$v_{ss} = \frac{1}{b} (\bar{F} + \Delta F)$$

Using Equation (5.37), the step response of velocity is found

$$v_{1g}(t) = \frac{(\bar{F} + \Delta F)}{b} - \left[\frac{(\bar{F} + \Delta F)}{b} - \bar{v}_{1g}\right] e^{-(b/m)(t-t_0)} U_s(t - t_0)$$

which can be further simplified using $\bar{v}_{1g} = \bar{F}/b$

$$v_{1g}(t) = \bar{v}_{1g} + \left(\frac{\Delta F}{b}\right) [1 - e^{-(t-t_0)/\tau}] U_s(t - t_0)$$

The response of acceleration can be found by differentiating the velocity step response

$$a_{1g}(t) = \left(\frac{\Delta F}{m}\right) [e^{-(t-t_0)/\tau}] U_s(t - t_0)$$

Note that the use of $U_s(t - t_0)$ in the last three equations serves the purpose of synchronizing the solution at $t = t_0$ or, in other words, preventing the occurrence of a fallacious output before the step input occurs at $t = t_0$.

Both the velocity and acceleration step response curves are plotted in Figure 5.7.

A conclusion from these considerations is that a step response of a first-order model is always an exponential function of time involving three parameters: initial value, steady-state value, and time constant. If the initial value is specified in the problem statement, the other two parameters can be found simply by inspection of the model input-output equation, using relations (5.28) and (5.39). A step response of a first order model can thus be found in a very simple manner, without solving the differential equation. ■

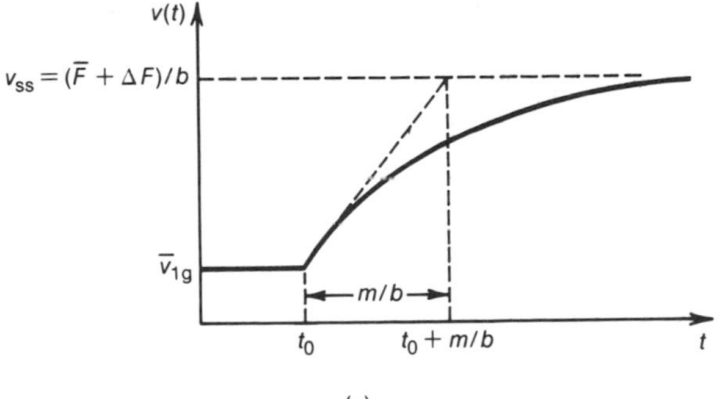

(a)

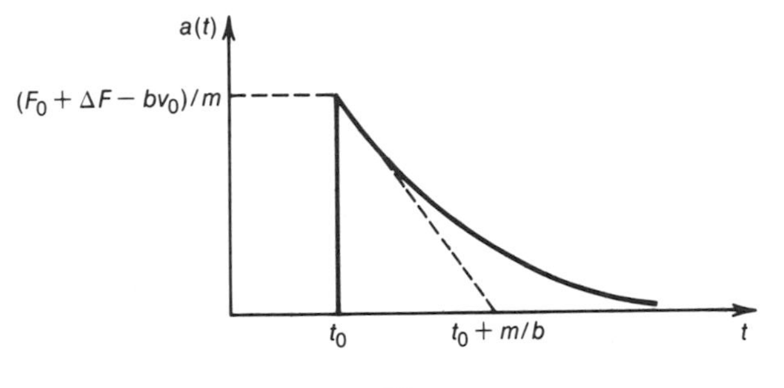

(b)

FIGURE 5.7 (a) Velocity and (b) acceleration step response curves of the system considered in Example 5.1.

5.5 Second-Order Models

An input-output equation for a stationary, linear second-order model is

$$a_2\ddot{y} + a_1\dot{y} + a_0y = u(t) \tag{5.53}$$

The initial conditions are

$$y(0^+) = y_0 \quad \text{and} \quad \dot{y}(0^+) = \dot{y}_0 \tag{5.54}$$

The characteristic equation is

$$a_2p^2 + a_1p + a_0 = 0 \tag{5.55}$$

5.5.1 Free Response

The form of the homogenous solution representing the system free response depends on whether the two roots of the characteristic Equation (5.55), p_1 and p_2, are distinct or identical. If the roots are distinct, $p_1 \neq p_2$, the homogenous solution is of the form

$$y_h(t) = K_1e^{p_1t} + K_2e^{p_2t} \tag{5.56}$$

If the roots are identical, $p_1 = p_2$, the free response is

$$y_h(t) = K_1 e^{p_1 t} + K_2 t e^{p_1 t} \tag{5.57}$$

If the roots are complex, they occur as pairs of complex conjugate numbers, i.e.

$$p_1 = a + jb \quad \text{and} \quad p_2 = a - jb \tag{5.58}$$

Substitution of the expressions (5.58) for the complex roots into Equation (5.56) gives the homogenous solution for this case.

$$y_h(t) = K_1 e^{(a+jb)t} + K_2 e^{(a-jb)t} \tag{5.59}$$

Or, using the trigonometric forms of the complex numbers,

$$\begin{aligned} y_h(t) &= K_1 e^{at}(\cos bt + j \sin bt) + K_2 e^{at}(\cos bt - j \sin bt) \\ &= e^{at}(K_3 \cos bt + K_4 \sin bt) \end{aligned} \tag{5.60}$$

where the constants K_3 and K_4 are a different, but corresponding, set of integration constants

$$K_3 = K_1 + K_2 \tag{5.61}$$

$$K_4 = j(K_1 - K_2) \tag{5.62}$$

Note that if the roots of the characteristic equation are complex, K_1 and K_2 are also complex conjugate numbers, but K_3 and K_4 are real constants.

The model free response has been shown to depend on the type of roots of the model characteristic equation, and for a second-order model the free response takes the form of Equation (5.56), (5.57), or [(5.59) or (5.60)] if the roots are real and distinct, real and identical, or complex conjugate, respectively. Moreover, the character of the free response depends on whether real roots or real parts of complex roots are positive or negative. Table 5.2 illustrates the effect of location of the roots in a complex plane on the model impulse response.

Several general observations can be made based on Table 5.2. First, the impulse response of a second-order model is seen to be oscillatory if the roots of the model characteristic equation are complex and nonoscillatory when the roots are real. Furthermore, the impulse response approaches zero as time approaches infinity only if the roots are either real and negative or complex and have negative real parts. The systems that have all the roots of their characteristic equation located in the left-hand side of a complex plane are referred to as stable systems. However, if at least one root of the model characteristic equation lies in the right-hand side of a complex plane, the model impulse response grows without bound with time; such a system is considered unstable. Marginal stability occurs when there are no roots in the right-hand side of a complex plane and at least one root is located on the imaginary axis. System stability constitutes one of the most important problems in analysis and design of feedback systems and will be treated more thoroughly in Chapter 13.

Two important parameters widely used in characterizing the response of second-order systems are the damping ratio, ζ, and the natural frequency, ω_n. These two parameters appear in the modified input-output equation (5.63).

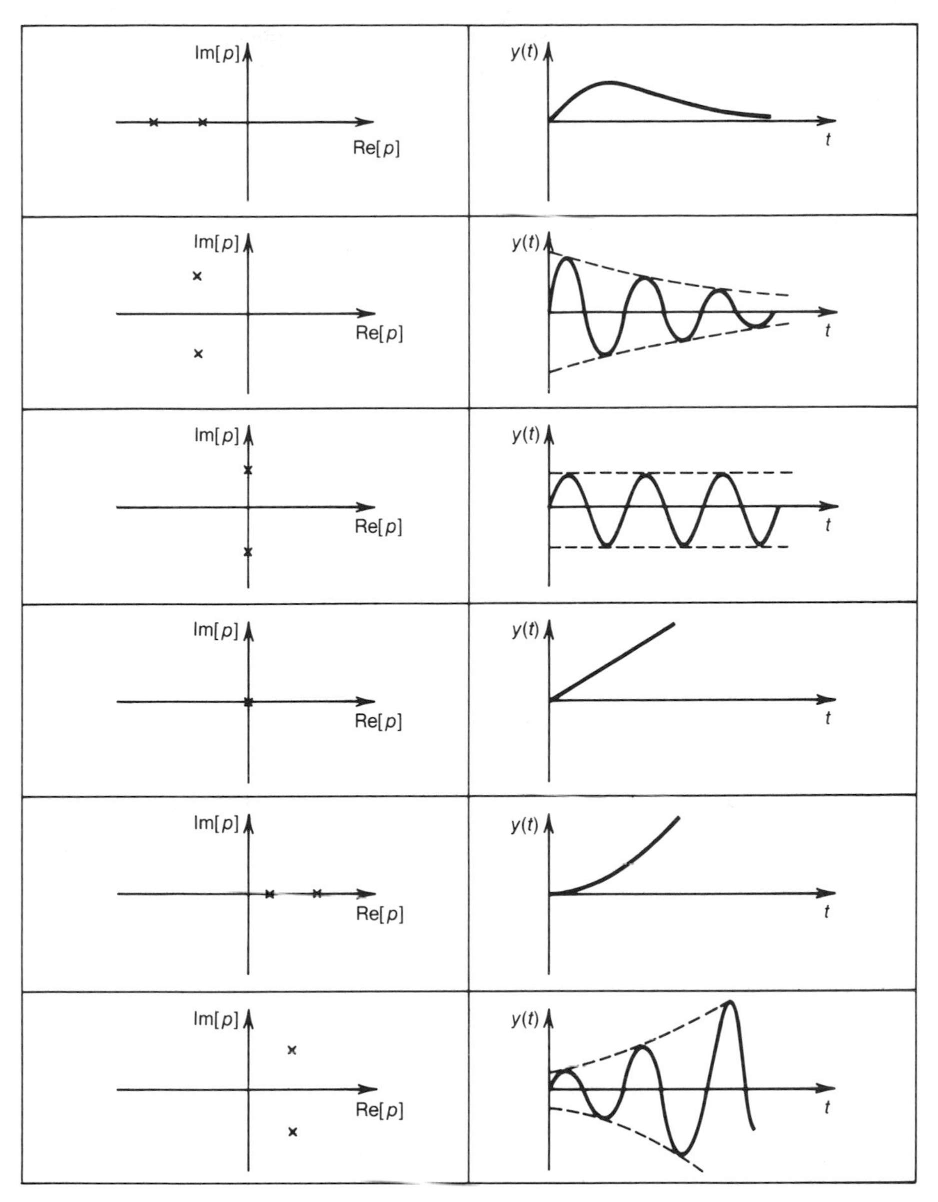

TABLE 5.2 Locations of roots of characteristic equation and the corresponding impulse response curves.

$$\ddot{y} + 2\zeta\omega_n\dot{y} + \omega_n^2 y = f(t) \tag{5.63}$$

The damping ratio ζ represents the amount of damping in a system, whereas the natural frequency ω_n is a frequency of oscillations in an idealized system with zero damping. In other words, since the amount of damping is related to the rate of dissipation of energy in the system, ω_n represents the frequency of oscillations in an idealized system that does not dissipate energy.

The input-output Equation (5.53) can be rewritten in the following form

$$\ddot{y} + (a_1/a_2)\dot{y} + (a_0/a_2)y = (1/a_2)u(t) \tag{5.64}$$

By comparing Equations (5.63) and (5.64), the natural frequency and the damping ratio can be expressed in terms of the coefficients of the input-output equation

$$\omega_n = \sqrt{\frac{a_0}{a_2}} \tag{5.65}$$

$$\zeta = \frac{a_1}{2\sqrt{a_0 a_2}} \tag{5.66}$$

Note also that $u(t)$ from Equation (5.53) is now replaced by $a_2 f(t)$. The system characteristic equation is

$$p^2 + 2\zeta\omega_n p + \omega_n^2 = 0 \tag{5.67}$$

The roots of the characteristic equation can be expressed in terms of the coefficients a_0, a_1, and a_2

$$p_1, p_2 = \frac{\left[-a_1 \pm \sqrt{(a_1^2 - 4a_0 a_2)}\right]}{2a_2} \tag{5.68}$$

or, equivalently, in terms of ζ and ω_n for the underdamped case,

$$p_1, p_2 = -\zeta\omega_n \pm j\omega_n\sqrt{1 - \zeta^2} \tag{5.69}$$

or, similarly, in terms of τ_1 and τ_2 for the overdamped case,

$$p_1, p_2 = -1/\tau_1, \ -1/\tau_2 \tag{5.70}$$

where

$$\tau_1, \tau_2 = \frac{[a_1 \pm \sqrt{a_1^2 - 4a_0 a_2}]}{2a_0} \tag{5.71}$$

It was pointed out earlier that a system is stable if real parts of the complex roots of the characteristic equation are negative, that is if

$$\text{Re}[p_1] = \text{Re}[p_2] = -\zeta\omega_n < 0 \tag{5.72}$$

Since the natural frequency is not negative, for stability of a second-order system the damping ratio must be positive so that $\zeta > 0$. From Equations (5.66) and (5.69) it can be deduced that the roots are complex: thus the system response is oscillatory if

$$0 < \zeta < 1 \tag{5.73}$$

A system is said to be underdamped when the damping ratio satisfies (5.73). The frequency of oscillation of an underdamped system is called a damped natural frequency and is equal to

$$\omega_d = \omega_n\sqrt{1 - \zeta^2} \tag{5.74}$$

If the damping ratio is equal to or greater than one, the expressions on the right-hand sides of Equation (5.69) become real, and thus the system free response is nonoscillatory. A system for which the damping ratio is greater than one is referred to as an overdamped system. The damping is said to be critical if the damping ratio is equal to one.

Table 5.2 shows how the nature of the system impulse response depends on the location of roots of the characteristic equation. In addition, the dynamics of a second order model can be uniquely described in terms of the natural frequency and the damping ratio. Therefore, the conclusion can be drawn that there must exist a unique relationship between pairs of roots of the characteristic equation and pairs of ω_n and ζ values. To determine this relationship, consider a second order model having two complex roots, p_1 and p_2, located as shown in Figure 5.8.

The real parts of both roots are

$$\text{Re}\,[p_1] = \text{Re}\,[p_2] = -\zeta\omega_n \tag{5.75}$$

and the imaginary parts are

$$\text{Im}\,[p_1] = \omega_n\sqrt{1 - \zeta^2} = \omega_d \tag{5.76}$$

$$\text{Im}\,[p_2] = -\omega_n\sqrt{1 - \zeta^2} = -\omega_d \tag{5.77}$$

The damped natural frequency is therefore equal to the ordinates of the points p_1 and p_2 with a plus or minus sign, respectively. In order to identify a corresponding natural frequency ω_n, consider the distance d between the points p_1 and p_2 and the origin. In terms of the real and imaginary parts of the two roots, d can be calculated as

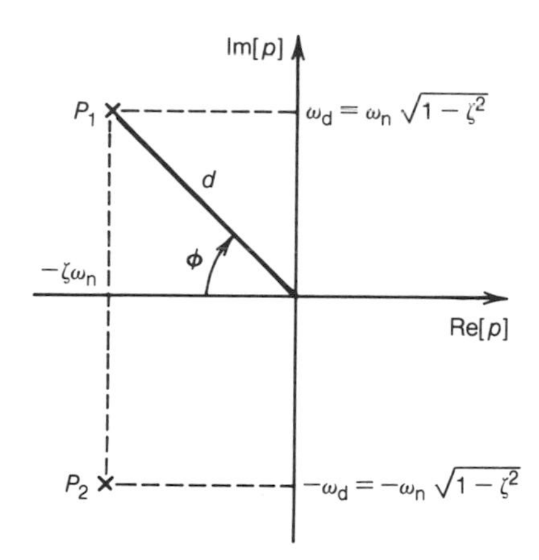

FIGURE 5.8 Pair of complex roots and corresponding values of ζ, ω_n, ω_d.

$$d = \sqrt{(\text{Re}\,[p_1])^2 + (\text{Im}\,[p_1])^2} = \sqrt{(\text{Re}\,[p_2])^2 + (\text{Im}\,[p_2])^2} \tag{5.78}$$

Substituting the expressions for the real and imaginary parts of p_1 and p_2 from Equations (5.75), (5.76), and (5.77), yields

$$d = \omega_n \tag{5.79}$$

The natural frequency of a second-order underdamped system is thus equal to the distance between the locations of the system characteristic roots and the origin of the coordinate system in the complex plane.

Finally, by comparing Equations (5.75) and (5.79), the damping ratio can be expressed as

$$\zeta = -\text{Re}\,[p_1]/d \tag{5.80}$$

which, after inspection of Figure 5.8, can be rewritten as

$$\zeta = \cos\varphi \tag{5.81}$$

where φ is the acute angle measured from the negative real axis.

The results of the above considerations are presented graphically in Figures 5.9, 5.10, and 5.11. Figure 5.9 shows loci of constant natural frequency. The loci are concentric circles with radii proportional to ω_n. The farther from the origin of the coordinate system are the roots of the characteristic equation, the higher is the value of the corresponding natural frequency.

In Figure 5.10 the horizontal lines represent loci of constant damped natural frequency. The greater the distance between the roots of the characteristic equation and the real axis, the higher the value of ω_d.

The loci of constant damping ratio for an underdamped system, shown in Figure 5.11, take the form of straight lines described by Equation (5.81). Practically, the use of the damping ratio is limited to stable systems having both roots of the

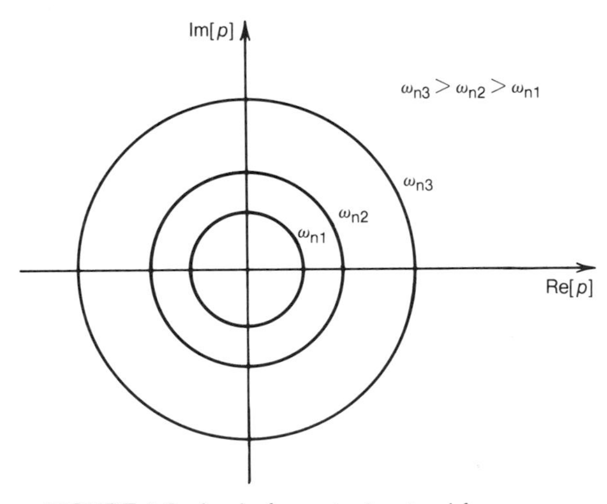

FIGURE 5.9 Loci of constant natural frequency.

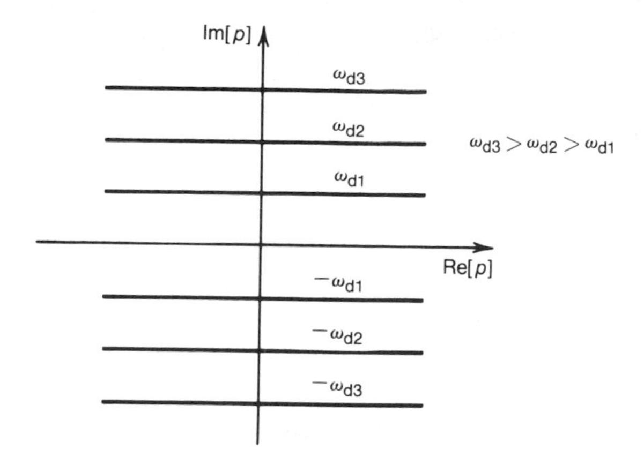

FIGURE 5.10 Loci of constant damped natural frequency.

characteristic equation in the left-hand side of a complex plane, and the loci in Figure 5.11 represent only stable underdamped systems. When the system is overdamped the roots lie along the real axis, and there is no oscillation in the response.

5.5.2 Step Response

A step response equation for a second order model will now be derived for the three cases of an underdamped, critically damped, and overdamped system.

Underdamped Case, $0 < \zeta < 1$. Consider the second-order model equation in another modified form

$$\ddot{y} + 2\zeta\omega_n\dot{y} + \omega_n^2 y = \omega_n^2 u(t) \tag{5.82}$$

Both initial conditions are zero, $y(0) = 0$, and $\dot{y}(0) = 0$.

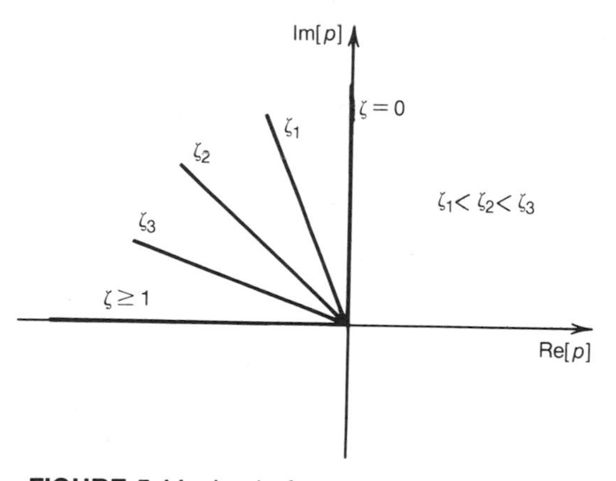

FIGURE 5.11 Loci of constant damping ratio.

Let $u(t)$ be a unit step function

$$u(t) = U_s(t)$$

The characteristic equation is

$$p^2 + 2\zeta\omega_n p + \omega_n^2 = 0 \tag{5.83}$$

For $0 < \zeta < 1$, Equation (5.83) has two complex conjugate roots

$$p_1 = -\zeta\omega_n - j\omega_d \tag{5.84}$$

$$p_2 = -\zeta\omega_n + j\omega_d \tag{5.85}$$

where ω_d is defined by Equation (5.74). The general form of the homogenous solution is given by Equation (5.60). Substituting real and imaginary parts of p_1 and p_2 from Equations (5.84) and (5.85) yields

$$y_h(t) = e^{-\zeta\omega_n t}(K_3 \cos \omega_d t + K_4 \sin \omega_d t) \tag{5.86}$$

where K_3 and K_4 are unknown real constants. The particular solution for a unit step input is, in accordance with Equation (5.12),

$$y_p(t) = 1 \tag{5.87}$$

The complete solution is a sum of the right-hand sides of (5.86) and (5.87).

$$y(t) = 1 + e^{-\zeta\omega_n t}(K_3 \cos \omega_d t + K_4 \sin \omega_d t) \tag{5.88}$$

Using the initial conditions, the two constants K_3 and K_4 are found.

$$K_3 = -1 \tag{5.89}$$

$$K_4 = -\frac{\zeta}{\sqrt{1 - \zeta^2}} \tag{5.90}$$

Hence the step response of an underdamped second order model is

$$y(t) = 1 - e^{-\zeta\omega_n t}\left(\cos \omega_d t + \frac{\zeta}{\sqrt{1 - \zeta^2}} \sin \omega_d t\right) \tag{5.91}$$

Critically Damped Case, $\zeta = 1$. The step response in this case can be found simply by taking a limit of the right-hand side of Equation (5.91) for ζ approaching unity, which yields

$$y(t) = 1 - e^{-\omega_n t}(1 + \omega_n t) \tag{5.92}$$

The same result can also be obtained by following the general procedure for solving linear differential equations presented in Section 5.2. The characteristic equation in this case has a double root.

$$p_1 = p_2 = -\omega_n \tag{5.93}$$

The general form of the homogenous solution is as given by Equation (5.10).

$$y_h(t) = K_1 e^{-\omega_n t} + K_2 t e^{-\omega_n t} \tag{5.94}$$

The particular solution for a unit step input is as before

$$y_p(t) = 1 \tag{5.95}$$

and hence the complete solution is

$$y(t) = 1 + K_1 e^{-\omega_n t} + K_2 t e^{-\omega_n t} \tag{5.96}$$

Using the initial conditions, $y(0) = 0$ and $\dot{y}(0) = 0$, the integration constants are found to be

$$K_1 = -1 \tag{5.97}$$

$$K_2 = -\omega_n \tag{5.98}$$

Substitution of the expressions for the constants K_1 and K_2 into Equation (5.94) yields the complete solution as obtained earlier, Equation (5.92).

Overdamped Case, $\zeta > 1$. When $\zeta > 1$, the characteristic equation (5.83) has two distinct real roots given by Equation (5.70). The homogeneous solution takes the following form

$$y_h(t) = K_1 e^{-t/\tau_1} + K_2 e^{-t/\tau_2} \tag{5.99}$$

The particular solution for a unit step input again is unity

$$y_p(t) = 1 \tag{5.100}$$

Combining (5.99) and (5.100) gives the complete solution

$$y(t) = 1 + K_1 e^{-t/\tau_1} + K_2 e^{-t/\tau_2} \tag{5.101}$$

The integration constants are found by using the zero initial conditions

$$K_1 = \frac{-\tau_1}{(\tau_1 - \tau_2)} \tag{5.102}$$

$$K_2 = \frac{\tau_2}{(\tau_1 - \tau_2)} \tag{5.103}$$

The complete solution for a unit step response of an overdamped second order model is

$$y(t) = 1 - \frac{1}{\tau_1 - \tau_2}(\tau_1 e^{-t/\tau_1} - \tau_2 e^{-t/\tau_2}) \tag{5.104}$$

A family of unit step response curves for a second order model with zero initial conditions is plotted in Figure 5.12 for different values of the damping ratio and the same natural frequency.

A dynamic behavior of a second order model described by Equation (5.82) can also be described in terms of selected specifications of the model step response. Figure 5.13 shows the most common specifications of a step response of an underdamped second order model. Although the mathematical relationships that fol-

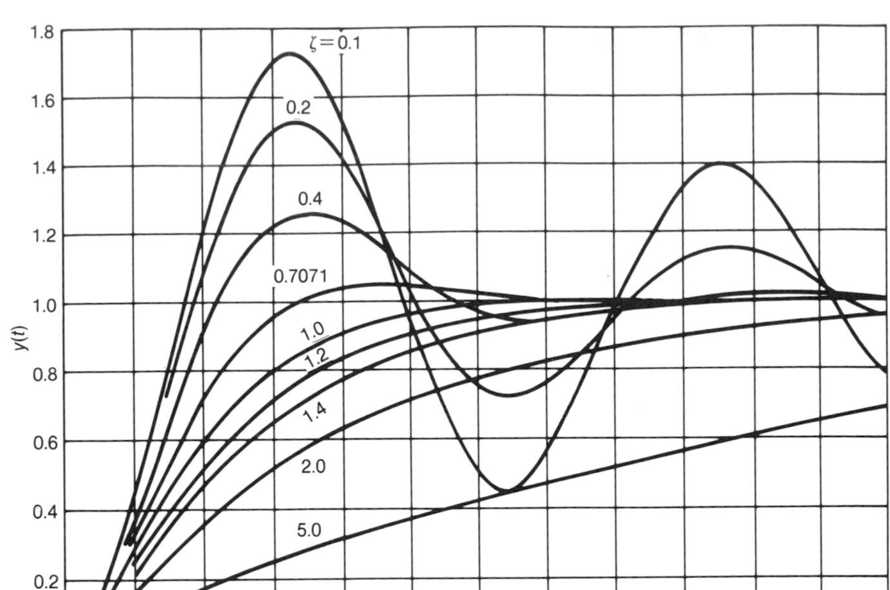
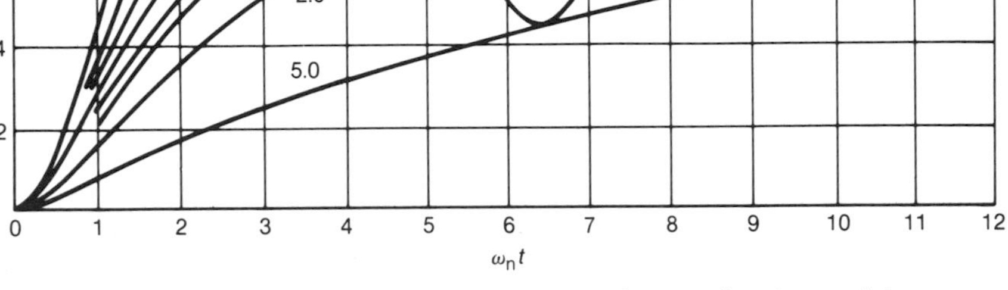

FIGURE 5.12 Unit step response curves of second-order model.

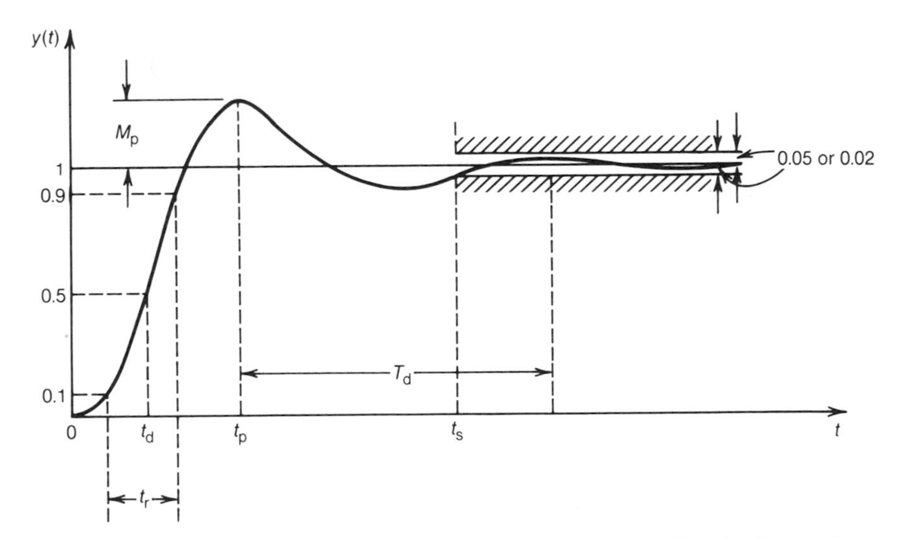

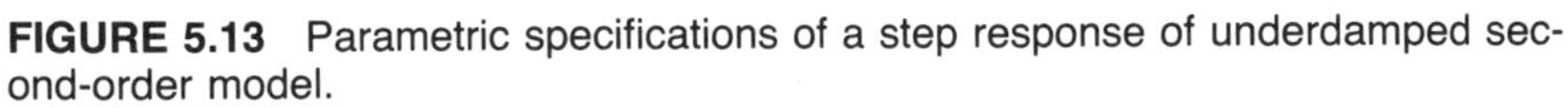
FIGURE 5.13 Parametric specifications of a step response of underdamped second-order model.

low apply only to a second order model with zero initial conditions, the use of the presented step response specifications is not limited to a single type of model.

The period of oscillations T_d is related to the damped natural frequency by the following equation

$$T_d = \frac{2\pi}{\omega_d} \tag{5.105}$$

The peak time t_p is the time between the start of the step response and its first maximum. It is equal to half of the period of oscillations.

$$t_p = \frac{\pi}{\omega_d} \tag{5.106}$$

The peak time is a measure of the speed of response of an underdamped system. For a critically damped or overdamped system the speed of response is usually represented by a delay time or a rise time.

The delay time t_d is the time necessary for the step response to reach a point half way between the initial value and the steady-state value, which can be expressed mathematically as

$$y(t)|_{t=td} = 0.5[y_{ss} - y(0^+)] + y(0^+) \tag{5.107}$$

The rise time t_r is the time necessary for the step response to rise from 10 percent to 90 percent of the difference between the initial value and the steady-state value.

Settling time t_s is defined as the time required for the step response to settle within a specified percentage of the steady-state value. A 2 percent settling time is the time for which the following occurs

$$|y(t) - y_{ss}| \leq 0.02(y_{ss} - y(0^+)), \text{ for } t \geq t_s \tag{5.108}$$

An oscillatory character of a system step response is represented by a maximum overshoot, M_p, defined as

$$M_p = y_{max} - y_{ss} \tag{5.109}$$

A percent maximum overshoot, $M_p^{\%}$ is used more often.

$$M_p^{\%} = \left(\frac{y_{max} - y_{ss}}{y_{ss} - y(0^+)}\right) 100\% \tag{5.110}$$

For a second-order model, the percent maximum overshoot can be expressed as a function of the damping ratio

$$M_p^{\%} = 100e^{-\left(\frac{\pi\zeta}{\sqrt{1-\zeta^2}}\right)} \tag{5.111}$$

Relationship (5.111) is presented in a graphical form in Figure 5.14.

Another useful specification of the transient response of an underdamped second-order system is a decay ratio DR, defined as the ratio of successive amplitudes of

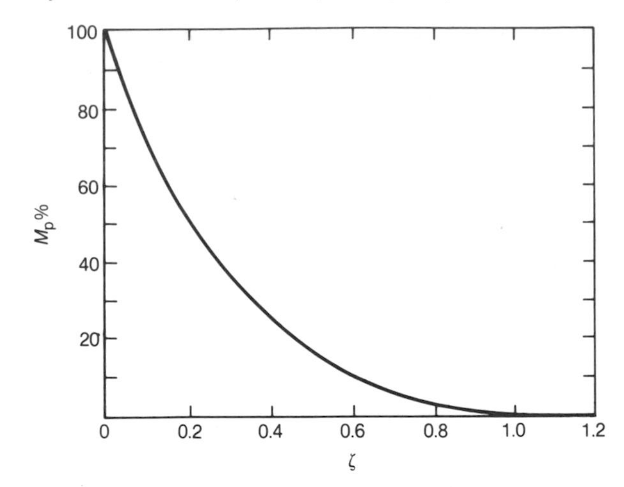

FIGURE 5.14 Maximum percent overshoot versus damping ratio.

the system step response. Referring to Figure 5.13, we find the decay ratio can be expressed as

$$\text{DR} = \frac{[y(t_p + T_d) - y_{ss}]}{[y(t_p) - y_{ss}]} \tag{5.112}$$

For a system described by Equation (5.82), the decay ratio can be related to the damping ratio by the following formula.

$$\text{DR} = e^{-\frac{2\pi\zeta}{\sqrt{1-\zeta^2}}} \tag{5.113}$$

A logarithmic decay ratio LDR is sometimes used instead of the decay ratio DR, where

$$\text{LDR} = \ln(\text{DR}) = \frac{-2\pi\zeta}{\sqrt{1 - \zeta^2}} \tag{5.114}$$

The decay ratio is useful in determining system damping ratio from the system oscillatory step response.

EXAMPLE 5.2

In the system shown in Figure 5.15a, mass $m = 9$ kg is subjected to force $F(t)$ acting vertically and undergoing a step change from 0 to 1.0 N at time $t = 0$. The mass, suspended on a spring of constant $k = 4.0$ N/m, is moving inside an enclosure with a coefficient of friction between the surfaces $b = 4.0$ N-sec/m. Using force $F(t)$ as the input variable and the position of mass $x(t)$ as the output variable, sketch an approximate step response of the system. If this response is oscillatory, determine the necessary modification to make the system critically damped.

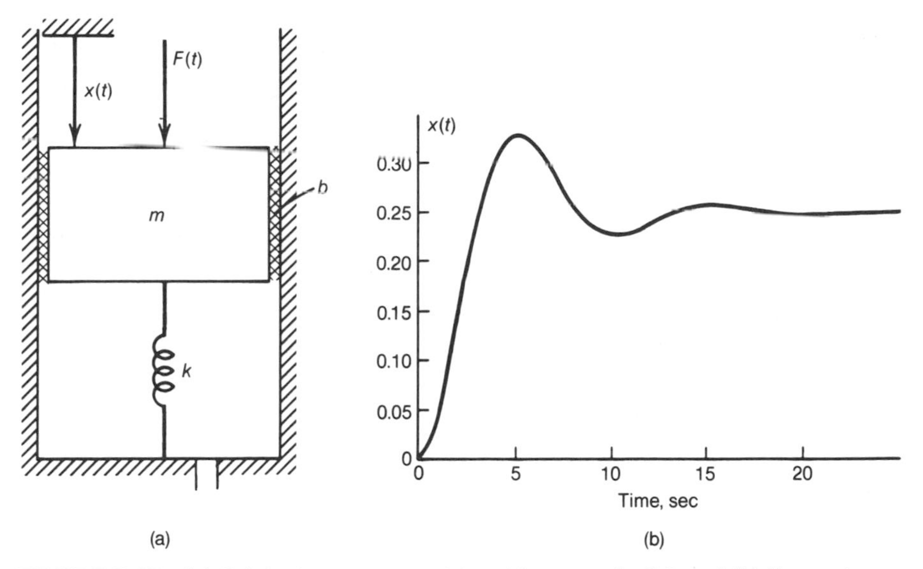

FIGURE 5.15 (a) Original system considered in example 5.2 and (b) the system unit step response.

Solution

The input-output equation of the system is

$$m\ddot{x} + b\dot{x} + kx = F(t)$$

Substitution of the numerical values for the system parameters yields

$$9\ddot{x} + 4\,\dot{x} + 4x = F(t)$$

From Equation (5.66), the damping ratio is

$$\zeta = \frac{4}{2\sqrt{9 \cdot 4}} = 0.3333$$

Since $\zeta < 1$, the system step response will be oscillatory. Other step response specifications useful in sketching the step response curve can be found from Equations (5.65), (5.74), (5.105), and (5.111) as follows

$$\omega_n = \sqrt{4/9} = 0.6667 \text{ rad/sec}$$

$$\omega_d = 0.6667\sqrt{1 - 0.3333^2} = 0.6286 \text{ rad/sec}$$

$$T_d = 10.0 \text{ sec}$$

$$M_p^{\%} = 100e^{-\frac{\pi \cdot 0.3333}{\sqrt{1-0.3333^2}}} = 32.94\%$$

Given these parameters the step response can be sketched as shown in Figure 5.15b.

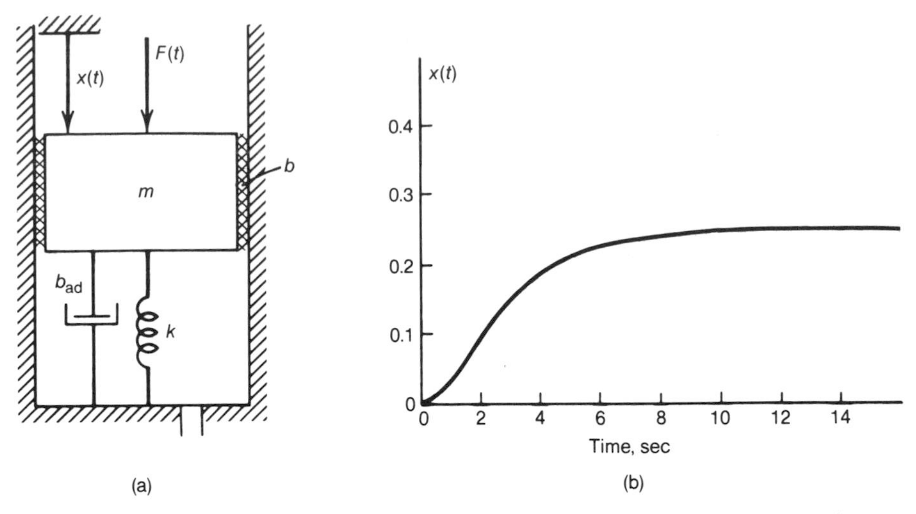

FIGURE 5.16 (a) Modified system from Example 5.2 and (b) the system unit step response.

In order to make the system critically damped, another damper, b_{ad}, is added as shown in Figure 5.16a. The input-output equation of the modified system is

$$9\ddot{x} + (4 + b_{ad})\dot{x} + 4x = F(t)$$

The damping ratio is now given by the expression

$$\zeta = \frac{(4 + b_{ad})}{2\sqrt{4 \cdot 9}}$$

For critical damping the term on the right-hand side of the preceding equation must be equal to 1.0, which yields the value of additional damping, $b_{ad} = 8$ N-sec/m. The step response curve of the system with additional damping is shown in Figure 5.16b. ■

5.6 Third- and Higher-order Models

Theoretically, the analysis of linear models of third and higher order should be nothing more than a simple extension of the methods developed for first- and second-order models, presented in earlier sections. Practically, however, methods that are fast and easy to use with lower-order models become excessively involved and cumbersome in applications involving third- and higher-order models. An analytical solution of an nth-order input-output equation for a linear, stationary model can be obtained by simply following the same general procedure described in Section 5.2; however, the algebra involved in the solution is considerably more complex. First- and second-order linear differential equations are not only easy to solve analytically but in addition, the parameters used in these equations have a straightforward physical meaning, which allows evaluation of general system characteristics even without solving the model equation. The physical meaning of the parameters in third- and higher-order model equations is considerably less clear and more difficult to interpret in terms of the system dynamic properties.

Another apparently insignificant problem occurs in obtaining the homogenous solution, which requires determining the roots of a third- or higher-order characteristic equation. There are many computer programs for solving higher-order algebraic equations; however, in such situations the following question arises: If it becomes necessary to use a computer to solve part of a problem (find roots of a characteristic equation), wouldn't it be worthwhile to use the computer to solve the entire problem (find the solution of an input-output equation)?

Computer programs for solving linear differential equations, or rather, for solving equivalent sets of first-order equations are almost as readily available and easy to use as the programs for finding roots of a characteristic equation. Some of these programs will be described in Chapter 6. Generally speaking, the higher the order of the input-output equation, the more justified and more efficient is the use of a computer. As has been and still will be stressed throughout this text, however, it is always necessary to verify the computer-generated results with a simple approximate analytical solution. In this section a concept of so-called dominant roots of a characteristic equation will be presented. The dominant roots are very useful in obtaining approximate solutions of third- or higher-order differential equations.

The general form of a homogenous solution of a nth-order differential equation having m distinct and $n - m$ multiple roots, presented in Chapter 4, is

$$y_h(t) = \sum_{i=1}^{m} K_i e^{p_i t} + \sum_{i=m+1}^{n} K_i t^{i-m-1} e^{p_i t} \tag{5.115}$$

If a system is stable, all roots, $p_1, p_2, \ldots, p_n$, are real and negative or complex and have negative real parts. The rate at which the exponential components on the right-hand side of Equation (5.115) decay depends in the magnitudes of the real parts of $p_1, p_2, \ldots, p_n$. The larger the magnitude of a negative real root and/or the larger the magnitude of a negative real part of a complex root, the faster the decay of the corresponding exponential terms in the model free response. In other words, the roots of the characteristic equation located farther away from the imaginary axis in the left half of the complex plane affect the model free response relatively less than the roots closer to the imaginary axis. The roots of the characteristic equation nearest to the imaginary axis in the complex plane are called the dominant roots.

The concept of dominant roots will now be illustrated in an example of a third-order model. The model is assumed to have distinct, real and negative roots, p_1, p_2, p_3, with the last root, p_3, being much farther away from the imaginary axis of the complex plane than p_1 and p_2, so that

$$|p_3| >> |p_1| \quad \text{and} \quad |p_3| >> |p_2| \tag{5.116}$$

The model differential equation is

$$\frac{d^3y}{dt^3} - (p_1 + p_2 + p_3)\frac{d^2y}{dt^2} + (p_1p_2 + p_1p_3 + p_2p_3)\frac{dy}{dt} - p_1p_2p_3y = 0 \tag{5.117}$$

Hence, the characteristic equation is

$$p^3 - (p_1 + p_2 + p_3)p^2 + (p_1p_2 + p_1p_3 + p_2p_3)p - p_1p_2p_3 = 0 \tag{5.118}$$

Dividing both sides of the characteristic equation by p_3 gives

$$\left(\frac{1}{p_3}\right)p^3 - \left(\frac{p_1}{p_3} + \frac{p_2}{p_3} + 1\right)p^2 + \left(\frac{p_1p_2}{p_3} + p_1 + p_2\right)p - p_1p_2 = 0 \tag{5.119}$$

The terms having p_3 in denominator can be neglected, based on assumptions (5.116), to yield the following approximation of the characteristic equation.

$$p^2 - (p_1 + p_2)p + p_1p_2 = 0 \tag{5.120}$$

The corresponding differential equation is

$$\frac{d^2y}{dt^2} - (p_1 + p_2)\frac{dy}{dt} + p_1p_2y = 0 \tag{5.121}$$

It should be noted that both p_1 and p_2 are assumed to be real and negative. These are the dominant roots of the model Equation (5.117). The solution of Equation (5.121) will thus approximate the solution of Equation (5.117), with the approximation being better the larger is the distance between the dominant roots, p_1 and p_2, and the other root, p_3, measured in a horizontal direction in the complex plane.

The simplification procedure just described should be used with caution because it leads to a system model that may be deficient in some applications. For instance, the small time constant associated with the rejected root p_3, $\tau_3 = -1/p_3$, may still be significant in some closed-loop systems employing high gain feedback.

In addition to simplifying the analysis, the rejection of a nondominant root (or roots) makes it easier to run a check solution on the computer and verify it before proceeding with the more complete model when it is required.

5.7 Synopsis

This chapter has reviewed the classical methods frequently employed to solve for the dynamic response of linear systems modeled by ordinary differential equations. Greatest emphasis has been placed on finding the responses of first- and second-order systems to step inputs; however responses to non-equilibrium initial conditions (so-called free response) and impulse response have also been covered.

In each case the solution was shown to be of the form

$$y(t) = y_p(t) + y_h(t)$$

where $y_p(t)$ is the particular or forced part of the response, usually having the same form as the input and/or its derivatives, and $y_h(t)$ is the homogeneous or natural part of the response consisting of exponential terms employing the roots of the system characteristic equation.

Thus the solution for a first-order system is

$$y(t) = y_p(t) + K_1 e^{pt} = y_p(t) + K_1 e^{-t/\tau}$$

where $p = -1/\tau$ is the single root of the first-order characteristic equation. The constant K_1 is then found from using the initial condition equation with the values of the particular part, the homogeneous part, and the output at $t = 0^+$

$$y(0^+) = y_p(0^+) + K_1 e^0$$

or

$$K_1 = y(0^+) - y_p(0^+)$$

For a second-order characteristic equation of the form

$$a_2 p^2 + a_1 p + a_0 = 0$$

the roots may be real or conjugate complex. For the underdamped case, the conjugate complex roots, involving the undamped natural frequency and damping ratio, are expressed by Equation (5.68). For the overdamped and critically damped cases, the real roots, involving two time constant τ_1 and τ_2, are expressed by Equation (5.71).

The solution for a second-order system is of the form

$$y(t) = y_p(t) + y_h(t) = \begin{cases} K_a e^{-\zeta\omega_n t} \cos(\omega_d t + \psi) & , \quad 0 < \zeta < 1 \\ y_p(t) + K_1 e^{p_1 t} + K_2 t e^{p_1 t} & , \quad \zeta = 1 \\ K_1 e^{p_1 t} + K_2 e^{p_2 t} & , \quad \zeta > 1 \end{cases}$$

where ω_n, ζ, and $\omega_d = \omega_n\sqrt{1 - \zeta^2}$ are the undamped natural frequency, the damping ratio, and the damped natural frequency respectively of an underdamped ($0 < \zeta < 1$) system having complex conjugate roots, p_1 repeated are the identical roots of a critically damped ($\zeta = 1$) system, and $p_1 = -1/\tau_1$ and $p_2 = -1/\tau_2$ are the real roots of an overdamped ($\zeta > 1$) system.

In order to assess each new situation as it arises, solving first for $\zeta = a_1/(2\sqrt{a_2 a_0})$ determines immediately which case is to be dealt with. Then when $y_p(t)$ has been determined from the form of the input, the constants K_1 and K_2, or the constant K_a and phase angle ψ, are readily determined from the two initial condition Equations (5.54).

Although the use of the classical methods for the finding responses of higher-order systems becomes tedious, for the case of a third-order system where one of the roots must be real, it is relatively easy to find that first root by using a simple iteration starting with an educated guess, employing a hand-held calculator. The remaining second-order polynomial, factored out by long division, is then solved for the remaining roots as before. The constants K_1, K_2, and K_3 or K_1, K_a, and ψ are then found through the use of the three initial condition equations,

$$y(0^+) = y_p(0^+) + y_h(0^+)$$
$$\dot{y}(0^+) = \dot{y}_p(0^+) + \dot{y}_h(0^+)$$
$$\ddot{y}(0^+) = \ddot{y}_p(0^+) + \ddot{y}_h(0^+)$$

Greatest emphasis has been placed on first- and second-order systems in order to simplify the illustration of the classical method of solving differential equations. In addition, first- and second-order models exhibit the major features encountered in the response of higher-order systems: exponential decay with time constant/s, and/or oscillatory response with decaying amplitude of oscillation, which are also present in the responses of third- and higher-order systems. Having a good understanding of first- and second-order system response behaviour facilitates the verification of computer simulations and the ''debugging'' of computer programs when they are being developed for computer simulation.

Using a simplified model employing only the dominant roots of a higher-order system makes it possible to uncover the most significant dynamic response characteristics of the higher-order model through the use of a lower order model. Also, running a reduced-order check solution as part of a simulation study helps to find program errors and ''glitches'' that seem to plague even the most experienced programmers. Once this simplified model is running properly on the computer, it is usually a simple matter to reinsert the less dominant roots and to then produce a ''full-blown'' solution which includes all of the higher order effects inherent in the higher-order system differential equation model of the system.

Problems

5.1. A first-order model of a dynamic system is

$$2\dot{y} + 5y = 5u(t)$$

- **(a)** Find and sketch the response of this system to the unit step input signal $u(t) = U_s(t)$, for $y(0^+) = 2$.
- **(b)** Repeat part (a) for zero initial condition, $y(0^+) = 0$.
- **(c)** Repeat part (a) for a unit impulse input $u(t) = U_i(t)$.
- **(d)** Repeat part (b) for a unit impulse input, $u(t) = U_i(t)$.

5.2. The roots of a second-order model are $p_1 = -1 + j$ and $p_2 = -1 - j$.

- **(a)** Find and sketch the system unit step response assuming zero initial conditions, $y(0^+) = 0$ and $\dot{y}(0^+) = 0$.
- **(b)** Repeat part (a) for the roots of the characteristic equation $p_1 = 1 + j$ and $p_2 = 1 - j$. Explain the major difference between the step responses found in parts (a) and (b).

5.3. A rotational mechanical system has been modeled by the equation

$$J\dot{\Omega} + B\Omega = T(t)$$

Determine the values of J and B for which the following conditions are met:

- Steady-state rotational velocity for a constant torque, $T(t) = 10$ N-m, is 50 rpm (revolutions per minute).
- The speed drops below 5 percent of its steady-state value within 160 msec after the input torque is removed, $T(t) = 0$.

5.4. Output voltage signals $y_1(t)$ and $y_2(t)$ of two linear first-order electrical circuits were measured as shown in Figures P5.4a and b. Write analytical expressions describing the two signals.

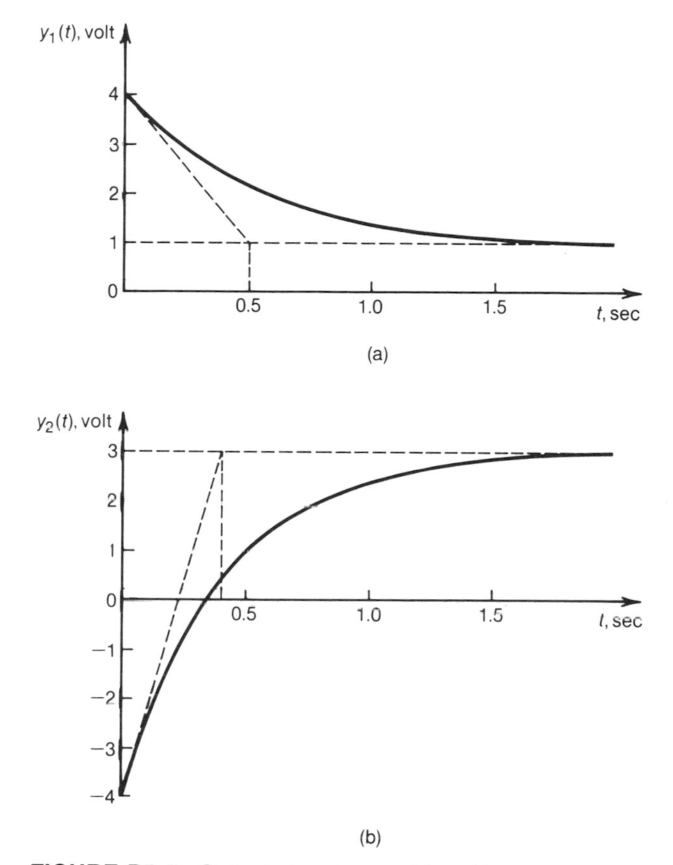

FIGURE P5.4 Output signals considered in Problem 5.4.

5.5. A mass $m = 1.5$ lb-sec^2/ft sliding on a fixed guideway is subjected to a suddenly applied constant force $F(t) = 100$ lb at time $t = 0$. The coefficient of linear friction between the mass and the guideway is $b = 300$ lb-sec/ft. Find the system time constant. Write the system model equation and solve it for the response of mass velocity v as a function of time, assuming $v(0^-) = 0$. Sketch and label the system response versus time.

5.6. The electric generator in the steam turbine drive system shown in Figure P5.6 has been running steadily at speed $(\Omega_{1g})_0$ with a constant input steam torque, T_{steam}. At time $t = t_0$ the shaft, which has a developing fatigue crack, breaks suddenly. This problem is concerned with how the speed Ω_{1g} varies with time after the shaft breaks (what happens to the generator is of no concern here).

(a) Find the steady torque $(T_g)_0$ in the shaft before the shaft breaks.

(b) Find the time constant of the remaining part of the system to the left of the crack for $t > t_0$.

(c) Determine the initial condition $\Omega_{1g}(t_0^+)$

(d) Solve for the response $\Omega_{1g}(t)$ for $t > t_0$ and sketch the response.

(e) Determine how long it takes (in terms of the system time constant) for $\Omega_{1g}(t)$ to reach a 5% overspeed condition.

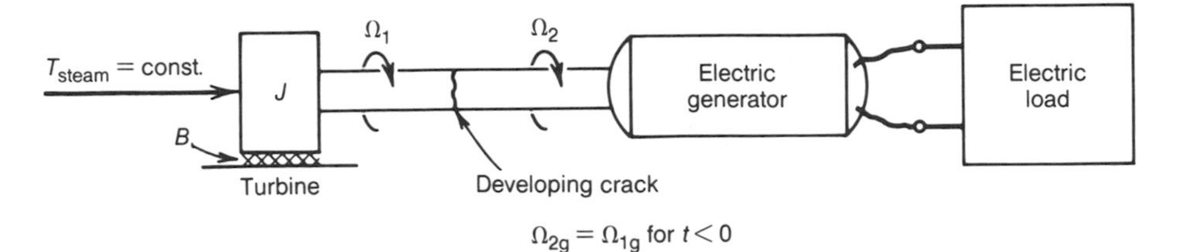

FIGURE P5.6 Steam turbine drive system considered in Problem 5.6.

5.7. A first-order system has been modeled by the following equation

$$\dot{y}(t) = ay(t) + bu(t)$$

With the system initially at rest, a unit step input $u(t) = U_s(t)$ was applied. Two measurements of the output signal were taken

$$y(0.5) = 1.2 \quad \text{and} \quad \lim_{t \to \infty} y(t) = 2.0$$

Find a and b.

5.8. A mechanical system is described by the following set of state variable equations:

$$\dot{q}_1 = -6q_1 + 2q_2$$
$$\dot{q}_2 = -6q_2 + 5u$$

The state variable q_1 is also the output variable. Sketch a unit step response of this system.

5.9. A mechanical device shown schematically in Figure P5.9 is used to measure a coefficient of friction between a rubber shoe and a pavement surface. The value of mass m is 4 kg, but the value of the spring constant is not exactly known. In a measuring procedure mass m is subjected to force $F(t)$, which changes suddenly from 0 to 10 N at time $t = 0$. The position of the mass $x(t)$ is recorded for $t \geq 0$. Based on the system step response curve recorded as shown in Figure P5.9b, find the unknown system parameters, k and b.

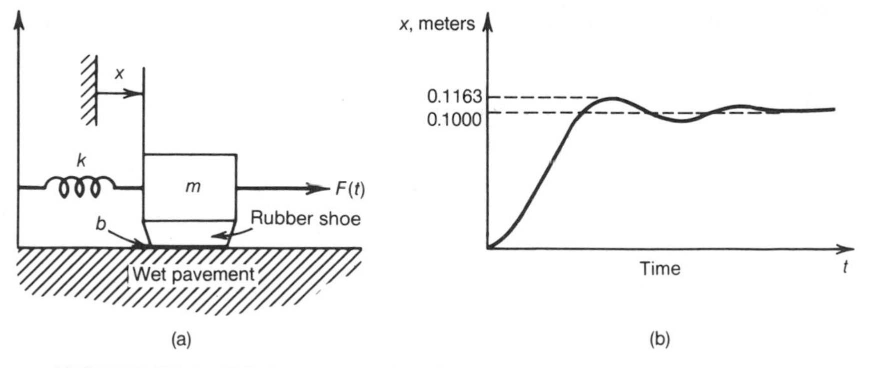

FIGURE P5.9 Friction-measuring device and its step response curve.

5.10. Consider again the mechanical device shown in Figure P5.9 but with mass m of 5 kg. The response of this system to a step change in force $F(t)$ was found to be very oscillatory (Figure P5.10). The only measurements obtained were two successive amplitudes, A_1 and A_2, equal to 55 cm and 16.5 cm, respectively, and the period of oscillation T_d equal to 1 sec. Determine the values of the spring constant k and the coefficient of friction b in this case.

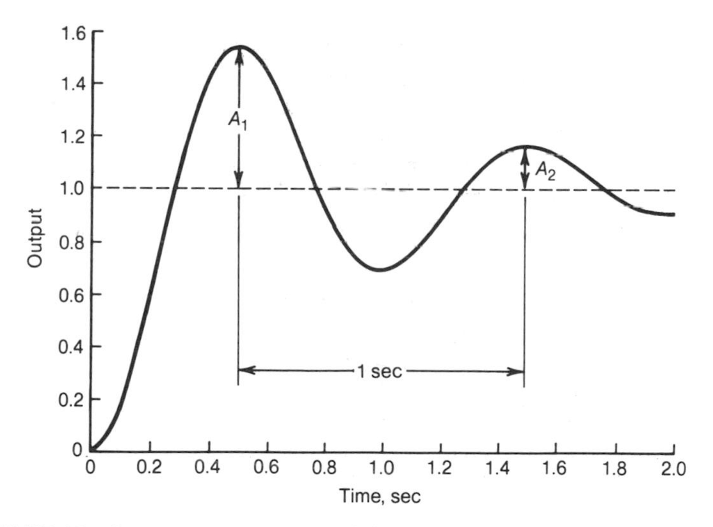

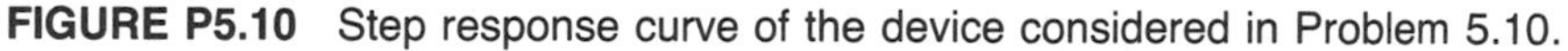

FIGURE P5.10 Step response curve of the device considered in Problem 5.10.

5.11. For each of the three mechanical systems shown in Figures P5.11a, b, and c, do the following:

(a) Derive an input-output equation.

(b) Write an expression for the unit step response.

(c) Sketch and carefully label the unit step response curve.

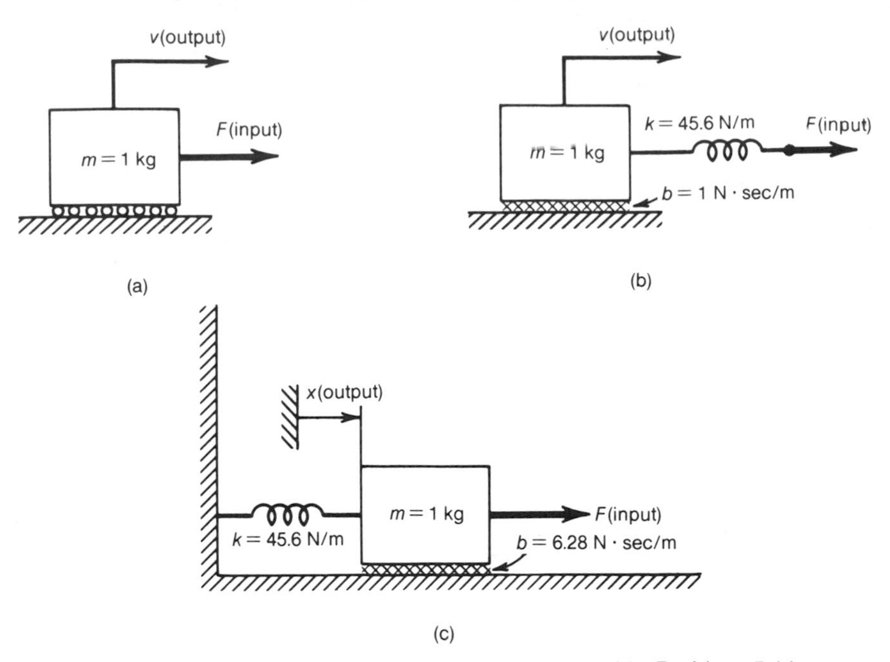

FIGURE P5.11 Mechanical systems considered in Problem 5.11.

5.12. An input-output model of a third-order system was found to be

$$\dddot{y} + 12\ddot{y} + 25\dot{y} + 50y = u(t)$$

The system step response for $u(t) = 50U_s(t)$ is plotted in Figure P5.12. Compute the roots for this third-order system and then find an approximating second-order model for this system using the dominant roots of the third-order model. Sketch the response of the approximating second-order model

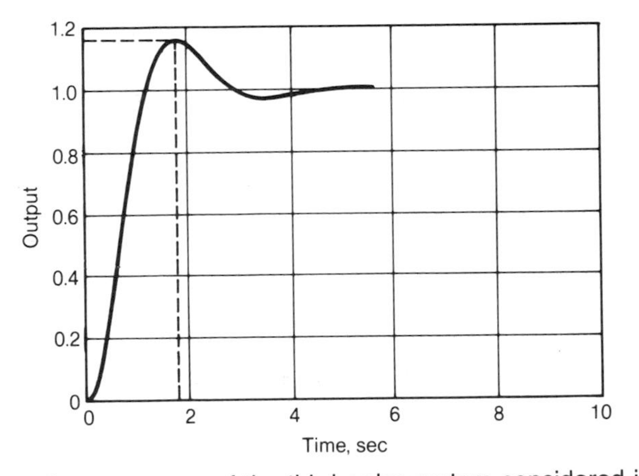

FIGURE P5.12 Step response of the third-order system considered in Problem 5.12.

to input $u(t) = 50U_s(t)$ and compare it with the curve shown in Figure P5.12.

5.13. A schematic of the mechanical part of a drive system designed for use in a drilling machine is shown in Figure P5.13. The driving torque T_m supplied by an electric motor is applied through a gear reduction unit having ratio R_1/R_2 to drive a drilling spindle represented here by mass m_s. The gears moments of inertia are J_1 and J_2 and their equivalent coefficient of rotational friction is B_{eq}. The spindle is suspended on air bearings of negligible friction, and it is pulled using a steel cable, which is assumed massless and has a spring constant k_s.

In order to examine the basic dynamic characteristics of the mechanical part of the system, the motor was shut off, $T_m = 0$, and an impact force F_{imp} was applied to the spindle. The response of the system, measured as the position $x(t)$ of the spindle, was found to be excessively oscillatory. In order to identify the source of this oscillation, determine the locations of the roots of the system characteristic equation and suggest how the system parameters should be changed, relative to their values during the test, to provide more damping.

Hint: Find approximate analytical expressions for the real roots and/or ζ and ω_n associated with complex conjugate roots. Assume that the one real root of the equation

$$p^3 + a_2p^2 + a_1p + a_0 = 0$$

is approximated by $p_1 = -a_0/a_1$.

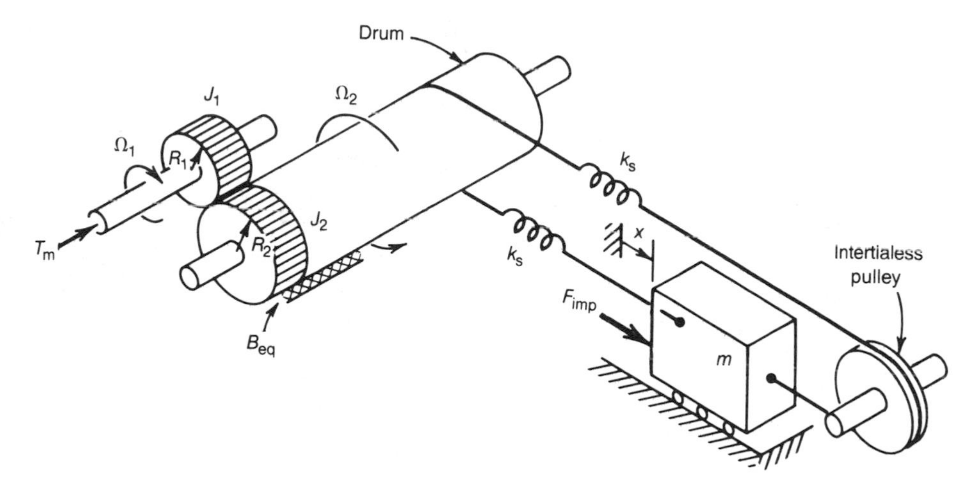

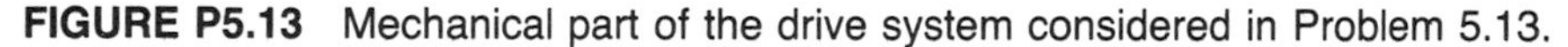
FIGURE P5.13 Mechanical part of the drive system considered in Problem 5.13.

6

Digital Computer Simulation

6.1 Introduction

For centuries engineers and scientists have sought help from calculating machines of all kinds in solving mathematical equations modeling dynamic systems. A digital computer, the most recent version of the calculating machine, has come a long way from an over 5000 year old Babylonian abacus. Most engineers would probably prefer a computer to an abacus because of its superior computational power, but both devices are only capable of performing those tasks that engineers already know how to perform but either choose not to do, for some reason, or can't do because of lack of sufficient speed or memory or both. Despite today's fascination with modern digital computers, it is important to remember that they can only do what they are programmed for within their vast yet finite performance limits.

Generally speaking, correctness of computer results depends on two conditions: the correctness of the formulation of the problem and the computational capability of the computer to solve it. It does not seem widely recognized how often at least one of the two conditions is not met, leading to worthless computer results. Great care must therefore always be taken in considering computer output. These authors' advice is never to accept computer results unless they can be fully understood and verified against the basic laws of physics.

In the next two sections selected computer methods for solving ordinary differential equations are presented. In Section 6.2 a classical Euler's method is described. The method of Euler has more historical than practical significance today due to its large computational error. More accurate methods, including an improved Euler, fourth-order Runge-Kutta, and exponential Euler method are introduced in Section 6.3. In Section 6.4 the effect of the size of the integration step on accuracy of numerical integration is discussed.

6.2 Euler's Method

In 1768 Leonhard Euler, the most prolific mathematician of the eighteenth century, and perhaps of all time, published the first numerical method for solving first-order differential equations of the following general form

$$\frac{dx}{dt} = f(x, t) \tag{6.1}$$

with an initial condition, $x(t_0) = x_0$. Note that system state variable Equations (4.6) as well as first-order input-output equations (5.21) can be presented in the form of Equation (6.1). In the case of the input-output model the dependent variable is y and Equation (6.1) becomes

$$\frac{dy}{dt} = f(y, t) = -\left(\frac{a_0}{a_1}\right) y + \left(\frac{1}{a_1}\right) u(t) \tag{6.2}$$

The method proposed by Euler was based on a finite difference approximation of a continuous first derivative dx/dt using a formula derived by another great mathematician, contemporary with Euler, Brook Taylor. Taylor's approximation of a first-order continuous derivative defined as

$$\frac{dx}{dt} = \lim_{\Delta t \to 0} \frac{[x(t_0 + \Delta t) - x(t_0)]}{\Delta t} \tag{6.3}$$

is

$$\frac{dx}{dt} \approx \frac{[x(t_0 + \Delta t) - x(t_0)]}{\Delta t} \tag{6.4}$$

Hence the estimate of the value of function x at time $t_0 + \Delta t$ is

$$x(t_0 + \Delta t) \approx x(t_0) + \left.\frac{dx}{dt}\right|_{t_0} \Delta t \tag{6.5}$$

Substituting for dx/dt from Equation (6.1) gives the solution equation

$$x(t_0 + \Delta t) \approx x(t_0) + f(x, t_0)\,\Delta t \tag{6.6}$$

Figure 6.1 shows a geometric interpretation of the Euler's method.

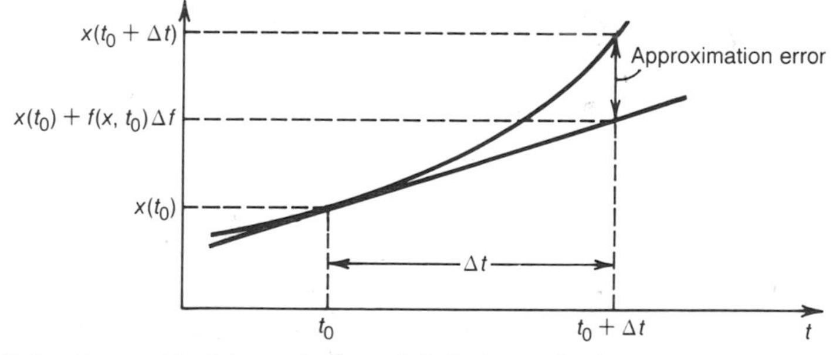

FIGURE 6.1 Geometric interpretation of Euler's method.

Equation (6.6) provides a recursive algorithm for computation of $x(t_0 + k\Delta t)$, $k = 1, 2, \ldots, N$. The Euler's solution procedure is marching from the initial time, t_0, to the final time, $t_f = t_0 + N\Delta t$ with a constant time step, Δt. The following equations are solved successively in the computational process

$$\begin{aligned} x(t_0 + \Delta t) &= x_0 + f(x, t_0)\Delta t \\ x(t_0 + 2\Delta t) &= x(t_0 + \Delta t) + f[x, (t_0 + \Delta t)]\Delta t \\ &\vdots \\ x(t_0 + N\Delta t) &= x(t_0 + (N - 1)\Delta t) + f[x, (t_0 + (N - 1)\Delta t)]\Delta t \end{aligned} \tag{6.7}$$

In general, in the case of an n-th order system represented by a set of n state variable Equations (4.6), a corresponding set of difference equations, one for each state variable, similar to Equation (6.6) has to be provided. The state variable equations can be rearranged into the form of Equation (6.1) as follows

$$\frac{dq_i}{dt} = f_i(\mathbf{q}, t) \qquad i = 1, 2, \ldots, n \tag{6.8}$$

where the state vector $\mathbf{q}$ is a column vector containing $q_1, q_2, \ldots, q_n$, and $f_i(\mathbf{q}, t)$ includes the inputs $u_j(t)$, $(j = 1, 2, \ldots, 1)$, as well as the state and input matrix coefficients. At each step of the numerical solution process a set of Equations (6.9) will be solved

$$q_i(t_0 + \Delta t) = q_i(t_0) + f_i(\mathbf{q}, t_0)\Delta t \qquad i = 1, 2, \ldots, n \tag{6.9}$$

Euler's method will be illustrated by Example 6.1.

EXAMPLE 6.1

Use Euler's method to obtain a numerical solution of the differential equation

$$4\frac{dx}{dt} + x = 4$$

over a period of time from 0 to 12 sec. The initial condition is $x(0^-) = 10$, and the system time constant is 4 sec.

Solution
First rewrite the differential equation in the same form as Equation (6.1).

$$\frac{dx}{dt} = -0.25x + 1$$

Using Euler's formula (6.6), the successive values of x are calculated as follows

$$x(1) = x(0) + (-0.25x(0) + 1)\Delta t$$
$$x(2) = x(1) + (-0.25x(1) + 1)\Delta t$$
$$\vdots$$

To find the numerical values of x, the size of the time step, Δt, must be selected. Selection of Δt has a great impact on the accuracy and speed of numerical solution procedure and will be discussed more in detail in Section 6.4. A rule of thumb for selection of the time step is to use $\Delta t = \tau/4$, where τ is the system time constant. In this example the time constant is 4 sec and thus Δt is chosen to be equal to 1 sec. The recursive numerical solution formula becomes

$$x(k) = 0.75x(k - 1) + 1 \qquad \text{for } k = 1, 2, \ldots, 12$$

The numerical results are given in Table 6.1 together with the values of the exact analytical solution of the original differential equation, which was found to be

$$x(t) = 6e^{-t/4} + 4$$

TABLE 6.1. Euler Solution Versus Exact Solution of the Equation from Example 6.1.

t	Euler	Exact
0	10.000000	10.000000
1	8.500000	8.672805
2	7.375000	7.639184
3	6.531250	6.834199
4	5.898438	6.207277
5	5.423828	5.719029
6	5.067871	5.338781
7	4.800904	5.042644
8	4.600678	4.812012
9	4.450508	4.632395
10	4.337881	4.492510
11	4.253411	4.383567
12	4.190058	4.298722

A simple BASIC program using Euler's method to solve the first order differential equation in Example 6.1 is listed in Table 6.2.

TABLE 6.2. Computer Program in BASIC for Solving Differential Equation $dx/dt = -0.25x + 1$

```
10 REM Numerical solution of first order differential equation
20 REM dx/dt=f(x, t) using Euler's method.
```

```
30 INPUT"Initial independent variable = ",T0
40 INPUT"Final independent variable = ",TF
50 INPUT"Number of integration steps = ",N
60 DELT=(TF-T0)/N
70 PRINT"Integration step = ";DELT
80 DIM X(N+1),T(N+1)
90 INPUT"Initial dependent variable = ",X0
100 T(0)=T0
110 X(0)=X0
120 FOR I=1 TO N
130 XDOT=-.25*X(I-1)+1
140 X(I)=X(I-1)+XDOT*DELT
150 T(I)=T(I-1)+DELT
160 NEXT I
170 PRINT"TIME                    OUTPUT"
180 FOR I=0 TO N
190 PRINT T(I),X(I)
200 NEXT I
210 END
```

■

The program can be modified for solving any first-order differential equation by changing line 130, which should have the following form

```
130 XDOT=f(X(I-1),T(I-1))
```

where f(X,T) represents the right hand side in Equation (6.1).

6.3 More Accurate Methods

The Euler's method presented in the previous section is very simple and easy to use, but it is seldom used in serious computation because of its poor accuracy. In general, the accuracy of a numerical integration method can be improved in two ways: by using a more sophisticated algorithm for numerical approximation of continuous derivatives or by reducing the size of the integration step. This section addresses the former approach; the effect of the numerical integration step on accuracy will be discussed in Section 6.4.

The accuracy of the classical Euler's method presented in Section 6.2 can be considerably improved by a relatively simple modification of the finite difference formula (6.6). According to that equation, the value of x at the end of the integration time step is calculated assuming that the dependent function x varies over the duration of the time step in a linear fashion at a constant rate determined by the value of dx/dt at the beginning of the time step. In other words, the value of $x(t_0 + \Delta t)$ is estimated as a linear projection from $x(t_0)$ with the projection rate equal to dx/dt at time t_0. In reality the rate of change of x, determined by the continuous derivative of x with respect to time, varies during the duration of the time step, which results in a computation error. The magnitude of this error can be reduced by using a better approximation of dx/dt than its initial value. In the improved Euler's method, known also as Heun's method[1], presented here, the continuous derivatives are estimated as an average of the estimates at the beginning and at the end of the integration step. In the improved algorithm the number of computations of function $f(x, t)$ on the right hand side of Equation (6.1) is doubled, compared with the classical Euler's method, but the computational process remains uncomplicated and the improvement of the accuracy of the numerical solution is definitely worth the additional computational effort. The listing of the BASIC computer program for solution of the first-order differential equation considered in Example 6.1 is given in Table 6.3. The results shown in Table 6.4 indicate a significant improvement in the accuracy of the improved method over the classical Euler's method. The maximum error which for both methods occurs at $t = 4$ sec is reduced more than ten times when the improved algorithm is used.

TABLE 6.3. Computer Program Using an Improved Euler's Method for Solving Differential Equation $dx/dt = -.25x + 1$

```
10 REM Numerical solution of first order differential equation
20 REM dx/dt=f(x,t) using improved Euler's method
30 INPUT"Initial independent variable = ",T0
40 INPUT"Final independent variable = ",TF
50 INPUT"Number of integration steps = ",N
60 DELT=(TF-T0)/N
70 PRINT"Integration step = ";DELT
80 DIM X(N+1), T(N+1)
90 INPUT"Initial dependent variable = ",X0
100 T(0)=T0
110 X(0)=X0
120 FOR I=1 TO N
130 XDOT1=-.25*X(I-1)+1
140 X(I)=X(I-1)+XDOT1*DELT
150 XDOT2=-.25*X(I)+1
160 XDOT=(XDOT1+XDOT2)/2
170 X(I)=X(I-1)+XDOT*DELT
180 T(I)=T(I-1)+DELT
190 NEXT I
200 PRINT"TIME                    OUTPUT"
210 FOR I=0 TO N
220 PRINT T(I),X(I)
230 NEXT I
240 END
```

[1]W. S. Dorn and D D. McCracken, *Numerical Methods with Fortran IV Case Studies*, John Wiley & Sons, Inc., New York 1972, pp. 368–372.

TABLE 6.4. Improved Euler Solution Versus Exact Solution of the Equation Considered in Example 6.1

t	Improved Euler	Exact
0	10.0	10.0
1	8.687500	8.672805
2	7.662110	7.639184
3	6.861023	6.834199
4	6.235174	6.207277
5	5.746230	5.719029
6	5.364242	5.338781
7	5.065814	5.042644
8	4.832668	4.812012
9	4.650522	4.632395
10	4.508220	4.492510
11	4.397047	4.383567
12	4.310193	4.298722

The improved Euler's method is still relatively simple. There are many more accurate and also more complex methods available for solving first-order ordinary differential equations. Probably the most popular are Runge-Kutta type methods. A very thorough description of those methods can be found in Yakowitz and Szidarovszky.[2] A brief description of the classical fourth order Runge-Kutta method will now be given.

In the classical Euler's method a continuous derivative is approximated using the first order term from the Taylor's series. The accuracy of the approximation of the first derivative, and eventually of the estimate of $x(t_0 + \Delta t)$, can be improved by including higher order terms from the Taylor's series, and this is essentially the idea behind most of the modern numerical integration techniques. The trick in the Runge-Kutta method is that the higher derivatives are approximated by finite difference expressions and thus do not have to be calculated from the original differential equation. The approximating expressions are calculated using data obtained from tentative steps taken from t_0 towards $t_0 + \Delta t$. The number of steps used to estimate $x(t_0 + \Delta t)$ determines the order of the Runge-Kutta method. In the most common version of the method four tentative steps are made within each time step, and the successive value of the dependent variable is calculated as

$$x(t_0 + \Delta t) = x(t_0) + \left(\frac{\Delta t}{6}\right)(k_1 + 2k_2 + 2k_3 + k_4) \tag{6.10}$$

where

$$k_1 = f[x(t_0), t_0] \tag{6.11}$$

[2]S. Yakowitz and F. Szidarovszky, *An Introduction to Numerical Computation*, Macmillan Publishing Company, New York, 1986.

$$k_2 = f\left[\left(x(t_0) + \Delta t\,\frac{k_1}{2}\right), \left(t_0 + \frac{\Delta t}{2}\right)\right] \tag{6.12}$$

$$k_3 = f\left[\left(x(t_0) + \Delta t\,\frac{k_2}{2}\right), \left(t_0 + \frac{\Delta t}{2}\right)\right] \tag{6.13}$$

$$k_4 = f[(x(t_0) + \Delta t\,k_3), (t_0 + \Delta t)] \tag{6.14}$$

For higher-order systems a set of n equations, one for each state variable, of the form (6.10) is solved at every time step of the numerical solution.

A BASIC computer program using fourth-order Runge-Kutta method solving the differential equation from Example 6.1 is listed in Table 6.5, and Table 6.6 shows the results of the calculations.

TABLE 6.5. Computer Program Using Runge-Kutta Method for Solving Equation dx/dt = −0.25x + 1

```
10 REM Numerical solution of first order differential equation
20 REM dx/dt=f(x,t) over the interval [T0,TF] using Runge-Kutta method
30 INPUT "Initial independent variable = ",T0
40 INPUT"Final value of independent variable = ",TF
50 INPUT"Number of integration steps = ",N
60 DELT=(TF-T0)/N
70 PRINT"Integration step = ";DELT
80 DIM X(N+1),T(N+1)
90 INPUT"Initial dependent variable = ",X0
100 T(0)=T0
110 X(0)=X0
120 FOR I=1 TO N
130 T(I)=T(I-1)+DELT
140 T1=T(I-1)
150 X1=X(I-1)
160 XDOT1=-.25*X1+1
170 T2=T1+.5*DELT
180 X2=X1+.5*DELT*XDOT1
190 XDOT2=-.25*X2+1
200 X3=X1+.5*DELT*XDOT2
210 XDOT3=-.25*X3+1
220 T3=T2+.5*DELT
230 X4=X1+DELT*XDOT3
240 XDOT4=-.25*X4+1
250 X(I)=X(I-1)+DELT*(XDOT1+2*XDOT2+2*XDOT3+XDOT4)/6.
260 NEXT I
270 PRINT"TIME                OUTPUT"
280 FOR I=0 TO N
290 PRINT T(I),X(I)
300 NEXT I
310 END
```

TABLE 6.6. Runge-Kutta Solution Versus Exact Solution of Equation $dx/dt = -0.25x + 1$.

t	Runge-Kutta	Exact
0	10.0	10.0
1	8.672852	8.672805
2	7.639257	7.639184
3	6.834285	6.834199
4	6.207366	6.207277
5	5.719116	5.719029
6	5.338862	5.338781
7	5.042717	5.042644
8	4.812077	4.812012
9	4.632453	4.632395
10	4.492560	4.492510
11	4.383610	4.383567
12	4.298759	4.298722

The improvement in accuracy in terms of the maximum error achieved with the Runge-Kutta method is over 300 times compared with the improved Euler method, while the computation time was increased about five times.

In order to use the program listed in Table 6.5 to solve a different first order differential equation of the form

$$\frac{dx}{dt} = f(x, t)$$

lines 160, 190, 210, and 240 have to be changed to represent the particular form of the function f(x,t) as follow

```
160 XDOT1=f(X1,T1)
190 XDOT2=f(X2,T2)
210 XDOT3=f(X3,T2)
240 XDOT4=f(X4,T3)
```

It has been shown (see footnote 1) that a fourth-order Runge-Kutta formulation is equivalent to using the first five terms of the Taylor series expansion for $e^{-\Delta t/\tau}$ when computing a free response, that is, when solving the homogenous equation

$$\frac{dx}{dt} = \frac{-x}{\tau} \tag{6.15}$$

The Runge-Kutta formulation in this case is

$$x(t_0 + \Delta t) = x(t_0)\left[1 + (-\Delta t/\tau) + \frac{(-\Delta t/\tau)^2}{2} + \frac{(-\Delta t/\tau)^3}{6} + \frac{(-\Delta t/\tau)^4}{24}\right] \tag{6.16}$$

where the series for $e^{(-\Delta t/\tau)}$ is

$$e^{(-\Delta t/\tau)} = 1 + (-\Delta t/\tau) + \frac{(-\Delta t/\tau)^2}{2} + \frac{(-\Delta t/\tau)^3}{6} + \frac{(-\Delta t/\tau)^4}{24} + \frac{(-\Delta t/\tau)^5}{120} + \cdots \tag{6.17}$$

In other words, for the fourth-order Runge-Kutta formulation, the series for $e^{(-\Delta t/\tau)}$ is truncated by leaving off terms beyond $(-\Delta t/\tau)^4$. This suggests that an even better way to compute successive values $x(t_0 + \Delta t)$ from $x(t_0)$ for the input-output equation

$$\frac{dx}{dt} + \frac{x}{\tau} = u(t) \tag{6.18}$$

is to use the exponential

$$x(t_0 + \Delta t) = x(t_0)e^{-\Delta t/\tau} + \tau u_{ave}(1 - e^{-\Delta t/\tau}) \tag{6.19}$$

where u_{ave} is the average value of the input $u(t)$ during the interval from t_0 to $(t_0 + \Delta t)$. It is assumed that the accuracy of the computer algorithm that generates the exponential is good to at least the first term beyond the $(-\Delta t/\tau)^4$ term.

A BASIC computer program using the exponential Euler method based on Equation (6.19) is listed in Table 6.7 and the solution of the differential equation from Example 6.1 is shown in Table 6.8. Computational errors of numerical solutions of the equation from Example 6.1 for classical Euler, improved Euler, Runge-Kutta, and exponential Euler methods are shown in Table 6.9.

TABLE 6.7. Computer Program Using Exponential-Euler Method for Solving Equation $dx/dt = -0.25x + 1$

```
10 REM Numerical solution of first order differential equation
20 REM with constant input, dx/dt + x/tau = u, where tau = 4 sec
30 REM u = 1.0 so that dx/dt = -0.25x + 1, carried out over the
40 REM interval [T0,TF] using the exponential Euler method.
50 INPUT"INITIAL INDEPENDENT VARIABLE = ", T0
60 INPUT"FINAL VALUE OF INDEPENDENT VARIABLE = ",TF
70 INPUT"SYSTEM TIME CONSTANT = ",TAU
80 INPUT"INPUT U = ",U
90 DELT=TAU/4.
100 DT=DELT/TAU
110 N=(TF-T0)/DELT
120 DIM X(N+1),T(N+1)
130 INPUT"INITIAL VALUE OF DEPENDENT VARIABLE = ",X0
140 T(0)=T0
150 X(0)=X0
160 FOR I=1 TO N
170 T(I)=T(I-1)+DELT
180 T1=T(I-1)
190 X(I)=X(I-1)*EXP(-DT)+TAU*U*(1-EXP(-DT))
```

```
200 NEXT I
210 PRINT"TIME                OUTPUT"
220 FOR I=0 TO N
230 PRINT T(I),X(I)
240 NEXT I
250 END
```

TABLE 6.8. Exponential-Euler Solution Versus Exact Solution of Equation $dx/dt = -0.25x + 1$

t	Exponential Euler	Exact
0	10.0	10.0
1	8.672805	8.672805
2	7.639184	7.639184
3	6.834199	6.834199
4	6.207277	6.207277
5	5.719029	5.719029
6	5.338781	5.338781
7	5.042644	5.042644
8	4.812012	4.812012
9	4.632395	4.632395
10	4.492510	4.492510
11	4.383567	4.383567
12	4.298722	4.298722

TABLE 6.9. Computational Errors of Numerical Solutions of Equation $dx/dt = -0.25x + 1$ for Different Numerical Integration Methods.

t	Euler	Improved Euler	Runge-Kutta	Exponential Euler
0	0.0	0.0	0.0	0.0
1	−0.172805	0.014695	0.000047	0.000000
2	−0.264184	0.022926	0.000073	0.000000
3	−0.302949	0.026824	0.000086	0.000000
4	−0.308839	0.027997	0.000089	0.000000
5	−0.295201	0.027201	0.000087	0.000000
6	−0.270910	0.025461	0.000081	0.000000
7	−0.241740	0.023170	0.000073	0.000000
8	−0.211334	0.020656	0.000065	0.000000
9	−0.181887	0.018127	0.000058	0.000000
10	−0.154629	0.015710	0.000050	0.000000
11	−0.130156	0.013480	0.000043	0.000000
12	−0.108664	0.011471	0.000037	0.000000

It is interesting to note that the computation time required when using the exponential Euler method is not significantly greater than when using the improved Euler method. Although the exponential Euler method gives exact results for a first-order system with a constant input, the results from using it, or any other method, on a system of n state variable equations with a time-varying input are not as accurate as for a single state variable equation. Again, the best results for an n-th order system are obtained when the smallest possible integration step is used for the computation.

For the student wanting to use a personal computer to experiment with digital simulation, a program written in BASIC, together with a completely worked-out check solution for a third-order linear system is given in Appendix III. Different subroutines (Euler, Runge-Kutta, and so forth) may be substituted for the exponential Euler subroutine provided in this program in order to get a feeling for the relative merits of each method. This program prompts the user for all initial conditions and all matrix coefficients. It computes an initial estimated time step based on system loop gains which can be over-ridden by a prompt, and it prompts for such things as the number of initial waiting period steps, the frequency of printing or plotting, and the number of simulation steps to be computed. Output is available for printing and/or plotting.

6.4 Effect of Integration Step Size on Accuracy

As pointed out in Section 6.3, the accuracy of numerical methods for solving ordinary differential equations is strongly affected by the size of the integration step. It can be seen from Equation (6.2) that the first-order finite difference approximation of a continuous derivative approaches the exact value as the time step approaches zero. It can, therefore, be expected that the smaller time step will result in higher accuracy of the numerical solution.[3] This fact is clearly illustrated by the results of the solution of the differential equation from Example 6.1 obtained using Euler's method for time steps of 1.0, 0.5, 0.2, given in Table 6.10. The plot of the maximum computational error versus the size of the time step is shown in Figure 6.2.

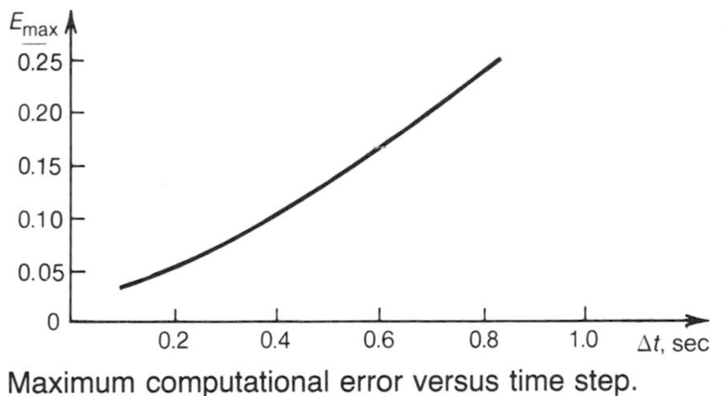

FIGURE 6.2 Maximum computational error versus time step.

[3]Actually, there is another type of error, called round-off error, that begins to become important when working with very small numbers on a floating point digital machine, so that when very, very small values of time step are used, the results may become questionable.

TABLE 6.10. Euler's Solution of Equation $dx/dt = -0.25x + 1$ for Different Time Steps.

t	$\Delta t = 1.0$	$\Delta t = 0.5$	$\Delta t = 0.2$	Exact
0	10.0	10.0	10.0	10.0
1	8.500000	8.593750	8.642685	8.672805
2	7.375000	7.517090	7.592421	7.639184
3	6.531250	6.692772	6.779747	6.834199
4	5.898438	6.061654	6.150915	6.207277
5	5.423828	5.578454	5.664337	5.719029
6	5.067871	5.208504	5.287832	5.338781
7	4.800904	4.925261	4.996500	5.042644
8	4.600678	4.708403	4.771072	4.812012
9	4.450508	4.542371	4.596641	4.632395
10	4.337881	4.415253	4.461669	4.492510
11	4.253411	4.317928	4.357230	4.383567
12	4.190058	4.243414	4.276418	4.298722

It can be concluded from the above considerations that in order to achieve a required accuracy of numerical results the integration step must be small enough. The initial selection of the size of the time step is usually made in accordance with the rule of thumb used in Example 6.1, that is Δt is chosen equal to one fourth of the time constant (for higher-order systems the selection of the initial time step is based on the smallest time constant or largest natural frequency of the linearized model). After the first computer run with $\Delta t = \tau/4$, several additional runs are performed with decreasing time step until little difference is achieved between results from successive runs. Although that sounds simple, very often the additional cost (in terms of computer time) involved in reducing the size of the integration step to a necessarily low value is exceedingly high.

In all methods considered so far, a fixed integration step was assumed. There are two problems associated with a fixed integration step in numerical computation. First, it is difficult to determine a proper size of the step for a particular problem (especially if the problem is nonlinear), and second, the computer time necessary to perform the computation with the selected integration step may be very high, as indicated above. Both these problems can be resolved if the size of the integration step is controlled during computer simulation. The size of the integration step necessary to maintain some specified accuracy during a simulation run varies from relatively small values, when rapid changes of the model variables occur or when model nonlinearities, such as Coulomb friction, dominate the system response, to relatively large steps, when the model is approaching steady state. Special algorithms have been developed that can be incorporated in the computer simulation program in order to determine the integration step necessary to maintain a desired accuracy throughout the duration of the simulation process. One such algorithm is used in subroutine DASCRU described in Christiansen[4]. In this subroutine a method of Merson is used to obtain a predetermined accuracy.

To explain the method of Merson consider a single integration step from t_0 to $t_0 + \Delta t$. The form of the differential equation is

$$\frac{dx}{dt} = f(x, t) \tag{6.1}$$

Four intermediate points A, B, C, D, are selected within the time step such that

$$\begin{aligned} t_A &= t_B = t_0 + \Delta t/3 \\ t_C &= t_0 + \Delta t/2 \\ t_D &= t_0 + \Delta t \end{aligned} \tag{6.20}$$

The values of x at these points are estimated using the following formulas

$$\begin{aligned} x(t_A) &= x(t_0) + (\Delta t/3)f(x(t_0), t_0) \\ x(t_B) &= x(t_0) + (\Delta t/6)[f(x(t_0), t_0) + f(x(t_A), t_A)] \\ x(t_C) &= x(t_0) + (\Delta t/8)[f(x(t_0), t_0) + 3f(x(t_B), t_B)] \\ x(t_D) &= x(t_0) + (\Delta t/2)[f(x(t_0), t_0) - 3f(x(t_B), t_B) + 4f(x(t_C), t_C)] \end{aligned} \tag{6.21}$$

Using these estimates, the value of $x(t_0 + \Delta t)$ is calculated as

$$x(t_0 + \Delta t) = x(t_0) + (\Delta t/6)[f(x(t_0), t_0) + 4f(x(t_C), t_C) + f(x(t_D), t_D)] \tag{6.22}$$

The estimated error resulting from the selected time step Δt is

$$E(\Delta t) = \frac{\Delta t}{5}\left|\frac{\Delta t}{3} f(x(t_0), t_0) - \left(3\frac{\Delta t}{2}\right) f(x(t_B), t_B) + 4\frac{\Delta t}{3} f(x(t_C), t_C) - \frac{\Delta t}{6} f(x(t_D), t_D)\right| \tag{6.23}$$

The size of the time step is then adjusted based on the estimate of the computation error given by Equation (6.23) in the following manner

If $(E_{max}/32) < E(\Delta t) < E_{max}$	then Δt remains unchanged
If $E(\Delta t) < (E_{max}/32)$	then use $2\Delta t$
If $E(\Delta t) > E_{max}$	then use $\Delta t/2$

where E_{max} is an initially chosen maximum acceptable error. In addition to selecting the value of E_{max}, the user of the program has to specify only the initial value of

[4]J. Christiansen, "Numerical Solution of Ordinary Simultaneous Differential Equations of the 1st Order Using a Method for Automatic Step Change," *Numerical Mathematics*, 14, pp. 317–324, (1970).

the integration time step (usually $\tau/4$) which is subsequently adjusted by the procedure described.

A computer program using the method of Merson for automatic change of the integration step is as accurate as any Runge-Kutta type program with fixed integration step but much faster.

6.5 Synopsis

The numerical integration of a set of simultaneous linear state variable equations requires a careful initial selection of the time step Δt, hence the statement given in Section 6.4 that it should be based on the smallest system time constant (largest real root or largest natural frequency). Another basis for the rule of thumb "$t = \tau/4$" is to use the time constant τ_n of the state variable equation having the fastest individual response (largest $|a_{nn}|$ in the $\mathbb{A}$-matrix). If several or all of the a_{nn}'s are very small in magnitude or zero, then the roots of the system are likely to be lightly damped and then the rule of thumb is "$\Delta t = 1/(4\omega_n)$" where ω_n is the largest natural frequency of the linearized input-output equation for the system. When the system is nonlinear, it is always helpful to prepare a linearized version, even if it represents a very crude model, both to help in choosing the initial time step and to help in verifying the computer simulation. *If too large an initial value of Δt is used, the solution is likely to be unstable, giving meaningless results.*

Once a reasonable initial simulation is achieved, the accuracy of the simulation is usually improved with the use of smaller time steps. Use of a numerical integration routine incorporating an error estimator and automatic step change algorithm can be very helpful in reducing the time required to obtain a reasonably accurate simulation. Such a routine is even more helpful when the system is nonlinear.

Another helpful technique is to draw the complete simulation block diagram for the system revealing all of the inherently closed loops within the system. The loops with highest gains are the fastest-responding parts of the system. In some cases the fastest-responding parts of the system can be assumed to have instantaneous response so that they may be described by simple algebraic expressions (a process somewhat analogous to the selection of dominant roots for system simplification discussed in Chapter 5). This reduces the order of the system, allows the use of a larger time step, and reduces the computation time for the simulation.

For obtaining the most reliable results when simulating large systems, it is recommended that digital simulation be carried out on a high-speed data processor employing a well-supported proprietary numerical integration routine available in the system library of the main-frame computing system.

In every case, a known check solution should also be run to verify the correct operation of the whole system of hardware and software, and the results should be reviewed for reasonableness, satisfaction of known initial conditions, final steady-states, and so forth, before accepting the results as final.

Problems

6.1. Use the computer program listed in Table 6.2 (with necessary modifications) to obtain the numerical solution of Problem 5.1a. Use the integration time steps, $\Delta t = 0.1$ sec. and .02 sec.

6.2. Use the computer program with the Euler integration method, listed in Table 6.2, to obtain numerical solution of Problem 5.1a for the following values of the integration time step: $\tau/10$, $\tau/5$, $\tau/2$, τ, 2τ, 5τ, where τ is the time constant of the dynamic system considered in Problem 5.1. Compare the numerical solutions obtained for different sizes of the integration time step with the exact analytical solution.

6.3. Use the computer program with the improved Euler method, listed in Table 6.3, to obtain numerical solution of Problem 5.1a.

6.4. Use the computer program with the Runge-Kutta method, listed in Table 6.5, to obtain numerical solution of Problem 5.1a.

6.5. Use the improved Euler or Runge-Kutta method to verify the analytical solution of Problem 5.3.

6.6. Compare simulations using the exponential Euler method and the Runge-Kutta method to obtain the numerical solution of Problem 5.5.

6.7. Make the necessary changes in the improved Euler program listed in Table 6.3 for solving sets of up to four state-variable equations. Test the program by computing the step response of the original system considered in Example 5.2. Compare performance specifications (percent overshoot, t_p, DR) obtained in the analytical and numerical solutions.

6.8. Use the computer program written for Problem 6.7 to obtain a numerical solution of Problem 5.12.

6.9. A mechanical system has been modeled by the following nonlinear state-variable equations:

$$\dot{x} = v$$

$$\dot{v} = -\frac{k}{m}x - \frac{1}{m}F_{NLD} + \frac{1}{m}F_a(t)$$

The nonlinear friction force F_{NLD} is approximated by the expression

$$F_{NLD} = f_{NL}(v) = v^2 + v + 1$$

The input force $F_a(t)$ has a constant component $\bar{F}_a = 15$ N, which has been acting on the system for a very long time, and an incremental component $\hat{F}_a(t)$, equal to $\Delta F_a \cdot U_s(t)$. The values of the system parameters are $m = 10$ kg and $k = 5$ N/m.

(a) Determine the normal operating point values for the state variables $\overline{x}$ and $\overline{v}$, corresponding to $\overline{F}_a$.

(b) Linearize the system model in the vicinity of the normal operating point. Find natural frequency ω_n, damping ratio ζ, and period of damped oscillations T_d for the linearized model.

(c) Use the exponential Euler program in Appendix III to compute the nonlinear system response to $F_a(t)$ for the step change $\Delta F_a(t) = 1.5$ N. Suggested integration time step for the computer program is $\Delta t \leq 0.05T_d$. Compare the specifications of the computer-generated step response with those obtained analytically for the linearized model in part (b).

(d) Repeat part (c) for a magnitude of the input step change $\Delta F_a(t) = 15$ N.

(e) Repeat part (c) for the linearized model.

(f) Repeat part (d) for the linearized model.

(g) Compare the agreement between the results obtained for nonlinear and linearized models for small and large magnitude inputs.

6.10. **(a)** Use the exponential Euler method with the BASIC program in Appendix III to simulate the response of the system having the following input-output equation to a unit step change in $u(t)$ at $t = 0$

$$0.025\frac{d^3x}{dt^3} + 0.25\frac{d^2x}{dt^2} + 0.1\frac{dx}{dt} + 1 = 0.5u(t)$$

Assume all initial conditions at $t = 0^-$ are zero.

(b) Modify the subroutine for using the improved Euler method and run for comparison with (a) using the same time step as in (a).

(c) Compare both simulations with the exact solution obtained by the classical method described in Chapter 5.

7

Electrical Systems

7.1 Introduction

The A-type, T-type, and D-type elements employed in modeling electrical systems, which correspond to the mass, spring, and damper elements discussed in Chapter 2, are the capacitor, inductor, and resistor elements.

A capacitor, the electrical A-type element, stores energy in the electric field induced in an insulating medium between a closely-spaced pair of conducting elements, usually plates of metal, when opposite charges are applied to the plates. Capacitance is a measure of the ability of a capacitor to accept charge and hence its ability to store energy. It occurs naturally between the conductors of a coaxial cable, between closely spaced parallel cables, and in closely packed coils of wire. In these cases the capacitance is distributed, along with resistance and inductance, along the line, and the analysis of such situations is beyond the scope of this text. However, in some cases the resistance and inductance are negligible, making it possible to use a lumped-capacitance model. More frequently, specially designed off-the-shelf capacitors are employed that have negligible inductance and series resistance. In addition, the dielectric material between the plates is such a good insulator that the parallel leakage resistance between the plates is essentially infinite; thus it takes a very long time for a charge to leak away internally.

An inductor, the electrical T-type element, stores energy in the magnetic field surrounding a conductor or a set of conductors carrying electric current (i.e., flow of charge). Inductance is a measure of the ability of an inductor to store magnetic energy when a current flows through it. Like capacitance, inductance occurs naturally in coaxial cables, long transmission lines, and in coils of wire. Here it is usually distributed along with capacitance and resistance, requiring a level of modeling and analysis that is beyond the scope of this text. However, often the case is such that the capacitance and resistance are negligible and a lumped-inductor model will suffice. Even if resistance is distributed along with the inductance, a series lumped-inductor lumped-resistor model will be suitable.

A resistor, the electrical D-type element, dissipates energy, resulting in heat transfer to the environment equal to the electric energy supplied at its terminals. Resistance is a measure of how much voltage is required to drive one ampere (1 amp) of current through a resistor. It occurs naturally in all materials (excepting superconducting materials), including wires and metal structural elements. When resistance occurs in wires intended for conducting electricity, it is usually an unwanted phenomenon. In many other cases resistance is necessary to accomplish circuit requirements, and a wide variety of resistors is available as off-the-shelf components, each type specially designed to provide needed characteristics. Carbon in some form is often used as the resistive medium, but coils of wire often serve this purpose, especially when a great deal of heat must be dissipated to the environment. Incandescent light bulbs are sometimes used as resistors, but they are very nonlinear.

This chapter will deal only with lumped models made up only of combinations of ideal electrical elements, which constitute a great majority of the electrical control systems encountered by engineers and scientists.

7.2 Diagrams, Symbols, and Circuit Laws

All three of the basic types of circuit elements discussed in Section 7.1 are two-terminal elements, as shown schematically in Figure 7.1. The electric potential at each terminal is measured by its voltage with respect to ground or some local reference potential, such as a machine frame or chassis. The rate of flow of electrical charge through the element in coulombs per second is measured in terms of amperes. The elemental equation usually takes the form

$$e_{12} = f_1(i_A) \quad i_A = f_2(e_{12})$$

In addition there are two types of ideal source elements employed for driving circuits, shown in Figure 7.2. The ideal voltage source e_s is capable of delivering the designated voltage regardless of the amount of current i being drawn from it. The ideal current source i_s is capable of delivering the designated current, regardless of the voltage required to drive its load.

The two basic circuit laws needed to describe the interconnections between the elements of a circuit are known as Kirchhoff's voltage law and Kirchhoff's current law. The voltage law says that the sum of voltage drops around a loop must equal zero (Figure 7.3a). A corollary to the voltage law is that the total voltage drop across a series of elements is the sum of the individual voltage drops across each of the elements in series (Figure 7.3b).

The current law states that the sum of the currents at a node (junction of two or more elements) must be zero. This law is illustrated in Figure 7.4.

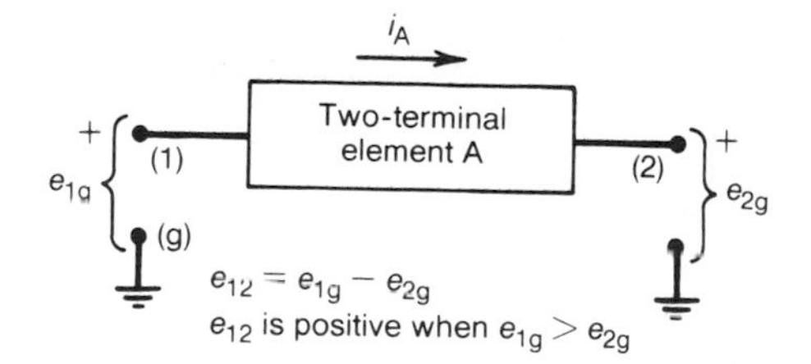

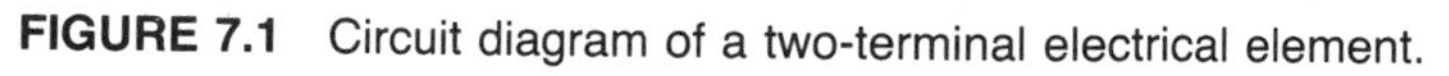

FIGURE 7.1 Circuit diagram of a two-terminal electrical element.

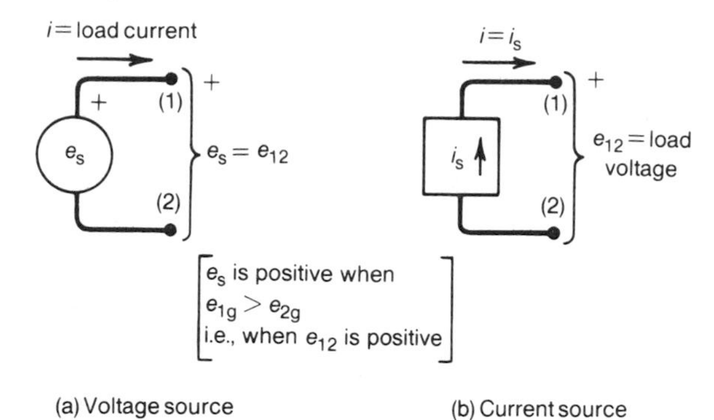

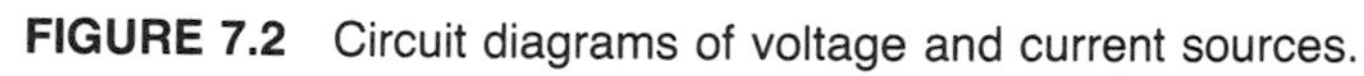

FIGURE 7.2 Circuit diagrams of voltage and current sources.

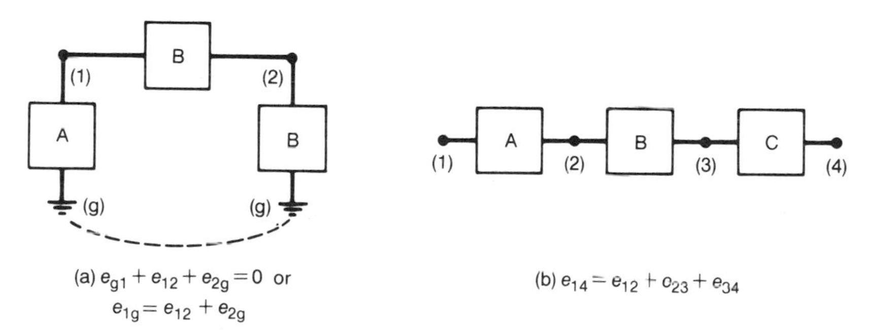

FIGURE 7.3 Illustration of Kirchhoff's voltage law for elements in a loop and for several elements in series.

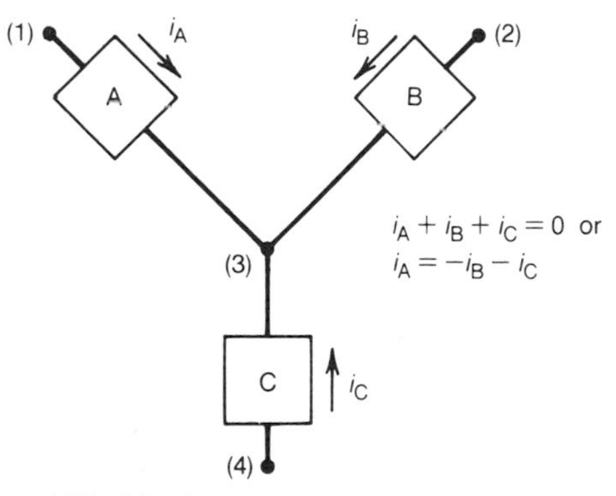

FIGURE 7.4 Illustration of Kirchhoff's current law.

7.3 Elemental Diagrams, Equations, and Energy Storage

Capacitors. The circuit diagram of an ideal capacitor is shown in Figure 7.5. The elemental equation in terms of the stored charge q_C of a capacitor is

$$q_C = C(e_{1g} - e_{2g}) = Ce_{12} \tag{7.1}$$

In terms of current i_C (rate of flow of charge q_C) the elemental equation is

$$i_C = C\frac{de_{12}}{dt} \tag{7.2}$$

Note the similarity to the elemental equation for an ideal mass given in Chapter 2. Since i_C is a through (T) variable and e_{12} is an across (A) variable, the capacitor C is said to be analogous to the mass m. Thus the state-variable for a capacitor is its voltage e_{12}. However, the capacitor is different in one respect: both of its terminals can "float above ground," that is, neither must be grounded, whereas the velocity for a mass must always be referred to a nonaccelerating reference such as the earth (ground).

The comments made in Chapter 2 about the response of a mass m to an input force apply here to the response of a capacitor C to an input current—it takes time for a finite current to change the charge stored in, and hence the voltage across the terminals of, a capacitor.

The energy stored in a capacitor resides mainly in the very closely confined static field between the plates of the element, and it is often referred to as static field energy. The dielectric material between the plates is a very good insulator, so that very, very little current can flow (i.e., leak) directly from one plate to the other. The current i_C simply represents the rate of flow of charge to and from the plates where the charge is stored.

The stored electric field energy is given by

$$\mathscr{E}_e = \frac{C}{2}e_{12}^2 \tag{7.3}$$

Here again it is apparent that imposing a sudden change in the voltage across a capacitor would not be realistic, as this would mean suddenly changing the stored energy, which would require an infinite current, i.e., an infinite power source.

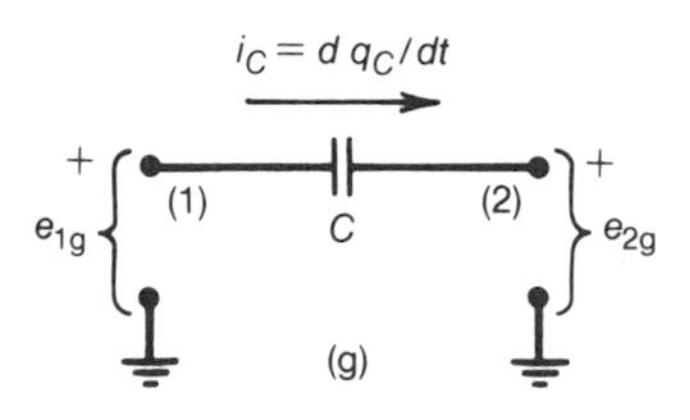

FIGURE 7.5 Circuit diagram of an ideal capacitor.

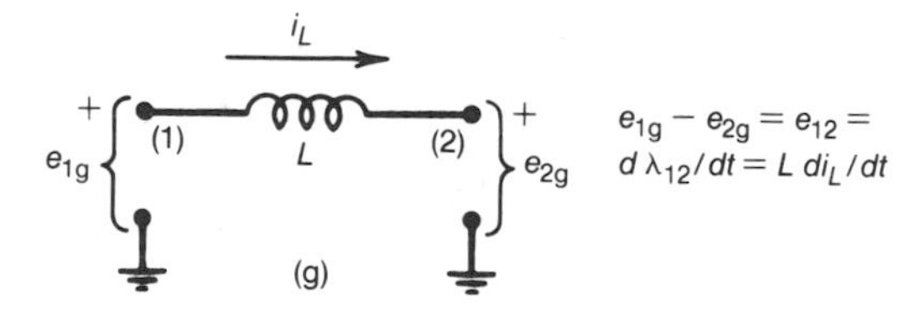

FIGURE 7.6 Circuit diagram of an ideal inductor.

Inductors. The circuit diagram of an ideal inductor is shown in Figure 7.6. The elemental equation in terms of the flux linkage λ_{12} of the coil is

$$\lambda_{12} = Li_L \tag{7.4}$$

In terms of the voltage $e_{12} = d\lambda_{12}/dt$, the elemental equation is

$$e_{12} = L\frac{di_L}{dt} \tag{7.5}$$

Note the similarity to the elemental equations for an ideal spring. Since i_L is a T variable and e_{12} is an A variable, an inductor is said to be analogous to a spring, with L corresponding to 1/k. Thus the state-variable for an inductor is its current i_L.

The energy stored in an inductor is stored in the magnetic field surrounding its conductors, and is known as magnetic field energy. The stored magnetic field energy is given by

$$\mathscr{E}_m = \frac{L}{2}i_L^2 \tag{7.6}$$

It can be seen from Equation (7.6) that to try to make a sudden change in the current through an inductor would not be realistic. Such a change would involve a sudden change in the stored energy, which would in turn require an infinite voltage, hence an infinite power source.

The magnetic field, which is induced by the current i_L flowing in many turns of a densely packed coil, is intensified if a ferromagnetic type metal core is installed within the coil. Air core inductors are usually linear, but contain the inherent resistance of the coil wire, which may not be negligible. This parasitic resistance is modeled by an ideal resistor in series with the inductance of the coil.

When the core is a ferromagnetic material, the inductance for a given coil current is much greater and the ability of the inductor to store energy is much greater, but a ferromagnetic core also introduces nonlinearity and hysteresis in the flux linkage versus current relationship so that Equation (7.4) becomes

$$\lambda_{12} = f_{\text{NL}}(i_{\text{NLI}}) \tag{7.7}$$

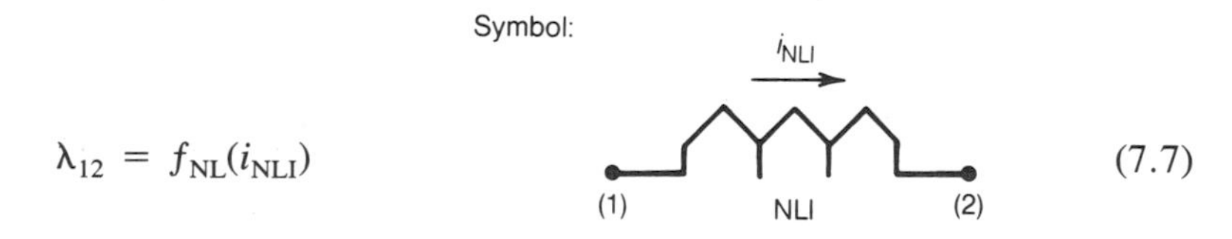

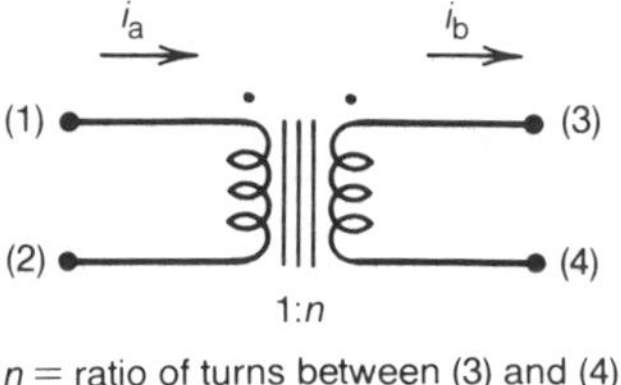

FIGURE 7.7 Circuit diagram of an ideal transformer.

Compared to the variety and quantities of capacitors and resistors available as off-the-shelf elements, the possibility of finding off-the-shelf inductors for specific applications is quite small. Thus inductors are usually custom designed and manufactured for each specific application.

Transformers. When two coils of wire are installed very close to each other so that they share the same core without flux leakage, an electric transformer results, shown schematically in Figure 7.7. Because a transformer is a four-terminal element, two elemental equations are needed to describe it.

The elemental equations are[1]

$$\lambda_{34} = n\lambda_{12} \quad \text{or} \quad e_{34} = ne_{12} \tag{7.8}$$

where n is the ratio of the number of turns between (3) and (4) to the number of turns between (1) and (2), and

$$i_b = \frac{1}{n} i_a \tag{7.9}$$

The dots appearing over each coil indicate that the direction of each winding is such that n is positive, in other words e_{34} has the same sign as e_{12} and i_a has the same sign as i_b.

Ideal transformers do not store energy, and are frequently used to couple circuits dynamically. Combining Equations (7.8) and (7.9) reveals the equality of energy influx and efflux.

$$e_{34}i_b = ne_{12} \frac{1}{n} i_a \tag{7.10}$$

Resistors. The circuit diagram for an ideal resistor is shown in Figure 7.8. The elemental equation for an ideal resistor can be written either as

$$e_{12} = Ri_R \tag{7.11}$$

[1]The ideal relations employed here are applicable only when fluctuating voltages and currents, such as AC signals occur. See R. E. Scott, *Linear Circuits II*, Addison-Wesley Publishing Co., Reading, MA, 1960, p. 594.

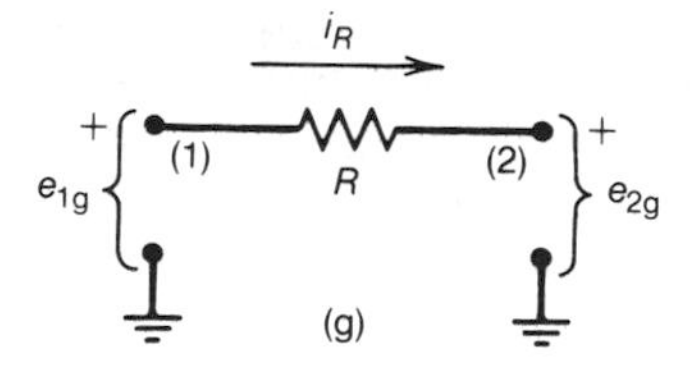

FIGURE 7.8 Circuit diagram of an ideal resistor.

or as

$$i_R = \frac{1}{R} e_{12} \tag{7.12}$$

Note that the voltage across and the current through a resistor are related 'instantaneously' to each other, since there is no energy storage, and no derivative of either e_{12} or i_R is involved in these equations.

Although many resistors are carefully designed and manufactured to be linear, many others are inherently or intentionally nonliner. The circuit diagram to be used here for a nonlinear resistor NLR is shown in Figure 7.9.

The elemental equation for such a nonlinear resistor is written either as

$$e_{12} = f_{\text{NL}}(i_{\text{NLR}}) \quad \text{or} \quad i_{\text{NLR}} = f_{\text{NL}}^{-1}(e_{12}) \tag{7.13}$$

When Equation (7.13) is linearized, for small perturbations about an operating point,

$$\hat{e}_{12} = R_{\text{inc}} \hat{i}_{\text{NLR}} \tag{7.13lin}$$

where the incremental resistance, R_{inc} is given by

$$R_{\text{inc}} = \left. \frac{df_{\text{NL}}}{di_{\text{NLR}}} \right|_{\bar{i}_{\text{NLR}}} \tag{7.13R}$$

7.4 Analysis of Systems of Interacting Electrical Elements

Electrical systems are usually referred to as electric circuits, and the techniques of analyzing electric circuits are very similar to those used in Chapter 2 for analyzing

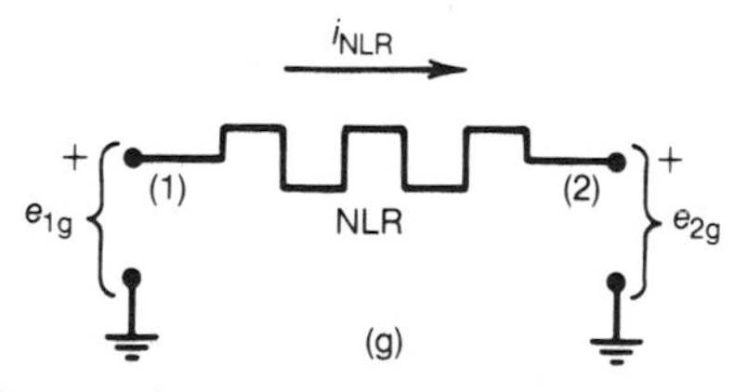

FIGURE 7.9 Circuit diagram for a nonlinear resistor.

mechanical systems: Simply write the equation for each element and employ the appropriate connecting laws (Kirchhoff's laws here) to obtain a complete set of n equations involving the n unknowns needed to describe the system completely; then combine (if everything is linear) to eliminate the unwanted variables to obtain a single input-output differential equation relating the desired output to the given input(s). If only a set of state-variable equations is needed, it is usually most convenient to "build" a state-variable equation around each energy-storage element, in which case nonlinearities are easily incorporated.

EXAMPLE 7.1

Find the input-output differential equation for the simple R, L, C circuit shown in Figure 7.10.

Solution

For the source

$$e_s = e_{1g} \tag{7.14}$$

For the resistor R_1

$$i_{R1} = \frac{1}{R_1}(e_{1g} - e_{2g}) \tag{7.15}$$

For the inductor L

$$e_{2g} - e_{3g} = L\frac{di_L}{dt} \tag{7.16}$$

For the resistor R_2

$$e_{3g} = R_2 i_L \tag{7.17}$$

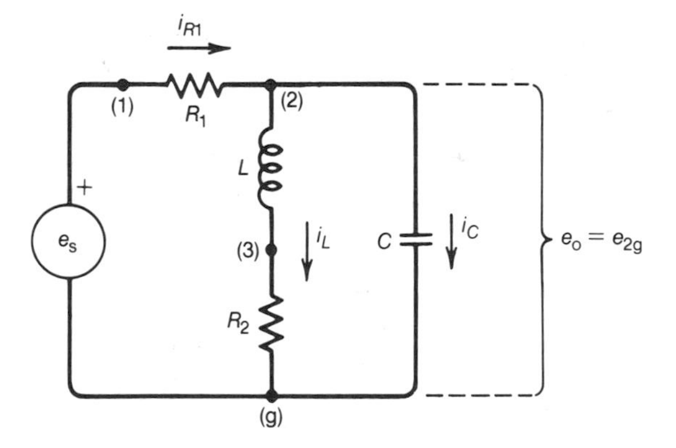

FIGURE 7.10 Circuit diagram of an R, L, C circuit.

For the capacitor C

$$i_C = C\,\frac{de_{2g}}{dt} \tag{7.18}$$

At node 2

$$i_{R1} = i_L + i_C \tag{7.19}$$

Equations (7.14) through (7.19) comprise a set of six equations involving six unknown variables: e_{1g}, e_{2g}, e_{3g}, i_{R1}, i_L, and i_C. (Note that using i_L to describe the current through both L and R_2 satisfies Kirchhoff's law at node 3 and eliminates one variable and one equation.)

The node method[2] may be applied to node 2 to eliminate the unwanted variables e_{1g}, e_{3g}, i_{R1}, i_L, and i_C. Substituting for i_{R1} from (7.15) and for i_C from (7.18) into (7.19)

$$\frac{1}{R_1}\,(e_{1g} - e_{2g}) = i_L + C\,\frac{de_{2g}}{dt} \tag{7.20}$$

Then using (7.14) for e_{1g} and rearranging (7.20)

$$i_L = \frac{1}{R_1}\,e_s - \frac{1}{R_1}\,e_{2g} - C\,\frac{de_{2g}}{dt} \tag{7.21}$$

Differentiating (7.21) with respect to time gives

$$\frac{di_L}{dt} = \left(\frac{1}{R_1}\right)\frac{de_s}{dt} - \left(\frac{1}{R_1}\right)\frac{de_{2g}}{dt} - C\,\frac{de_{2g}^2}{dt^2} \tag{7.22}$$

Combine Equations (7.16) and (7.17) to solve for e_{2g}.

$$e_{2g} = L\,\frac{di_L}{dt} + R_2 i_L \tag{7.23}$$

Substitution from (7.21) and (7.22) into (7.23) then eliminates i_L.

$$e_{2g} = \left(\frac{L}{R_1}\right)\left(\frac{de_s}{dt} - \frac{de_{2g}}{dt}\right) - LC\,\frac{d^2e_{2g}}{dt^2} + \left(\frac{R_2}{R_1}\right)(e_s - e_{2g}) - R_2C\,\frac{de_{2g}}{dt} \tag{7.24}$$

Collecting terms and noting that e_0 is the same as e_{2g},

$$LC\,\frac{d^2e_o}{dt^2} + \left(\frac{L}{R_1} + R_2C\right)\frac{de_o}{dt} + \left(1 + \frac{R_2}{R_1}\right)e_o = \frac{R_2}{R_1}\,e_s + \frac{L}{R_1}\frac{de_s}{dt} \tag{7.25}$$

■

[2]The node method involves starting by satisfying Kirchhoff's current law at each node, then using the elemental equation for each branch connecting to each node.

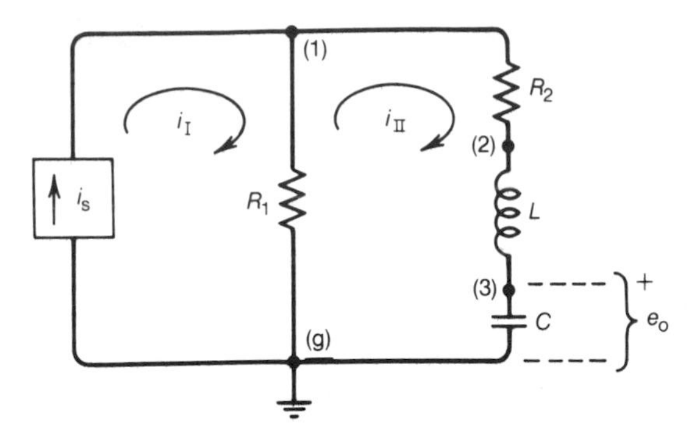

FIGURE 7.11 A different R, L, C circuit, driven by a current source.

EXAMPLE 7.2

Develop the input-output differential equation for the circuit shown in Figure 7.11.

Solution

Here the loop method[3] will be employed, and the elemental equations will be developed as needed for each loop. Like the use of the node method, the use of the loop method helps to eliminate quickly the unwanted variables. The two loops, I and II, are chosen as shown in Figure 7.11 carrying the loop currents i_I and i_{II}. Loop I is needed only to note that $i_I = i_s$. Using Kirchhoff's voltage law for loop II yields

$$R_2 i_{II} + L\frac{di_{II}}{dt} + \left(\frac{1}{C}\right)\int i_{II}\,dt + R_1(i_{II} - i_s) = 0 \tag{7.26}$$

Since $i_{II} = C\,de_o/dt$, substituting for i_{II} in (7.26) yields

$$(R_1 + R_2)C\frac{de_o}{dt} + LC\frac{d^2e_o}{dt^2} + e_o = R_1 i_s \tag{7.27}$$

Rearranging

$$LC\frac{d^2e_o}{dt^2} + (R_1 + R_2)C\frac{de_o}{dt} + e_o = R_1 i_s \tag{7.28}$$

■

EXAMPLE 7.3

(a) Write the state-variable equations based on the energy storage elements L and C for the circuit shown in Figure 7.12a.

(b) Linearize these equations for small perturbations of all variables.

[3]The loop method starts by satisfying Kirchhoff's voltage law for each independent loop and then uses the elemental equations for each part of each loop.

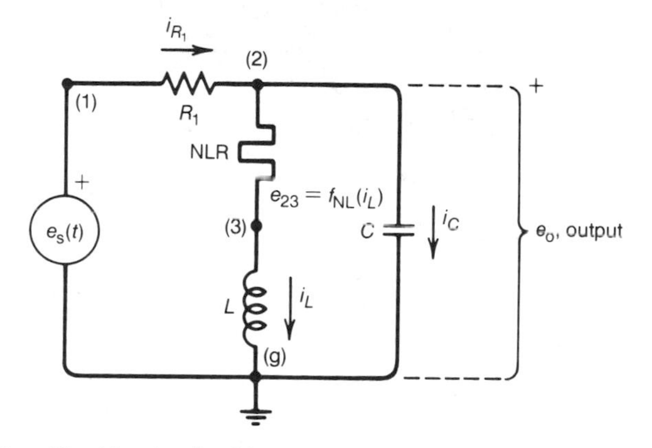

FIGURE 7.12a Electric circuit with nonlinear resistor.

(c) Combine to eliminate the unwanted variable and obtain the input-output system differential equation relating the output, e_{2g}, to the input $e_s(t)$.
(d) Draw the simulation block diagram for the nonlinear system.

Solution

(a) In general the state-variable for a capacitor is its voltage e_{2g}, and the state variable for an inductor is its current i_L. Thus for the inductor L,

$$\frac{di_L}{dt} = \frac{1}{L}(e_{2g} - e_{23})$$

or

$$\frac{di_L}{dt} = \frac{-1}{L} f_{NL}(i_L) + \frac{1}{L} e_{2g} \tag{7.29}$$

and for the capacitor C,

$$\frac{de_{2g}}{dt} = \frac{1}{C}(i_{R1} - i_L)$$

or

$$\frac{de_{2g}}{dt} = \frac{1}{C}\left(\frac{e_s - e_{2g}}{R_1} - i_L\right)$$

Rearranging,

$$\frac{de_{2g}}{dt} = \left(\frac{-1}{C}\right) i_L + \left(\frac{-1}{R_1C}\right) e_{2g} + \left(\frac{1}{R_1C}\right) e_s \tag{7.30}$$

Because of the nonlinear function $f_{NL}(i_L)$, Equations (7.29) and (7.30) cannot be combined to eliminate i_L until they are linearized for small perturbations of all variables.

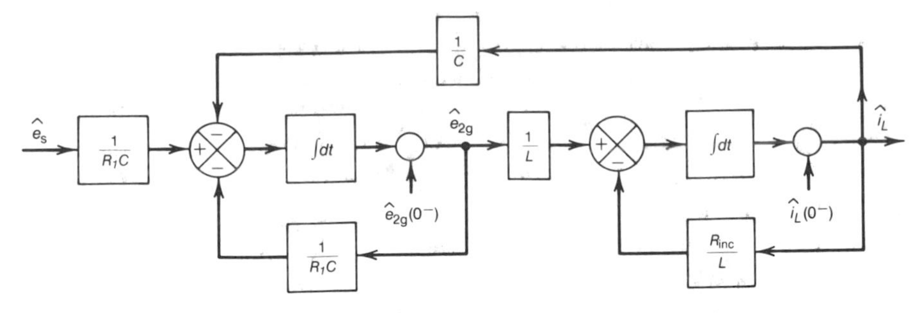

FIGURE 7.12b Simulation block diagram for the linearized system.

(b) After linearizing Equations (7.29) and (7.30) become

$$\frac{d\hat{\imath}_L}{dt} = \left(\frac{-1}{L}\right) \frac{de_{23}}{di_L}\bigg|_{\bar{\imath}_L} \hat{\imath}_L + \left(\frac{1}{L}\right) \hat{e}_{2g} \tag{7.31}$$

$$\frac{d\hat{e}_{2g}}{dt} = \left(\frac{-1}{C}\right) \hat{\imath}_L + \left(\frac{-1}{R_1C}\right) \hat{e}_{2g} + \left(\frac{1}{R_1C}\right) \hat{e}_s \tag{7.32}$$

(c) Equations (7.31) and (7.32) may now be combined as follows to eliminate i_L. From Equation (7.32)

$$\hat{\imath}_L = -C \frac{de_{2g}}{dt} - \left(\frac{1}{R_1}\right) \hat{e}_{2g} + \left(\frac{1}{R_1}\right) \hat{e}_s$$

Substituting in Equation (7.31) and using $R_{inc} = \dfrac{de_{23}}{di_L}\bigg|_{\bar{\imath}_L}$ or

$$-C \frac{d^2\hat{e}_{2g}}{dt^2} - \frac{1}{R_1}\frac{d\hat{e}_{2g}}{dt} + \frac{1}{R_1}\frac{d\hat{e}_s}{dt} = -\frac{R_{inc}}{L}\left(-C\frac{d\hat{e}_{2g}}{dt} - \frac{\hat{e}_{2g}}{R_1} + \frac{\hat{e}_s}{R_1}\right) + \frac{\hat{e}_{2g}}{L}$$

Collecting terms,

$$C\frac{d^2\hat{e}_{2g}}{dt^2} + \left(\frac{L + R_1R_{inc}C}{R_1L}\right)\frac{d\hat{e}_{2g}}{dt} + \left(\frac{R_{inc} + R_1}{R_1L}\right)\hat{e}_{2g} = \frac{1}{R_1}\frac{d\hat{e}_s}{dt} + \frac{R_{inc}}{LR_1}\hat{e}_s \tag{7.33}$$

(d) The simulation block diagram for the linearized system appears in Figure 7.12b. ■

EXAMPLE 7.4

The type of circuit shown in Figure 7.13 involves the use of a high-gain operational amplifier (OpAmp) with feedback to achieve desired dynamic response characteristics in automatic controllers. The gain, k_a, of the OpAmp is negative, and its input current i_a is so very small that it is negligible. The object of this example is to write the describing equations, eliminate unwanted variables, and develop the input-output system differential equation, revealing the dynamic characteristic achieved with capacitor feedback.

Solution

Employing Kirchhoff's current law at node 2

$$i_R = i_C$$

or

$$\frac{e_i - e_{2g}}{R} = C\,\frac{d(e_{2g} - e_{3g})}{dt} \tag{7.34}$$

And for the amplifier,

$$e_{3g} = k_a e_{2g}$$

or

$$e_{2g} = \frac{1}{k_a}\,e_{3g} \tag{7.35}$$

Combining (7.34) and (7.35) gives

$$e_i - \left(\frac{1}{k_a}\right) e_{3g} = RC\,\frac{d[(e_{3g}/k_a) - e_{3g}]}{dt}$$

Rearranging,

$$RC\,\frac{[1 - (1/k_a)]de_{3g}}{dt} - \left(\frac{1}{k_a}\right) e_{3g} = -e_i$$

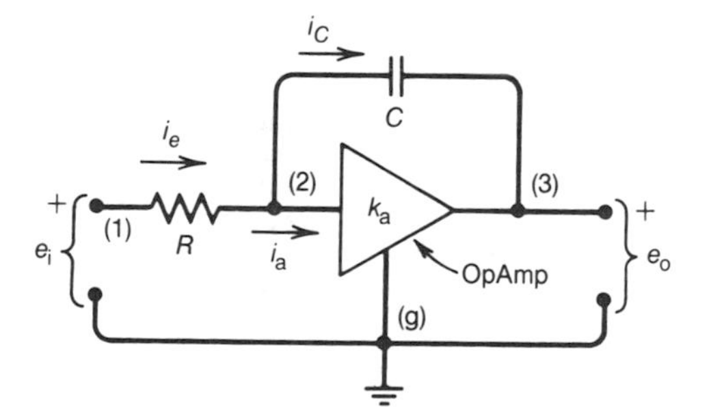

FIGURE 7.13 High gain OpAmp with capacitor feedback.

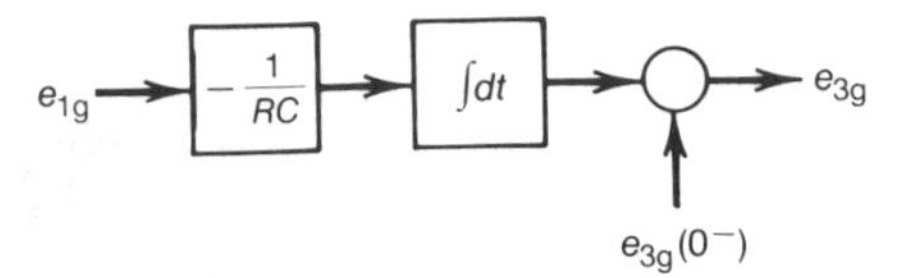

FIGURE 7.14 Simulation block diagram for integrator with time constant.

Now since the magnitude of k_a is very, very great,

$$\frac{de_{3g}}{dt} \approx -\frac{e_i}{RC}$$

or

$$e_{3g} = -\left(\frac{1}{RC}\right)\int_{0^-}^{t} e_i\, dt + e_{3g}(0^-)$$

Thus, use of a capacitor in the feedback with a resistor at the input results in an integrator with time constant RC, which is described by the simulation block diagram shown in Figure 7.14. ■

7.5 Linear Time-varying Electrical Elements

Many of the instrument and control systems used in engineering employ electrical elements that are time varying. The most common is the variable resistor, in which the resistance is caused to vary with time.

The elemental equation of this time-varying resistor is

$$e_{1g} - e_{2g} = e_{12} = R(t)i_R \qquad (7.36)$$

If such a resistor is supplied with a constant current, its voltage drop will have a variation with time that is proportional to the variation in R. Conversely, if the resistor is supplied with a constant voltage drop, it will have a current that varies inversely with the varying of R.

The capacitor may also be a time-varying element, having the time-varying elemental equations

$$q_C = C(t)e_{12} \qquad (7.37)$$

and because $i_C = dq_C/dt$,

$$i_C = C(t)\frac{de_{12}}{dt} + e_{12}\frac{dC}{dt} \qquad (7.38)$$

Similarly, the elemental equations for a time-varying inductor are

$$\lambda_{12} = L(t)i_L \qquad e_{12} = \frac{d\lambda_{12}}{dt} \tag{7.39}$$

and because $e_{12} = d\lambda_{12}/dt$,

$$e_{12} = L(t)\frac{di_L}{dt} + i_L\frac{dL}{dt} \tag{7.40}$$

In general, the state variable for a time-varying capacitor is its charge q_C, and the state variable for a time-varying inductor is its flux linkage λ_{12}.

EXAMPLE 7.5

A time-varying inductor in series with a voltage source and a fixed resistor is being used as part of a moving-metal detector. The circuit diagram for this system is shown in Figure 7.15. Find the input-output differential equation relating e_o to $L(t)$.

Solution

Employ Kirchhoff's voltage law to write

$$e_{12} = e_s - Ri_L \tag{7.41}$$

Substituting from Equation (7.40) for e_{12} in (7.41) gives

$$e_s - Ri_L - L(t)\frac{di_L}{dt} - i_L\frac{dL}{dt} = 0 \tag{7.42}$$

Employing small perturbations for all variables

$$\hat{e}_s - R\hat{i}_L - \bar{L}(t)\frac{d\hat{i}_L}{dt} - \bar{i}_L\frac{d\hat{L}}{dt} = 0 \tag{7.43}$$

Since e_s is constant, $\hat{e}_s = 0$, so that

$$\bar{L}\frac{d\hat{i}_L}{dt} + R\hat{i}_L = -\bar{i}_L\frac{d\hat{L}}{dt} \tag{7.44}$$

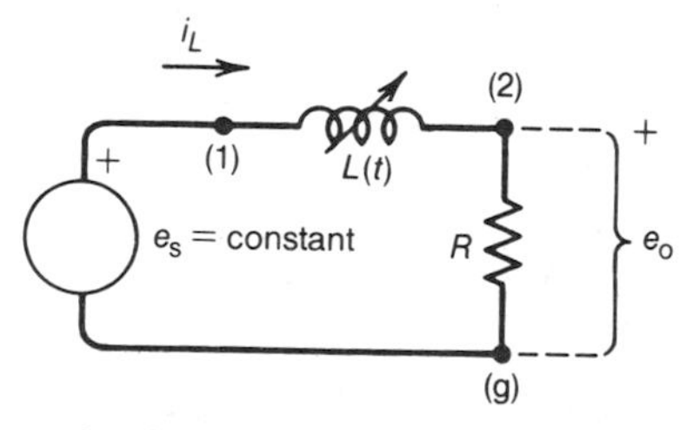

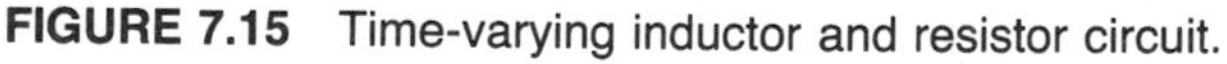

FIGURE 7.15 Time-varying inductor and resistor circuit.

Now $e_{2g} = Ri_L$ or $i_L = (1/R)e_{2g}$, making it possible to substitute for i_L in (7.44), yielding

$$\left(\frac{L}{R}\right)\frac{d\hat{e}_{2g}}{dt} + \hat{e}_{2g} = -\bar{\imath}_L \frac{d\hat{L}}{dt} \tag{7.45}$$

Since $e_{2g} = e_o$

$$\left(\frac{\bar{L}}{\bar{R}}\right)\frac{d\hat{e}_o}{dt} + \hat{e}_o = -\bar{\imath}_L \frac{d\hat{L}}{dt} \tag{7.46}$$

Note that to put this equation in state-variable form, the state variable must be λ_{12} as follows:

$$\frac{d\hat{\lambda}_{12}}{dt} = \left(-\frac{\bar{R}}{\bar{L}}\right)\hat{\lambda}_{12} + \left(\frac{-R\bar{\imath}_L}{\bar{L}}\right)\hat{L}(t) \tag{7.46'}$$

■

7.6 Synopsis

The basic physical characteristics of the linear, lumped-parameter electrical elements were discussed together with their mechanical system analogs: Type A, electrical capacitor and mechanical mass; Type T, electrical inductor and mechanical spring; Type D, electrical resistor and mechanical damper. Thus capacitance C is analogous to mass m, inductance L is analogous to inverse stiffness $\frac{1}{k}$, and resistance R is analogous to inverse damping $\frac{1}{b}$. Diagrams and describing equations were presented for the two-terminal elements including ideal voltage and ideal current sources. Kirchhoff's voltage and current laws were provided to serve as a means of describing the interactions occurring between the two-terminal elements in a system of interconnected elements (i.e., electrical circuits).

The dynamic analysis of electric circuits was then illustrated by means of several examples chosen to demonstrate different approaches to finding the input-output differential equation for a given circuit. In each case the procedure began with writing the elemental equations for all of the system elements and writing the required interconnection (Kirchhoff's Law) equations to form a necessary and sufficient set of n equations relating n unknown variables to the system input(s) and time. In order to then eliminate the unwanted unknown variables, one of four different methods was employed in each example: (a) step-by-step manipulation of the n equations to reduce the number of unwanted variables and reduce the number of equations one or two at a time; (b) the node method which eliminates many or all of the unwanted T-type variables (currents) in a few steps; (c) the loop method which eliminates many or all of the unwanted A-type variables (voltages) in a few steps; or (d) developing the set of state-variable equations based on each of the energy-storing elements, which eliminates all unwanted variables except the state-variables, so that only the voltages across capacitors and the currents through inductors remain as unknowns. In cases (b), (c), and (d), the elimination of the remaining unwanted variables involves the algebraic manipulation of a reduced set of only m equations containing only m unknowns to arrive at the desired input-output differential equation containing the one remaining unknown and its derivatives on the left-hand side. In the case when the system is to be simulated by computer, method (d) is preferred because it leads directly to the equations in proper form for programming on the computer, and the final reduction to one variable is needed only for the purposes of producing a check solution when needed. *Note that a properly constituted set of state variable equations must not have any derivative terms on the right-hand sides of any of its equations.*

Analyses of systems containing nonlinear elements and time-varying elements were carried out by the use of small perturbation techniques, and an analysis of a feedback system containing a high-gain OpAmp was included to illustrate the use of circuit analysis procedures for electronic circuits.

Problems

7.1. The source in the circuit shown in Figure P7.1 undergoes a step change so that e_s suddenly changes from 0 to 10 v at $t = 0$. Before the step change occurs, all variables are constant, i.e., e_s has been zero for a long time.

(a) Find $e_{32}(0^-)$ and $e_{32}(0^+)$.
(b) Find $e_{1g}(0^-)$ and $e_{1g}(0^+)$.

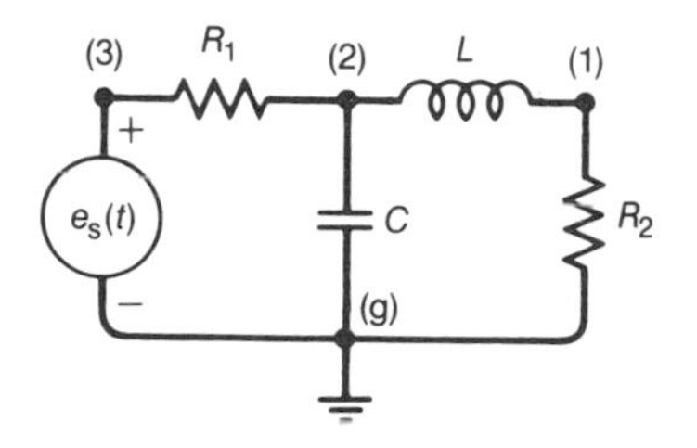

FIGURE P7.1

7.2. The circuit shown in Figure P7.2 is subjected to a step change in e_s from 5.0 to 7.0 v at $t = 0$. Before the step change occurs, all variables are constant, i.e., e_s has been 5.0 v for a very long time.

(a) Find $i_L(0^-)$, $i_L(0^+)$, and $i_L(\infty)$.
(b) Find $e_{3g}(0^-)$, $e_{3g}(0^+)$, and $e_{3g}(\infty)$.
(c) Find $e_{2g}(0^-)$, $e_{2g}(0^+)$, and $e_{2g}(\infty)$.

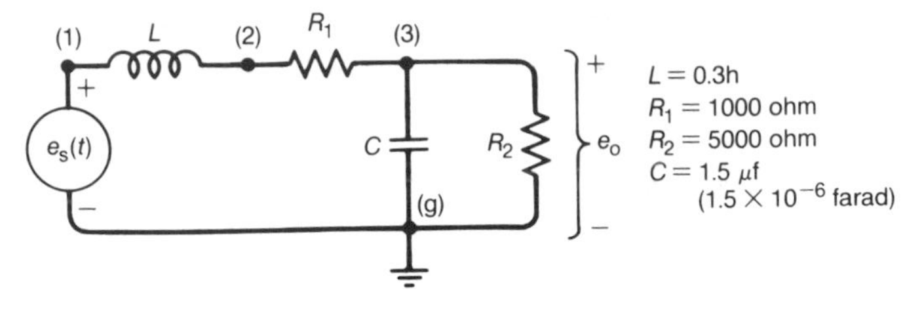

FIGURE P7.2

7.3. You have been asked if you can determine the capacitance of an unknown capacitor handed to you by a colleague. This capacitor is not an electrolytic device so that you do not have to worry about its polarity. Not having a capacitance meter, you decide to employ a battery that is close at hand, together with a decade resistor box, to run a simple test that will enable you to determine the capacitance. After the capacitor has been charged to 12.5 v from the battery and then disconnected from the battery, a high value of resistance, 100,000 ohm, is connected across the terminals of the capacitor, resulting in its slow discharge. The values of its voltage at succeeding intervals of time are recorded below.

Time, sec	Capacitor Voltage, e_{12}
0	12.5
10	9.3
20	7.1
30	5.3
40	3.9
50	2.9
60	2.2

(a) Write the system differential equation for the capacitor voltage e_{12} during the discharge interval. What is the input to this system? (*Hint*: What happens at $t = 0$?)

(b) Determine your estimated value for the capacitance, C.

7.4. This is a continuation of Problem 7.3. You have rearranged the components R and C as shown in Figure P7.4 and run another test by suddenly closing the switch and recording the capacitor voltage at successive increments of time. The results appear in the accompanying table.

Time, sec	e_{1g}, v
0	2.2
10	4.3
20	5.9
30	7.2
40	8.3
50	9.1
60	9.6

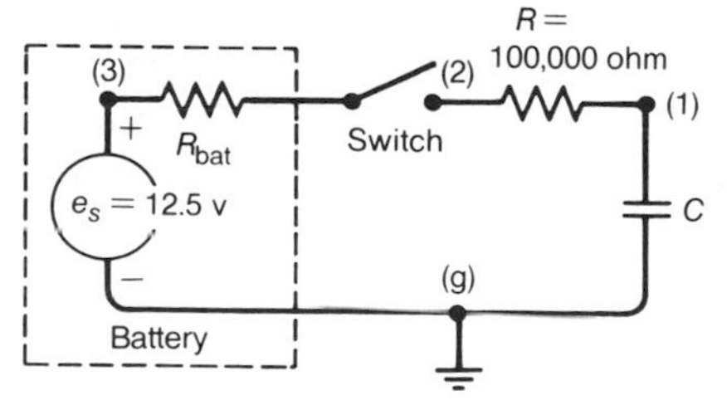

FIGURE P7.4

Employ the value of capacitance found in Problem 7.3b, to determine an estimated value for the internal resistance of the battery, R_{bat}.

7.5. **(a)** Develop the system differential equation relating the output voltage e_{3g} to the input voltage $e_g = e_{2g}$ for the circuit shown in Figure P7.5. (Use only the symbols in the figure.)

(b) The input voltage, which has been zero for a very long time, is suddenly increased to 5.0 v at $t = 0$. Find $e_{3g}(0^-)$, $e_{3g}(0^+)$, and $e_{3g}(\infty)$.

(c) Find the system time constant, and sketch the response of e_{3g} versus time t.

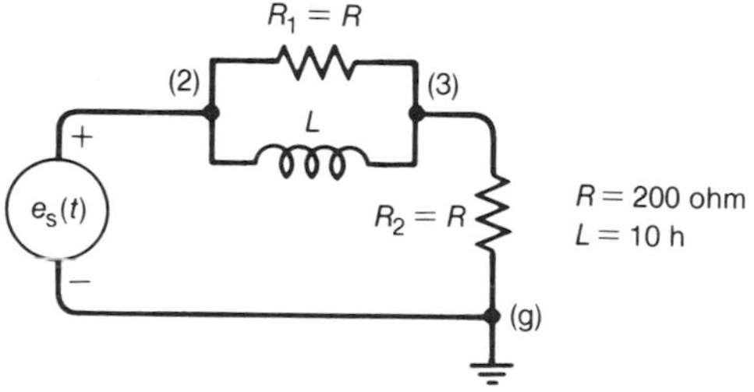

FIGURE P7.5

7.6. An electric circuit is being driven with a voltage source e_s as shown in Figure P7.6, and the output of primary interest is the capacitor voltage $e_o = e_{3g}$.

(a) Derive the system differential equation relating e_o to e_s.

(b) Find the undamped natural frequency and damping ratio for this system.

(c) Find $i_L(0^+)$, $e_o(0^+)$, and de_o/dt at $t = 0^+$, assuming that e_s has been equal to zero for a very long time and then suddently changes to 10 v at $t = 0$.

(d) Sketch the response of e_o versus time, indicating clearly the period of oscillation (if applicable) and the final steady-state value of e_o.

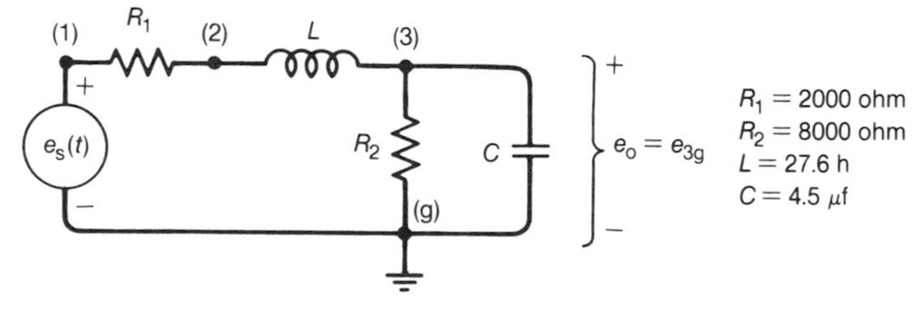

FIGURE P7.6

7.7. **(a)** Develop a complete set of state-variable equations for the circuit shown in Figure P7.7, employing e_{3g} and i_L as the state variables.

(b) Draw the simulation block diagram for this system, employing a separate block for each independent parameter and showing all system variables.

(c) Consider e_{3g} to be the output variable of primary interest, and develop the input-output system differential equation for this system.

(d) Having been equal to 5 v for a very long time, e_s is suddenly decreased to 2 v at time $t = 0$. Find $i_L(0^+)$, $i_C(0^+)$, $e_{3g}(0^+)$, and de_{3g}/dt at $t = 0^+$.

(e) Sketch the response of e_{3g} versus time, given the following system parameters: $R_1 = 80$ ohm; $R_2 = 320$ ohm; $L = 2$ h; $C = 50$ μf; $R_3 = 5$ ohm.

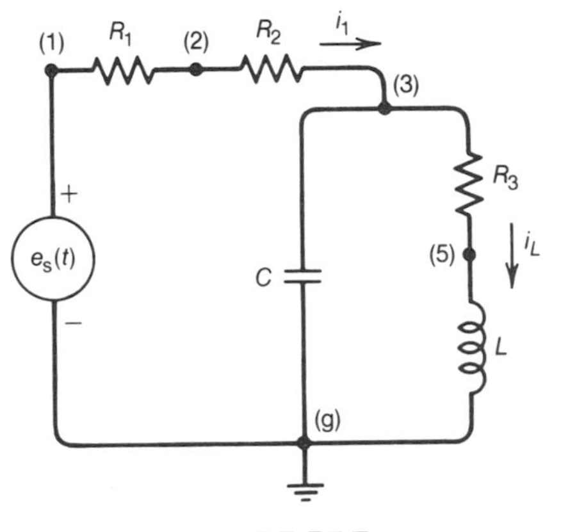

FIGURE P7.7

7.8. You have been given a field-controlled DC motor for use in a research project, but the values of the field resistance and field inductance are unknown. By conducting a simple test of the field with an initially charged capacitor of known capacitance, you should find it possible to estimate the values of R_f and L_f for this motor winding. The schematic diagram in Figure P7.8 shows such a test arrangement, the test being initiated by closing the switch at $t = 0$.

(a) Assuming a series *R-L* model for the field winding, draw the lumped-parameter circuit diagram for this system. (Why would a parallel *R-L* model for the field winding not be correct?)

(b) Derive the system differential equation for the voltage e_{21} after the switch is suddenly closed. (Note that the input for this system is a sudden change in the system at $t = 0$; after $t = 0$ this is a homogeneous system with no input variable.)

(c) After the switch is closed, the voltage e_{21} is displayed versus time on a CRO. A damped oscillation, having a period of 0.18 sec and a decay ratio of 0.8 per cycle is observed. Determine estimated values of R_f and L_f, based on your knowledge about the response of second-order systems.

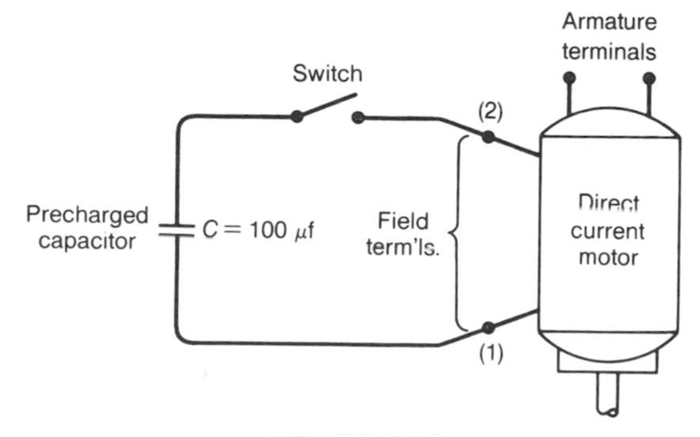

FIGURE P7.8

7.9. The nonlinear circuit shown in Figure P7.9 is to be subjected to an input voltage consisting of a constant normal operating value, $\bar{e}_s$ and an incremental portion, $\hat{e}_s$, which changes with time.

The system parameters are $R_1 = 400$ ohm; $L = 3.0$ h; $C = 5\ \mu f$; and for the nonlinear resistor NLR, $e_{23} = 3.0 \times 10^6\ (i_{NLR})^3$. The normal operating value of the source voltage $\bar{e}_s(t)$ is 5.0 v.

(a) Find the normal operating point value of the output voltage, $\bar{e}_{2g}$, and the incremental resistance, R_{inc}, of the nonlinear resistor.

(b) Develop the nonlinear state-variable equations for the circuit, employing e_{12} and i_{NLR} as the state variables, and linearize them for small perturbations from the normal operating point.

(c) Find the natural frequency, ω_n, and the damping ratio, ζ, of the linearized system, and carry out the analytical solution for e_{2g} when e_s suddenly increases by 0.5 v at $t = 0$. Sketch $e_{2g}(t)$.

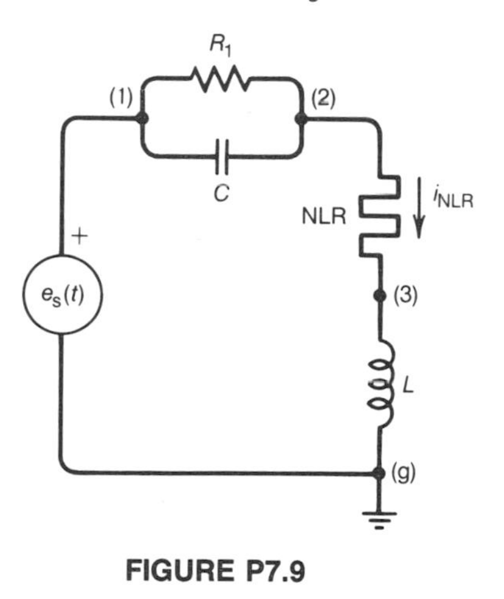

FIGURE P7.9

7.10. The nonlinear electric circuit shown in Figure P7.10 is subjected to a time-varying current source i_s, and its output is the voltage e_{2g}. The nonlinear resistor obeys the relation $e_{23} = 1.0 \times 10^7 (i_{NLR})^3$.

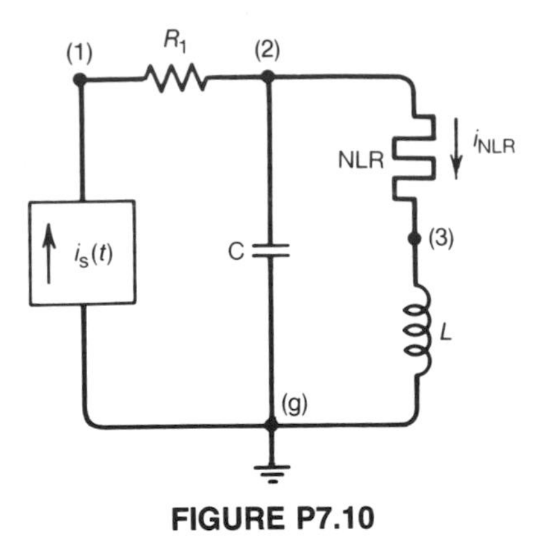

FIGURE P7.10

(a) Write the set of nonlinear state-variable equations employing i_{NLR} and e_{2g} as the state variables.

(b) Given the following normal operating point data, find the values of the resistance R_1 and the normal operating current $\bar{i}_{NLR}$.

$$\bar{e}_{12} = 1.2 \text{ v} \qquad \bar{e}_{23} = 3.75 \text{ v}$$

(c) Develop the set of linearized state-variable equations for this system.

(d) Draw the simulation block diagram for this linearized system, employing a separate block for each system parameter and showing all system variables.

7.11. The nonlinear electric circuit shown in Figure P7.11 is subjected to an input voltage e_s, which consists of a constant normal operating component $\bar{e}_s$ = 5.0 v and a time-varying component $\hat{e}_s(t)$. The nonlinear resistor NLR is a square-law device such that $e_{23} = K|i_{NRL}|i_{NLR}$.

(a) Develop the nonlinear state-variable equations for the circuit using e_{3g} and i_L as the state variables and using symbols for the system parameters.

(b) Determine the incremental resistance $R_{inc} = (de_{23}/di_{NLR})|_{\bar{i}_{NLR}}$ (i.e., the linearized resistance) for the resistor NLR. Use symbols only.

(c) Write the linearized state-variable equations for the system, using symbols only.

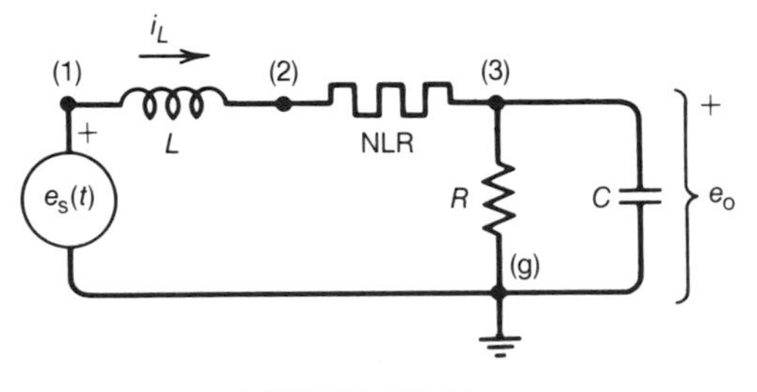

FIGURE P7.11

(d) Draw the simulation block diagram for the linearized system, using a separate block for each parameter and showing all system variables.

(e) Derive the input-output system differential equation for the linearized system, with $e_o = e_{3g}$ as the output.

7.12. In the variable reactance transducer circuit shown in Figure P7.12, the air gap varies with time producing a corresponding variation of inductance with time given by:

$$L(t) = 1.0 + 0.1 \sin 5t$$

The input voltage $e_s = 6.0$ v is constant, and the output signal is $e_{2g} = \bar{e}_{2g} + \hat{e}_{2g}(t)$. The values of the resistances are $R_1 = 1.2 \times 10^3$ ohm and $R_2 = 2.4 \times 10^3$ ohm.

(a) Derive the state-variable equation for this circuit using magnetic flux linkage λ_{2g} as the state variable.

(b) Determine the normal operating conditions (i.e., the values of $\bar{e}_{2g}$ and $\bar{i}_L$) for this system and linearize the state-variable equation for small perturbations about the normal operating point.

(c) Write the differential equation relating $\hat{e}_{2g}$ to $\hat{L}(t)$.

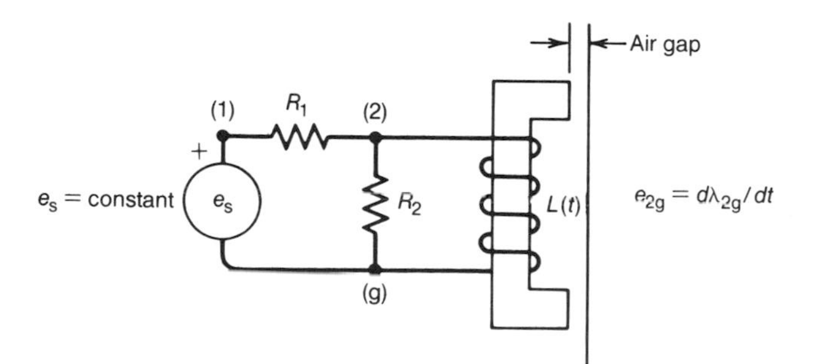

FIGURE P7.12

7.13. In the circuit shown in Figure P7.13, the value of the capacitance varies with time about its mean value, $\bar{C}$, and the voltage source e_s is constant. In other words

$$C = \bar{C} + \hat{C}(t)$$

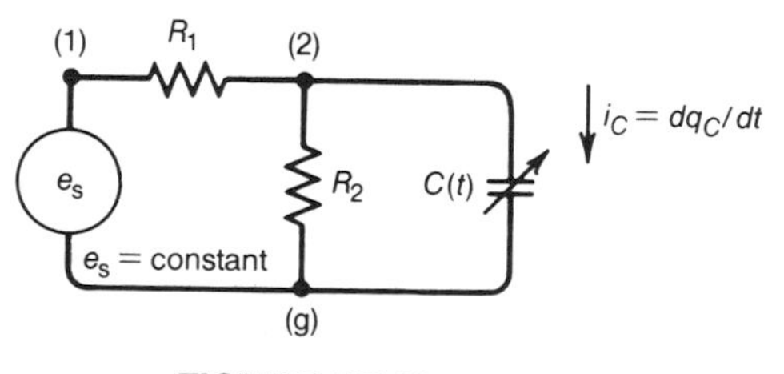

FIGURE P7.13

(a) Develop the state variable equation for this system, using charge q_C on the capacitor as the state-variable.

(b) Linearize this state-variable equation for small perturbations about the normal operating point.

(c) Write the linearized system differential equation relating $\hat{e}_{2g}$ to $\hat{C}(t)$.

7.14. The circuit shown in Figure P7.14 is often used to obtain desired dynamic characteristics for use in automatic control systems.

The operational amplifier has a very high negative gain k_a, which is of the order of -10^5 v/v. The input current i_a is very small, and may be considered negligible. The object here is to find how the output, $e_o = e_{3g}$, is related to the input signal, $e_i = e_{1g}$.

(a) Write the necessary and sufficient set of describing equations for this system.

(b) Eliminate unwanted variables to find the system input-output equation relating e_{3g} to e_{1g}.

(c) Considering the very large magnitude of k_a, eliminate very small terms to arrive at a simplified input-output differential equation for this system, using the following values for R_1, R_2, and L: $R_1 = 1200$ ohm; $R_2 = 2000$ ohm; $L = 1.5$ h.

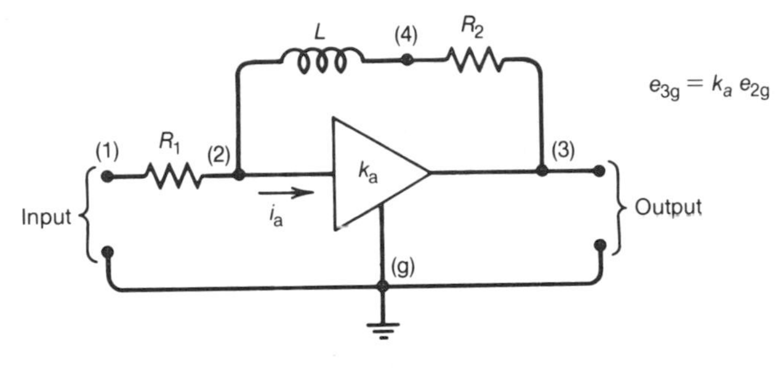

FIGURE P7.14

8

Thermal Systems

8.1 Introduction

Fundamentals of mathematical methods used today in modeling thermal systems were developed centuries ago by such great mathematicians and scientists as Laplace, Fourier, Poisson, and Stefan. Analytical solution of the equations describing the basic mechanisms of heat transfer—conduction, convection, and radiation — was always considered to represent an extremely challenging mathematical problem.

Temperature, which is an A-type variable in thermal systems, is usually dependent on spatial as well as temporal coordinates. As a result, the dynamics of thermal systems has to be described by partial differential equations, in order to assure high accuracy of calculations. Moreover, nonlinearities are often essential for describing dynamic performance of thermal systems. Very often nonlinear problems arise in modeling processes involving heat transfer by radiation or heat transfer by convection where the heat transfer coefficient is a function of a fluid flow rate, temperature, and/or other system variables. Very few problems described by nonlinear partial differential equations have been solved analytically. The availability of large digital computers together with fast, efficient finite difference algorithms for solving partial differential equations has opened new avenues in analysis of thermal systems.

The energy crisis that came upon the modern era in the late 1960s rekindled interest in thermal systems. Sudden shortage and dramatically rising cost of thermal energy inspired new designs and new energy-efficient processes. Waste heat recovery systems as well as solar and other alternative energy source systems have attracted engineers and scientists in recent years.

The material presented in this chapter covers the fundamentals of mathematical modeling of thermal systems. First, the equations describing the basic mechanisms of heat transfer by conduction, convection, and radiation are reviewed in Section 8.2. In Section 8.3 lumped models of thermal systems are introduced. Application of the lumped models leads to approximate solutions, and great care must be exercised in interpreting the results produced by this method. In many cases, however, especially in the early stages of system analysis, the lumped models are very useful because of their simplicity and easy solution methods.

8.2 Basic Mechanisms of Heat Transfer

In this section the classical mathematical equations describing heat transfer by conduction, convection, and radiation are reviewed. This part of the material will be presented in a rather condensed form since it is assumed that the reader is generally familiar with it.

Conduction. The net rate of heat transfer by conduction across the boundaries of a unit volume is equal to the rate of heat accumulation within the unit volume, which is mathematically expressed by the Laplace equation

$$\frac{\partial^2 T}{\partial x^2} + \frac{\partial^2 T}{\partial y^2} + \frac{\partial^2 T}{\partial z^2} = \frac{1}{\alpha}\frac{\partial T}{\partial t} \tag{8.1}$$

where T is temperature, α is thermal diffusivity, and x, y, z are the Cartesian space coordinates. The thermal diffusivity can be expressed as $\alpha = k/\rho c_p$, where k is thermal conductivity, ρ is density, and c_p is the specific heat of the material.

At steady state, that is, when temperature remains constant in time, the Laplace equation takes the form

$$\frac{\partial^2 T}{\partial x^2} + \frac{\partial^2 T}{\partial y^2} + \frac{\partial^2 T}{\partial z^2} = 0 \tag{8.2}$$

For one-dimensional heat conduction in the x direction, the rate of heat flow is determined by the Fourier equation

$$Q_{hk} = -kA\frac{dT}{dx} \tag{8.3}$$

where A is an area of heat transfer normal to x. Integrating with respect to x yields

$$\int_0^L Q_{hk}\,dx = -\int_{T_1}^{T_2} kA\,dT \tag{8.4}$$

and hence the rate of heat transfer is

$$Q_{hk} = -\left(\frac{A}{L}\right)\int_{T_1}^{T_2} k\,dT \tag{8.5}$$

The one-dimensional steady-state heat conduction is depicted in Figure 8.1. Note that the assumption of the temperature gradients in y and z directions being equal to zero implies that there is no heat loss through the sides, which are said to be

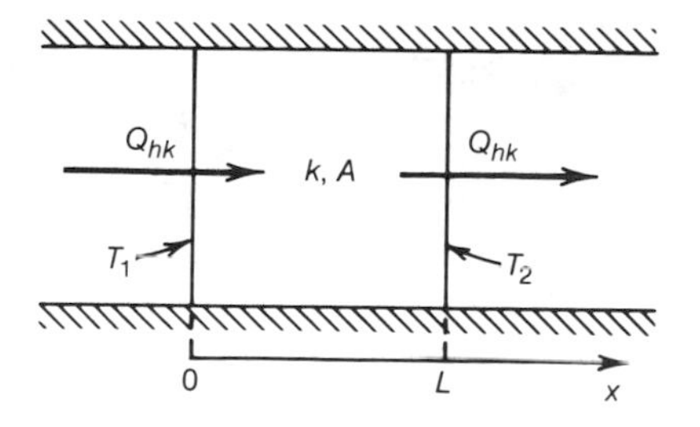

FIGURE 8.1 One-dimensional steady-state heat conduction.

perfectly insulated. At steady state the temperature is constant in time and thus there is no energy storage.

If thermal conductivity of the material does not depend on temperature, the rate of heat transfer can be expressed as

$$Q_{hk} = \left(\frac{kA}{L}\right)(T_1 - T_2) \tag{8.6}$$

In Equation (8.6) T_1 is the temperature at $x = 0$ and T_2 is the temperature at $x = L$. In general, if the value of the rate of heat transfer Q_{hk} is known, the value of temperature at any location between 0 and L, $0 < x < L$, can be calculated, substituting $T(x)$ for T_2 in Equation (8.6).

$$T(x) = T_1 - \left(\frac{x}{Ak}\right) Q_{hk} \tag{8.7}$$

Note that Q_{hk} is positive if heat is transfered in the direction of x and negative if heat is flowing in the direction opposite to x.

Convection. Convective heat transfer is usually associated with the transfer of mass in a boundary layer of a fluid over a fixed wall. In the system shown in Figure 8.2, the rate of heat transfer by convection Q_{hc} between a solid wall and a fluid flowing over it is given by Equation (8.8).

$$Q_{hc} = h_c A(T_w - T_f) \tag{8.8}$$

where h_c is a convective heat transfer coefficient, A is an area of heat transfer, and T_w and T_f represent wall and fluid temperature, respectively.

From the distribution of the fluid velocity within the boundary layer shown in Figure 8.2, it can be seen that the velocity is zero on the surface of the wall and thus no convective heat transfer can take place there. However, although heat is transfered across the wall–fluid boundary by conduction, it is carried away from there by convection with the flowing fluid. The rate at which the fluid carries heat from the wall surface is thus determined by the heat convection Equation (8.8). If the flow of fluid is caused by its density gradient in the gravity field, the heat convection is said to be free. The density gradient usually occurs due to temperature gradient. Forced convection, on the other hand, takes place when the flow of fluid is forced by an external energy source, such as a pump or a blower.

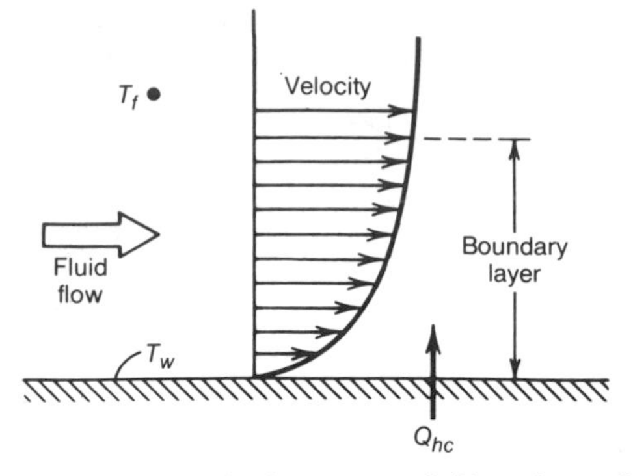

FIGURE 8.2 Convective heat transfer between a fluid and a wall.

Determination of the value of the convective heat transfer coefficient for an actual thermal system presents a challenging task. In many practical cases empirical models are developed that express a convective heat transfer coefficient as a function of other system variables for specified operating conditions. The applicability of such empirical formulas is usually limited to a narrow class of problems. Moreover, the mathematical forms used in convective heat transfer correlations are often nonlinear, which leads to nonlinear system-model equations, difficult to solve analytically.

Radiation. The rate of heat transfer by radiation between two separated bodies having temperatures T_1 and T_2, respectively, is determined by the Stefan-Boltzman law,

$$Q_{hr} = \sigma F_E F_A A(T_1^4 - T_2^4) \tag{8.9}$$

where $\sigma = 5.667 \times 10^{-8}$ W/m^2-deg^4 (the Stefan-Boltzman constant)
F_E is effective emissivity
F_A is the shape factor
A is the heat transfer area

The effective emissivity F_E accounts for the deviation of the radiating systems from black bodies. For instance, the effective emissivity of parallel planes having emissivities ϵ_1 and ϵ_2 is

$$F_E = \frac{1}{\dfrac{1}{\epsilon_1} + \dfrac{1}{\epsilon_2} - 1} \tag{8.10}$$

The values of the shape factor F_A range from 0 to 1 and represent the fraction of the radiative energy emitted by one body that reaches the other body. In a case when all radiation emitted by one body reaches the other body, for example in heat transfer by radiation between two parallel plane surfaces, the shape factor is equal to one.

The nonlinearity of Equation (8.9) constitutes a major difficulty in developing and solving mathematical model equations for thermal systems in which radiative heat transfer takes place. Determining values of model parameters also presents a challenging task, especially when radiation of gases in involved.

8.3 Lumped Models of Thermal Systems

Mathematical models of thermal systems are usually derived from the basic energy balance equations that follow the general form of Equation (8.11).

$$\begin{pmatrix}\text{rate of energy}\\ \text{stored}\\ \text{within system}\end{pmatrix} = \begin{pmatrix}\text{heat flow}\\ \text{rate}\\ \text{into system}\end{pmatrix} - \begin{pmatrix}\text{heat flow}\\ \text{rate}\\ \text{out of system}\end{pmatrix} + \begin{pmatrix}\text{rate of heat}\\ \text{generated}\\ \text{within system}\end{pmatrix} + \begin{pmatrix}\text{rate of work}\\ \text{done}\\ \text{upon system}\end{pmatrix} \tag{8.11}$$

For a stationary system composed of a material of density ρ, specific heat c_p, and constant volume $\mathcal{V}$, the energy balance equation takes the form

$$\rho c_p \mathcal{V} \frac{dT}{dt} = Q_{hin}(t) - Q_{hout}(t) + Q_{hgen} + \frac{d\mathcal{W}}{dt} \tag{8.12}$$

Equation (8.12) can only be used if the temperature distribution in the system, or in a part of the system is uniform, that is when the temperature is independent of spatial coordinates, $T(x, y, z, t) = T(t)$. The assumption about the uniformity of the temperature distribution also implies that the system physical properties, such as density and specific heat, are constant within the system boundaries.

Two basic parameters used in lumped models of thermal systems are thermal capacitance and thermal resistance.

Thermal capacitance of a thermal system of density ρ, specific heat c_p, and volume $\mathcal{V}$ is

$$C_h = \rho c_p \mathcal{V} \tag{8.13}$$

The rate of energy storage in a system of thermal capacitance C_h is

$$Q_{hstored} = C_h \frac{dT}{dt} \tag{8.14}$$

Note that a thermal capacitance is an A-type element because its stored energy is associated with an A-type variable, T. Physically, thermal capacitance represents a system ability to store thermal energy. It also provides a measure of the effect of energy storage on the system temperature. If the system thermal capacitance is large, the rate of temperature change owing to heat influx is relatively low. On the other hand, when the system thermal capacitance is small, the temperature increases more rapidly with the amount of energy stored in the system. For example, if the rate of thermal energy storage changes by ΔQ_h in a stepwise manner, the change in rate of change of temperature is, from Equation (8.14),

$$\frac{dT}{dt} = \frac{\Delta Q_h}{C_h} \tag{8.15}$$

Another parameter used in lumped models of thermal systems is thermal resistance, R_h. Thermal resistance to heat flow rate, Q_h, between two points having different temperatures, T_1 and T_2, is

$$R_h = \frac{T_1 - T_2}{Q_h} \tag{8.16}$$

The mathematical expressions for thermal resistance are different for the three different mechanisms of heat transfer, conduction, convection, and radiation. The conductive thermal resistance R_{hk} can be obtained from Equations (8.6) and (8.16),

$$R_{hk} = \frac{L}{kA} \tag{8.17}$$

From Equations (8.8) and (8.16), the convective thermal resistance is found to be

$$R_{hc} = \frac{1}{h_c A} \tag{8.18}$$

The ratio of the conductive thermal resistance over the convective thermal resistance yields a nondimensional constant known as the Biot number, Bi.[1]

$$\mathrm{Bi} = \frac{R_{hk}}{R_{hc}} \tag{8.19}$$

Using Equations (8.17) and (8.18), the Biot number can be expressed as

$$\mathrm{Bi} = \frac{h_c L}{k} \tag{8.20}$$

The value of the Biot number provides a measure of adequacy of lumped models to represent dynamics of thermal systems. At the beginning of this section lumped models were characterized as having a uniform temperature distribution, which occurs when an input heat flow (usually convective) encounters relatively low resistance within the system boundaries. The resistance to heat flow within the system is usually conductive in nature. Thermal systems represented by lumped models should therefore have relatively high thermal conductivity. It can thus be deduced that lumped models can be used when the ratio of convective thermal resistance over conductive thermal resistance is large or, equally, when the Biot number, which represents the reciprocal of that ratio, is small enough. The value

[1] J. P. Holman, "*Heat Transfer*," McGraw-Hill Book Company, 1976.

of 0.1 is usually accepted as the threshold for the Biot number below which lumped models can be used to describe actual thermal systems with sufficient accuracy. If $\text{Bi} > 0.1$, distributed models involving partial differential equations are necessary to adequately represent system dynamics.

In order to derive an expression for a radiative thermal resistance, the basic Equation (8.9), describing heat transfer by radiation, has to be linearized. Applying the general linearization procedure based on Taylor series expansion to the function of two variables $f_{NL}(T_1, T_2)$ yields

$$f_{NL}(T_1, T_2) \approx f_{NL}(\overline{T}_1, \overline{T}_2) + \hat{T}_1 \left.\frac{\partial f_{NL}}{\partial T_1}\right|_{\overline{T}_1, \overline{T}_2} + \hat{T}_2 \left.\frac{\partial f_{NL}}{\partial T_2}\right|_{\overline{T}_1, \overline{T}_2} \tag{8.21}$$

where $\overline{T}_1$ and $\overline{T}_2$ represent the normal operating point. The linearized expression for radiative heat transfer is then

$$Q_{hr} \approx \overline{Q}_{hr} + b_1\hat{T}_1 - b_2\hat{T}_2 \tag{8.22}$$

$$\text{where} \quad \overline{Q}_{hr} = \sigma F_E F_A A(\overline{T}_1^4 - \overline{T}_2^4)$$
$$b_1 = 4\sigma F_E F_A A \overline{T}_1^3$$
$$b_2 = 4\sigma F_E F_A A \overline{T}_2^3$$

The form of Equation (8.22) is not very convenient in modeling thermal systems because of its incompatibility with corresponding equations describing heat transfer by convention and conduction in lumped models. To achieve this compatibility a different linearization procedure can be used. First, rewrite Equation (8.9) in the following form.

$$Q_{hr} = \sigma F_E F_A A(T_1^2 + T_2^2)(T_1 + T_2)(T_1 - T_2) \tag{8.23}$$

Assuming that T_1 and T_2 in the first two parenthetical factors in Equation (8.23) represent normal operating point values, the approximating linear equation is

$$\hat{Q}_{hr} \approx \left(\frac{1}{R_{hr}}\right)[\hat{T}_1(t) - \hat{T}_2(t)] \tag{8.24}$$

where the radiative thermal resistance is

$$R_{hr} = \frac{1}{\sigma F_E F_A A(\overline{T}_1^2 + \overline{T}_2^2)(\overline{T}_1 + \overline{T}_2)} \tag{8.25}$$

A lumped linear model of combined heat transfer by convection and radiation in parallel can be approximated by Equation (8.26).

$$\hat{Q}_h = \left(\frac{1}{R_{hc}} + \frac{1}{R_{hr}}\right)(\hat{T}_1 - \hat{T}_2) \tag{8.26}$$

The lumped-model parameters, thermal capacitance and thermal resistance, will now be used in the following examples.

EXAMPLE 8.1

A slab of material of density ρ and specific heat c_p, shown in Figure 8.3, is perfectly insulated on all its sides except the top, which is in contact with fluid of temperature T_f in motion above the slab. The coefficient of heat transfer between the fluid and the top side of the slab was found to be h_c. The cross-sectional area of the slab is A, and its height is L.

Derive a state-variable model for the slab using its temperature, T_s, as the state variable and the temperature of the fluid, T_f, as the input variable.

Solution

First, assume that the temperature distribution within the slab is uniform, T_s. The energy balance equation for the system is

$$Q_{h\text{stored}} = Q_{h\text{conv.}}$$

where $Q_{h\text{stored}}$ represents the rate of heat storage in the slab, and $Q_{h\text{conv.}}$ is the rate of heat transferred by convection from the fluid.

$$Q_{h\text{stored}} = \rho c_p AL \left(\frac{dT_s}{dt} \right) \tag{8.27}$$

$$Q_{h\text{conv.}} = h_c A(T_f - T_s) \tag{8.28}$$

Introducing thermal capacitance C_h, defined earlier by Equation (8.13) and convective thermal resistance R_{hc}, given by Equation (8.18), and substituting the right-hand sides of equations (8.27) and (8.28) into the heat balance equation yields the state-variable equation

$$\dot{T}_s = -\frac{1}{R_{hc}C_h} T_s + \frac{1}{R_{hc}C_h} T_f$$

The system step response from an initial equilibrium condition $T_{s0} = T_{f0}$ is

$$T_s(t) = \Delta T_f(1 - e^{-t/\tau}) + T_{s0}$$

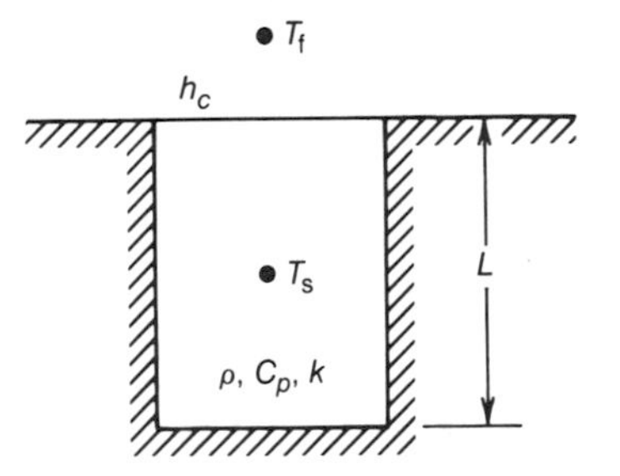

FIGURE 8.3 Slab represented by a lumped model in Example 8.1.

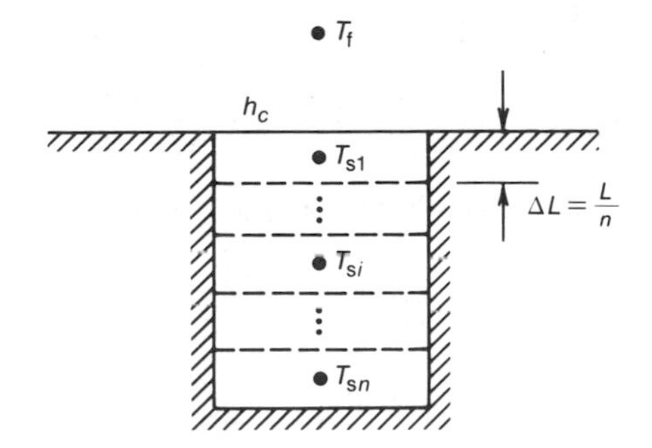

FIGURE 8.4 Slab represented by a multilayer model in Example 8.1.

where T_{s0} is the initial temperature of the slab, ΔT_f is the magnitude of the step change of the temperature of the fluid, and τ is the system time constant given by

$$\tau = R_{hc}C_h$$

To verify that a lumped model adequately represents the thermal system considered in this example, calculate the Biot number

$$\text{Bi} = \frac{R_{hk}}{R_{hc}} = h_c \frac{L}{k}$$

If the Biot number is smaller than 0.1, a lumped model is adequate.

If the Biot number is greater than 0.1, the slab has to be subdivided into layers of equal height ΔL, as shown in Figure 8.4, so that the Biot number calculated for the top layer, $h_c \Delta L/k$, is less than 0.1.

The energy balance equation for the top layer is then

$$\rho c_p \Delta L A \dot{T}_{s1} = h_c A(T_f - T_{s1}) - \frac{kA}{\Delta L}(T_{s1} - T_{s2})$$

where $T_{s1}, T_{s2}, \ldots, T_{sn}$ are the slab temperatures in layers 1, 2, . . . , n. Hence, the state variable equation in terms of the lumped parameters of the top layer is

$$\dot{T}_{s1} = -\frac{n}{C_h}\left(\frac{1}{R_{hc}} + \frac{n}{R_{hk}}\right) T_{s1} + \frac{n^2}{R_{hk}C_h} T_{s2} + \frac{n}{R_{hc}C_h} T_f$$

For an internal layer, that is for a layer located between two other slab layers, the state-variable equation is

$$\dot{T}_{si} = -\frac{2n^2}{R_{hk}C_h} T_{si} + \frac{n^2}{R_{hk}C_h} T_{s(i-1)} + \frac{n^2}{R_{hk}C_h} T_{s(i+1)}$$

where $i = 2, 3, \ldots, n-1$. Finally, the equation for the bottom layer is

$$\dot{T}_{sn} = -\frac{n^2}{R_{hk}C_h}T_{sn} + \frac{n^2}{R_{hk}C_h}T_{s(n-1)}$$

It should be noted that the slab is assumed to be perfectly insulated on all side and bottom faces and thus no heat is lost. ■

EXAMPLE 8.2

Consider a blending system shown schematically in Figure 8.5. Two identical liquids of different temperatures T_1 and T_2, flowing with different flow rates Q_1 and Q_2, are perfectly mixed in a blender of volume $\mathcal{V}$. The mixture of liquids is also heated in the blender by an electric heater supplying heat at a constant rate, Q_{hgen}. There are heat losses in the system and the coefficient of heat transfer between the blender and ambient air of temperature T_a is h_c. Although the mixing in the tank is assumed to be perfect, the work done by the mixer is negligible and the kinetic energies of the flows Q_1, Q_2, and Q_3 are very small. Derive a mathematical model of the blending process.

Solution

The unsteady flow energy balance equation includes the following terms.

$$\begin{pmatrix}\text{rate}\\\text{of}\\\text{enthalpy}\end{pmatrix}_1 + \begin{pmatrix}\text{rate}\\\text{of}\\\text{enthalpy}\end{pmatrix}_2 + \begin{pmatrix}\text{rate}\\\text{of}\\\text{heat}\end{pmatrix}_{\text{gen}}$$

$$= \begin{pmatrix}\text{rate}\\\text{of}\\\text{enthalpy}\end{pmatrix}_3 + \begin{pmatrix}\text{rate}\\\text{of}\\\text{heat}\end{pmatrix}_{\text{loss}} + \begin{pmatrix}\text{rate}\\\text{of}\\\text{change}\\\text{of}\\\text{energy}\end{pmatrix}_{\text{sto}}$$

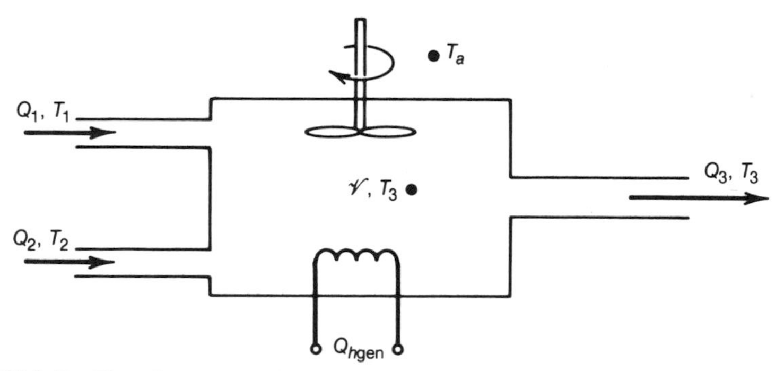

FIGURE 8.5 Blending system.

The first two components in this equation represent the rate of enthalpy supplied with the two incoming streams of liquids given by the equations

$$Q_{h1} = \rho c_p Q_1(T_1 - T_a)$$

$$Q_{h2} = \rho c_p Q_2(T_2 - T_a)$$

The next term in the energy balance equation represents the rate of heat generated by the heater, Q_{hgen}. The heat carried away from the tank is represented by the first two terms on the right-hand side of the energy balance equation. The rate of enthalpy carried with the outgoing stream of the mixture of the two input liquids is

$$Q_{h3} = \rho c_p(Q_1 + Q_2)(T_3 - T_a)$$

The rate of the heat lost by the liquid through the sides of the tank to the ambient air is

$$Q_{hloss} = h_c A(T_3 - T_a)$$

Finally, the rate of change of energy stored in the liquid contained in the tank is

$$\frac{d\mathcal{E}}{dt} = \rho c_p \mathcal{V} \frac{dT_3}{dt}$$

Substituting detailed mathematical expressions for the heat and enthalpy rates into the energy balance equation gives

$$\rho c_p Q_1(T_1 - T_a) + \rho c_p Q_2(T_2 - T_a) + Q_{hgen} = \rho c_p(Q_1 + Q_2)(T_3 - T_a) + h_c A(T_3 - T_a) + \rho c_p \mathcal{V} \frac{dT_3}{dt} \quad (8.29)$$

Equation (8.29) can be rearranged into a simpler form.

$$\frac{dT_3}{dt} = -\frac{1}{\mathcal{V}}\left(\frac{h_c A}{\rho c_p} + Q_1 + Q_2\right) T_3 + \frac{Q_1}{\mathcal{V}} T_1 + \frac{Q_2}{\mathcal{V}} T_2 + \frac{1}{\rho c_p \mathcal{V}} Q_{hgen} + \frac{h_c A}{\rho c_p \mathcal{V}} T_a \quad (8.30)$$

Equation (8.30) represents a first-order multidimensional model with six potential input signals, Q_1, T_1, Q_2, T_2, Q_{hgen}, and T_a. The system time constant is

$$\tau = \frac{\rho c_p \mathcal{V}}{h_c A + \rho c_p Q_3}$$

The model can be further simplified if the blender is assumed to be perfectly insulated, $h_c = 0$, and if there is no heat generation in the system, $Q_{hgen} = 0$. Under such conditions the system state equation becomes

$$\frac{dT_3}{dt} = -\left(\frac{1}{C}\right)\left(\frac{1}{R_1} + \frac{1}{R_2}\right) T_3 + \left(\frac{1}{R_1 C}\right) T_1 + \left(\frac{1}{R_2 C}\right) T_2$$

where the lumped-model parameters C, R_1, and R_2, are defined as follows.

$$C = \rho c_p \mathcal{V}$$

$$R_1 = \frac{1}{\rho\, c_p Q_1}$$

$$R_2 = \frac{1}{\rho c_p Q_2}$$

■

8.4 Synopsis

Exact models of thermal systems usually involve nonlinear partial differential equations. Deriving closed-form analytical solutions for such problems is often impossible and even obtaining valid computer models poses a very difficult task. Simplified lumped models of thermal systems were introduced in this chapter. Lumped models are very useful in early stages of system analysis and also in verifying more complex computer models. It is always very important, when simplified models are used, to make sure that such models retain the basic dynamic characteristics of the original systems. The criterion of applicability of lumped models in thermal problems is provided by the Biot number defined as the ratio of the conductive thermal resistance over the convective thermal resistance. Thermal systems can be represented by lumped models if the Biot number is small enough, usually less than one tenth. Basic elements of the lumped models, thermal capacitance and thermal resistance, are analogous to the corresponding elements in mechanical and electrical systems (although there is no thermal inductance). The state model equations are derived around the thermal system elements associated with the A-type variable, temperature, and the T-type variable, heat flowrate. The same analytical and numerical methods of solution of state-variable equations as those described in earlier chapters can be employed with thermal systems. Several examples of thermal systems are presented, including a liquid blending process, to illustrate thermal energy storage.

Problems

8.1. A slab of cross-sectional area A and length $L = L_1 + L_2$ is made of two materials having different thermal conductivities, k_1 and k_2, as shown in Figure P8.1. The left- and right-side surfaces of the slab are at constant temperatures, $T_1 = 200°C$ and $T_2 = 20°C$, while all other surfaces are perfectly insulated. The values of the system parameters are

$$k_1 = 0.05 \text{ W/m-°C} \qquad L_1 = 0.04 \text{ m}$$

$$k_2 = 0.7 \text{ W/m-°C} \qquad L_2 = 1.4 \text{ m}$$

$$A = 1 \; m^2$$

(a) Find the temperature at the interface of the two materials.
(b) Sketch the temperature distribution along the slab.
(c) Derive an expression for an equivalent thermal resistance of the entire slab in terms of the thermal resistances of the two parts.

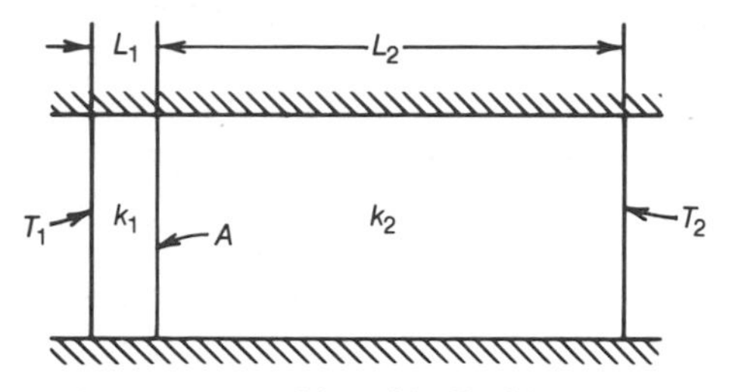

FIGURE P8.1 Thermal system considered in Problem 8.1.

8.2. The temperatures of the side surfaces of the composite slab shown in Figure P8.2 are T_1 and T_2. The other surfaces are perfectly insulated. The cross-sectional areas of the two parts of the slab are A_1 and A_2, and their conductivities are k_1 and k_2, respectively. The length of the slab is L.
(a) Find an equivalent thermal resistance of the slab and express it in terms of the thermal resistances of the two parts.
(b) Sketch the steady-state temperature distribution along the slab.

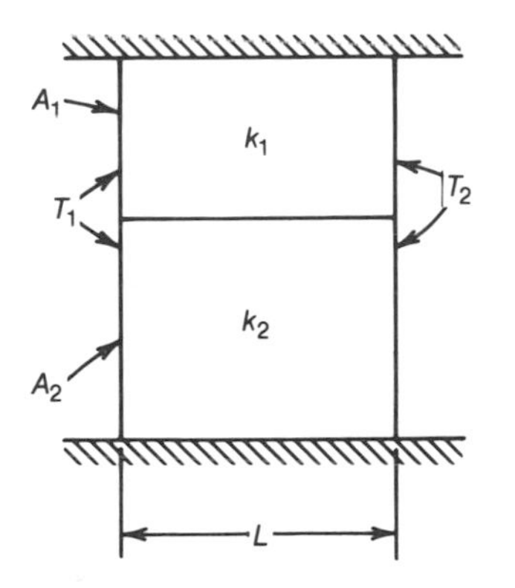

FIGURE P8.2 Thermal system considered in Problem 8.2.

8.3. A hollow cylinder is made of material of thermal conductivity k. The dimensions of the cylinder follow: inside diameter, D_i; outside diameter, D_0; length, L. The inside and outside surfaces are at constant temperatures T_i and T_0, respectively, whereas the top and bottom surfaces are both perfectly insulated. Find the expression for the lumped conductive thermal resistance of the cylinder for heat transfer in the radial direction only.

8.4. Consider again the cylinder from Problem 8.3 assuming that the inside diameter $D_i = 1m$, the outside diameter $D_0 = 2m$, and the thermal conductivity of the material $k = 54$ W/m °C. The outside surface of the cylinder is now exposed to a stream of air of temperature T_a and velocity v_a. The inside surface remains at constant temperature T_i. The convective heat transfer coefficient between the outside surface of the cylinder and the stream of air is approximated by the expression $h_c = 2.24v_a$. Determine the condition for the velocity of air under which the cylinder can be adequately represented by a lumped model.

8.5. A thermocouple circuit is used to measure the temperature of a perfectly mixed liquid, as shown in Figure P8.5 The hot junction of the thermocouple has the form of a small sphere of radius r_t. The density of the hot junction material is ρ_t, and the specific heat is c_t. Thermal capacitance of the thermocouple wire is negligible. The measuring voltage e_{21} is related to the hot junction temperature, T_t, by the equation $e_{21} = aT_t$.

(a) Derive a mathematical model for this system relating e_{21} to T_L.
(b) Sketch the response of e_{21} to a step change in the liquid temperature T_L.
(c) How long will it take for the measuring signal e_{21} to reach approximately 95 percent of the steady-state value after a step change of the liquid temperature. The thermocouple parameters are $\rho_t = 7800$ kg/m^3, $c_t = 0.4$ kJ/kg-°C, and $r_t = 0.2$ mm. The convective heat transfer between the liquid and the hot junction is $h_c = 150$ W/m^2-°C.

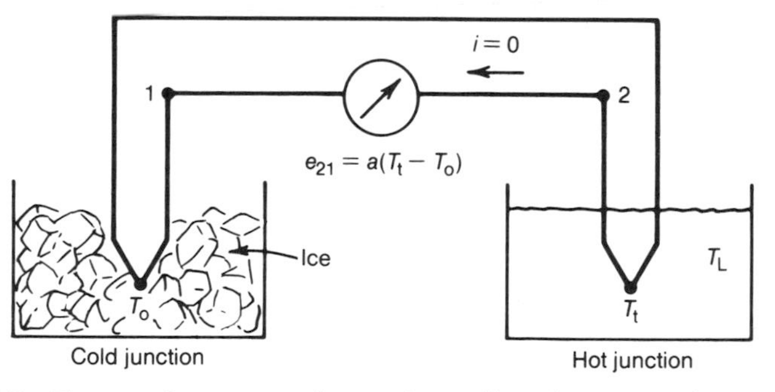

FIGURE P8.5 Temperature measuring system with a thermocouple.

8.6. Figure P8.6 shows a simple model of an industrial furnace. A packing of temperature T_1 is being heated in the furnace by an electric heater supplying heat at the rate $Q_{hi}(t)$. The temperature inside the furnace is T_2, the walls are at temperature T_3, and ambient temperature is T_a. The thermal capacitances of the packing, air inside the furnace, and the furnace walls are C_{h1}, C_{h2}, and C_{h3}, respectively. Derive state-variable equations for this system assuming that heat is transferred by convection only, with the convective heat transfer coefficients: h_{c1} (air-packing), h_{c2} (air-inside walls), and h_{c3} (outside walls-ambient air).

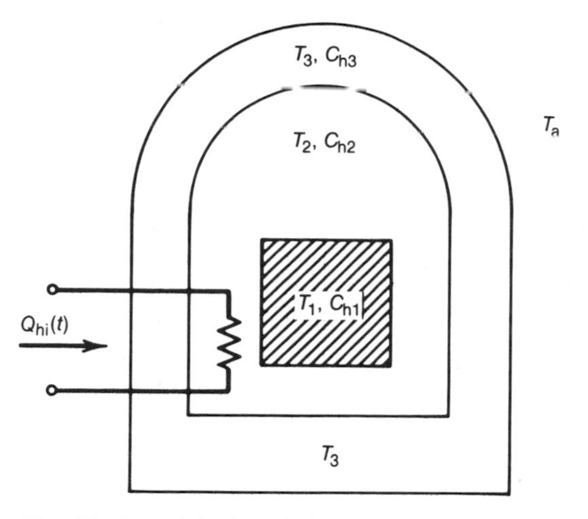

FIGURE P8.6 Simplified model of an industrial furnace.

8.7. A ceramic object shown in Figure P8.7 consists of two layers having different thermal capacitances, C_{h1} and C_{h2}. The top layer, having temperature T_1, is exposed to thermal radiation from a heater of temperature T_r and effective emissivity F_E. The area exposed to radiation is A_1 and the shape factor is F_{A1}. Both layers exchange heat by convection with ambient air of temperature T_a through their sides of areas A_s. The convective heat transfer coefficient is h_c.

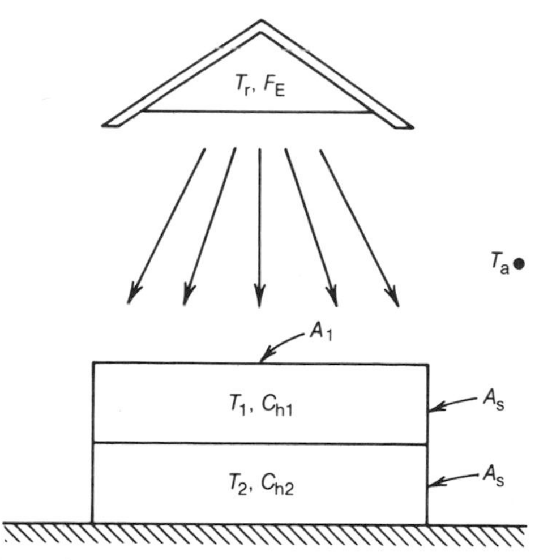

FIGURE P8.7 Radiative heating system considered in Problem 8.7.

Heat is also transfered between the two layers at the rate given by

$$Q_{h12} = \left(\frac{1}{R_{12}}\right)(T_1 - T_2)$$

where R_{12} represents the thermal resistance of the interface between the layers. Heat transfer through the bottom of the lower layer is negligible.

(a) Derive nonlinear state-variable equations for this system using temperatures T_1 and T_2 as the state variables.

(b) Obtain linearized state-model equations describing the system in the vicinity of the normal operating point given by $\overline{T}_r$, $\overline{T}_1$, $\overline{T}_2$.

8.8. A heat storage loop of a solar water heating system is shown schematically in Figure P8.8. The rate of solar energy incident per unit area of the collector is $S(t)$. The collector can be modeled by the Hottel-Whillier-Bliss equation, which gives the rate of heat absorbed by the collector, Q_{hcol}.

$$Q_{hcol}(t) = A_c F_R[S(t) - U_L(T_s - T_a)]$$

The values of the collector parameters are

Collector surface area, $A_c = 30 \text{ m}^2$
Heat removal factor, $F_R = 0.7$
Heat loss coefficient, $U_L = 4.0 \text{ W/m}^2\text{-°C}$

The rate of water flow in the loop is $Q_{mw} = 2160$ kg/hr and the specific heat of water is $c_w = 4180$ J/kg-°C. The water storage tank has capacity $m_w = 1800$ kg and is considered to be perfectly insulated. It is also assumed that there are no heat losses through the piping in the system. Derive a state-variable model for this system, assuming that the water in the storage tank is fully mixed.

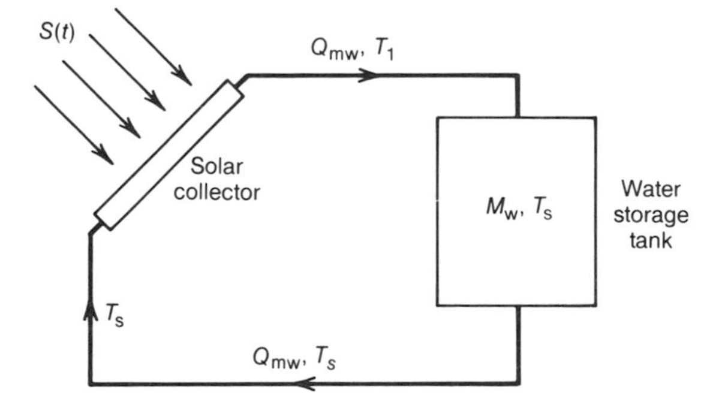

FIGURE P8.8 Heat storage loop in a solar water heating system.

8.9. Consider again the solar system shown in Figure P8.8 but now without assuming that the storage tank is fully mixed. In fact, stratification of a storage tank leads to enhanced performance of a solar heating system. Develop a mathematical model for the system shown in Figure P8.8 assuming that the water storage tank is made up of three isothermal segments of equal volume. Assume also that heat conduction between successive segments of water is negligible.

9

Fluid Systems

9.1 Introduction

Corresponding to the mass, spring, and damper elements discussed in Chapter 2, the A-type, T-type, and D-type elements employed in modeling fluid systems are the fluid capacitor, the inertor, and the fluid resistor elements.

Capacitance occurs as a result of elasticity or compliance in the fluid or its containing walls. Although liquids such as water and oil are often considered to be incompressible in many fluid flow situations, these hydraulic fluids are sometimes compressible enough to produce fluid capacitance. In other cases, the walls of the chambers or passages containing the fluid have enough compliance, when the fluid pressure changes rapidly, to produce fluid capacitance. In long lines and passages fluid capacitance is distributed along the line, together with inertance and resistance, and the analysis of such situations is beyond the scope of this text[1]. If both the resistance and inertance are negligible, however, it is possible to employ a lumped-capacitance model of the line.

When fluid capacitance is wanted for energy storage purposes, specially designed off-the-shelf capacitors with minimal resistance and inertance, called hydraulic accumulators, are employed. Storage tanks and reservoirs also serve as fluid capacitors. The energy stored in a capacitor is potential energy and is related to the work required to increase the pressure of the fluid filling the capacitor.

When the working fluid is a compressed gas such as air, fluid compliance is much more significant and must be accounted for even for small chambers and passages.

[1]For the case of a lossless line with distributed inertance and capacitance, see *Fluid Power Control*, by J. F. Blackburn, G. Reethof, and J. L. Shearer, M.I.T. Press, Cambridge, Mass., 1960, pp. 83–89 and 137–143.

Inertance is owing to the density of the working fluid when the acceleration of the fluid in a line or passage requires a significant pressure gradient to produce the rate of change of flowrate involved. It occurs mainly in long lines and passages, but it can be significant even in relatively short passages when the rate of change of flowrate is great enough. The energy stored in an inertor is kinetic, and it is related to the work done by the pressure forces at the terminals of the element to increase the momentum of the flowing fluid. If resistance (see below) is distributed along with the inertance in a passage, a series lumped-inertance, lumped-resistance model will be suitable, as long as the capacitance is negligible.

Fluid resistance is encountered in small passages and usually is owing to the effects of fluid viscosity, which impedes the flow and requires that significant pressure gradients be used to produce the viscous shearing of the fluid as it moves past the walls of the passage. Fluid resistance often becomes significant in long lines; it is then modeled in series with lumped inertance for the line when the rate of change of flowrate is large. A fluid resistor dissipates energy in the fluid, resulting in the increase of fluid temperature.

When fluid resistance results from turbulent flow in a passage, or from flow through an orifice, the kinetic effects of predominant inertia forces in the fluid flow result in nonlinear characteristics that can sometimes be linearized successfully.

9.2 Fluid System Elements

Fluid Capacitors. The symbolic diagram of a fluid capacitor is shown in Figure 9.1. Note that the pressure in a fluid capacitor must be referred to a reference pressure P_r. When the reference pressure is that of the surrounding atmosphere, it is the gage pressure; when the reference pressure is zero, i.e., a perfect vacuum, it is the absolute pressure. In this respect, a fluid capacitor is like a mass, that is, one of its across variables must be a reference.

The volume rate of flow Q_c is the through variable, even though no flow "comes out the other side," so to speak. The net flow into the capacitor is stored and corresponds somewhat to the energy that is stored in the charging process.

The elemental equation for an ideal fluid capacitor is

$$Q_c = C_f \frac{dP_{1r}}{dt} \tag{9.1}$$

where C_f is the fluid capacitance. Simplified expressions that may be used to evaluate the capacitance of a number of commonly encountered fluid capacitors are shown in Figure 9.2. Note that when dealing with compressed gases, the

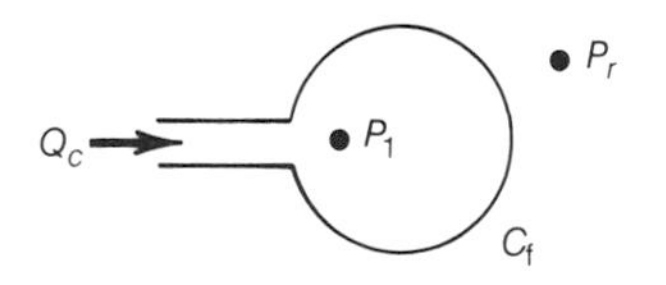

FIGURE 9.1 Symbolic diagram of a fluid capacitor.

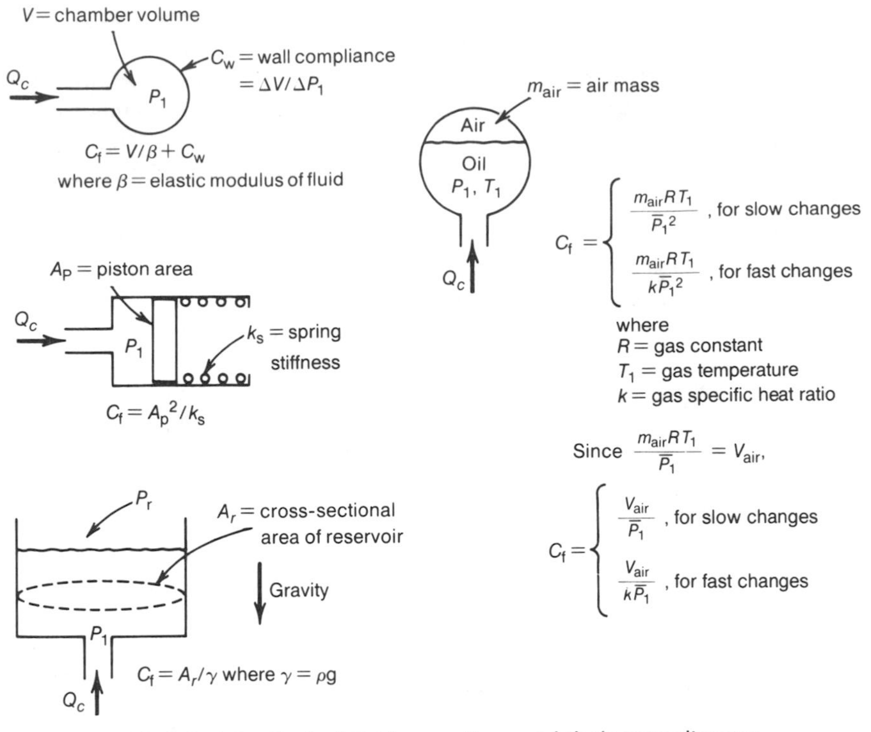

FIGURE 9.2 Typical fluid capacitors and their capacitances.

pressure in the capacitor must always be expressed as an absolute pressure; that is, the pressure must be referred to a perfect vacuum.

An alternative form of the elemental equation that is sometimes more useful, especially for time-varying or nonlinear fluid capacitors is given by

$$V_c = C_f P_{1r} \tag{9.2}$$

where V_c is the volume of the net flow, i.e., the time integral of Q_c.

The potential energy stored in an ideal fluid capacitor is given by

$$\mathscr{E}_p = \frac{C_f}{2} P_{1r}^2 \tag{9.3}$$

Hence the fluid capacitor is an A-type element, storing energy as the square of its across variable P_{1r}, and it would be unrealistic to try to suddenly change its pressure.

Fluid Inertors. The symbolic diagram of a fluid inertor is shown in Figure 9.3. The elemental equation for an inertor is

$$P_{12} = I \frac{dQ_I}{dt} \tag{9.4}$$

FIGURE 9.3 Symbolic diagram of a fluid inertor.

where I is the fluid inertance. For frictionless incompressible flow in a uniform passage having cross-sectional area A and length L, the inertance, $I = (\rho/A)L$, where ρ is the mass density of the fluid. For passages having nonuniform area, it is necessary to integrate $(\rho/A(x))dx$ over the length of the passage to determine I. The expression given above for I can be modified for a flow with a nonuniform velocity profile by applying the unsteady flow momentum equation to a small element of the passage and integrating across the passage area. The correction factor for a circular area with a parabolic velocity profile is 2.0; i.e., $I = (2\rho/A)L$. Since the nonuniform velocity profile usually results from viscosity effects, the accompanying fluid resistance would then be modeled as a series resistor.

The kinetic energy stored in an ideal inertor is given by

$$\mathscr{E}_K = \frac{I}{2} Q_I^2 \tag{9.5}$$

Hence the inertor is a T-type element, storing energy as a function of the square of its T-type variable; it would be unrealistic to try to suddenly change the flowrate through an inertor.

When enough fluid compressibility or wall compliance is also present, a lumped-parameter capacitance-inertance model might be suitable[2]. The justification for using this model would follow the lines discussed for modeling a mechanical spring in Chapter 2. Otherwise a distributed-parameter model is needed, which is beyond the scope of this book.

Fluid Resistors. The symbolic diagram for a fluid resistor is shown in Figure 9.4.

The elemental equation of an ideal fluid resistor is

$$P_{12} = R_f Q_R \tag{9.6}$$

where R_f is the fluid resistance, a measure of the pressure drop required to force a

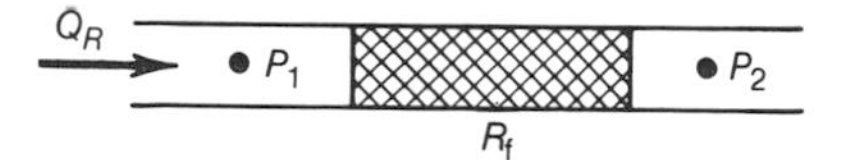

FIGURE 9.4 Symbolic diagram of a fluid resistor.

[2]For a brief discussion of lumped-parameter models of transmission lines, see *Handbook of Fluid Dynamics* edited by V. L. Streeter, McGraw-Hill Publishing Co., New York, 1961, pp. 21–24 through 21–28.

unit of flowrate through the resistor. Alternatively the elemental equation may be expressed by

$$Q_R = \frac{1}{R_f} P_{12} \tag{9.7}$$

Figure 9.5 shows some typically encountered linear fluid resistors, together with available expressions for their resistance.

When the flow in a passage becomes turbulent or when the flow is through an orifice, a nonlinear power-law relationship is used to express the pressure drop as a function of flow rate

$$P_{12} = C_R Q_{\text{NLR}}^n \tag{9.8}$$

where C_R is a flow constant, and n is either approximately 7/4 (from the Moody Diagram) for turbulent flow in a long, straight, smooth-walled passage or 2 (from Bernoulli's equation) for flow through a sharp-edged orifice[3]. For short tubes, often called "orificies", it is necessary to determine experimentally the nonlinear relationship between pressure drop and flowrate.

When a flow in a passage that undergoes transition from laminar-to-turbulent-to-laminar is to be modeled, hysteresis is likely to be present, as shown in Figure 9.6. Careful measurements of the actual line should be made to determine the transition points, and so on.

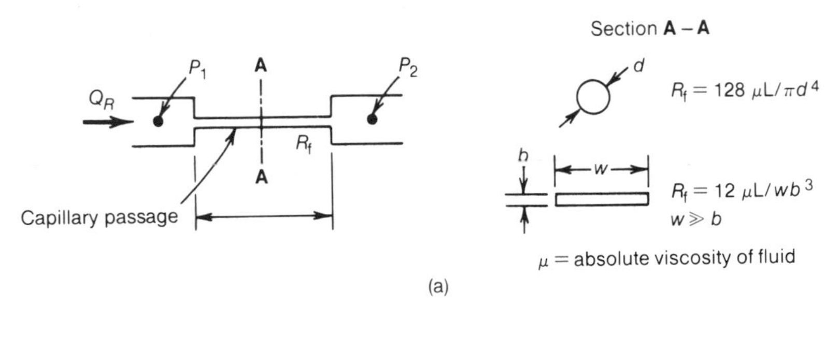

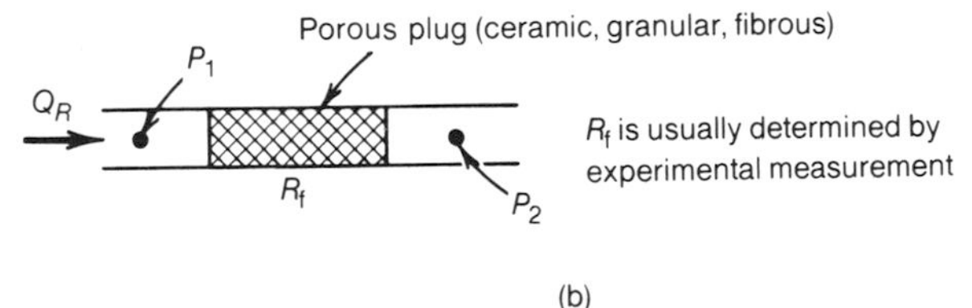

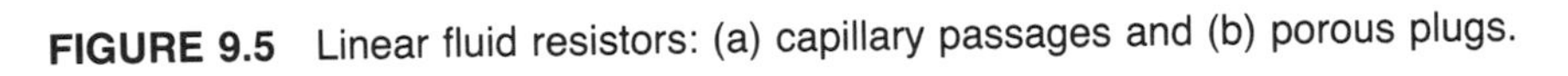
FIGURE 9.5 Linear fluid resistors: (a) capillary passages and (b) porous plugs.

[3]For such an orifice $Q_{\text{NLR}} = A_0 C_d (2P_{12}/\rho)^{.5}$, where A_0 = orifice area; C_d = discharge coefficient; ρ = mass density of fluid.

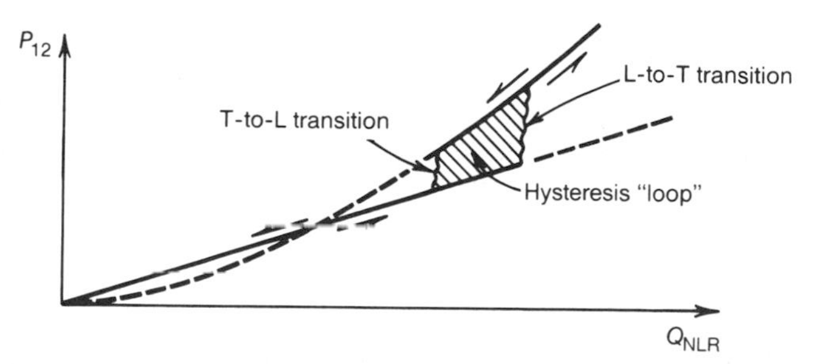

FIGURE 9.6 Steady-state pressure drop versus flowrate characteristics for a fluid line undergoing laminar-turbulent-laminar transition.

In pneumatic systems, where the fluid is very compressible, the Mach number of an orifice flow can easily exceed 0.2, and for cases involving such high local flow velocities, the compressible flow relations must be used or approximated[4]. The flowrate is then expressed in terms of mass or weight rate of flow because volume rate of flow of a given amount of fluid varies so greatly with local pressure. A modified form of fluid capacitance is then needed, unless the local volume rate of flow is computed at each capacitor.

Fluid Sources. The ideal sources employed in fluid system analysis are shown in Figure 9.7. An ideal pressure source is capable of delivering the indicated pressure, regardless of the flow required by what it is driving, whereas an ideal flow source is capable of delivering the indicated flowrate, regardless of the pressure required to drive its load.

Interconnection Laws. The two fluid system interconnection laws, corresponding to Kirchhoff's laws for electrical systems are the laws of continuity and compatibility. The continuity law says that the sum of the flowrates at a junction must be zero, and the compatibility law says that the sum of the pressure drops around a loop must be zero. These laws are illustrated in Figure 9.8.

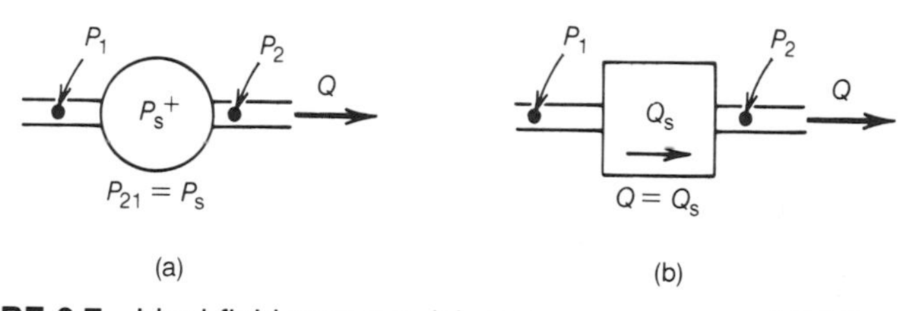

FIGURE 9.7 Ideal fluid sources: (a) pressure source and (b) flow source.

[4]For a discussion of compressible flow effects in orifices see *Fluid Power Control*, by J. F. Blackburn, G. Reethof, and J. L. Shearer, MIT Press, Cambridge, Mass., 1960, pp. 61–80 and 214–234.

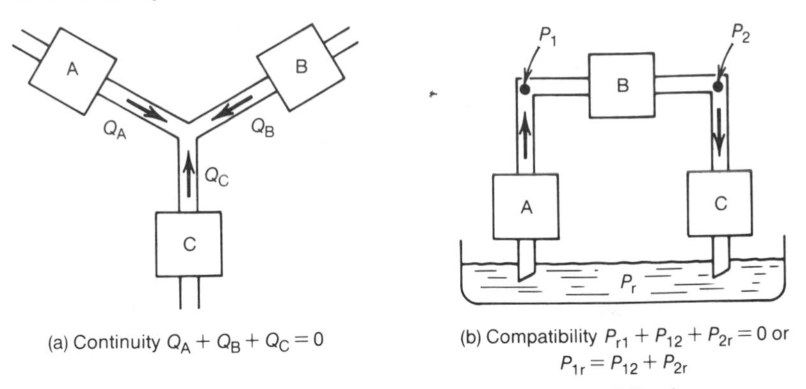

(a) Continuity $Q_A + Q_B + Q_C = 0$

(b) Compatibility $P_{r1} + P_{12} + P_{2r} = 0$ or $P_{1r} = P_{12} + P_{2r}$

FIGURE 9.8 Illustration of continuity and compatibility laws.

9.3 Analysis of Fluid Systems

The procedure followed in the analysis of fluid systems is similar to that followed earlier for mechanical and electrical systems: Write the elemental equations and the interconnection equations and then (a) combine, removing unwanted variables, to obtain the desired system input-output differential equation or (b) build a state-variable equation around each energy storage element by combining to remove all but the state variables for the energy storage elements. These state variables are: (a) Pressure for capacitor (A-Type) elements; and (b) Flowrate for inertor (T-type) elements.

EXAMPLE 9.1

Find a set of state-variable equations and develop the input-output differential equation relating the output pressure P_{3r} to the input pressure P_s for the fluid system shown in Figure 9.9.

Solution

The elemental equations are as follows: For the fluid resistor

$$P_{12} = R_f Q_R \tag{9.9}$$

For the inertor

$$P_{23} = I \frac{dQ_R}{dt} \tag{9.10}$$

For the fluid capacitor

$$Q_R = C_f \frac{dP_{3r}}{dt} \tag{9.11}$$

Continuity is satisfied by using Q_R for Q_I and Q_C. To satisfy compatibility

$$P_s = P_{1r} = P_{12} + P_{23} + P_{3r} \tag{9.12}$$

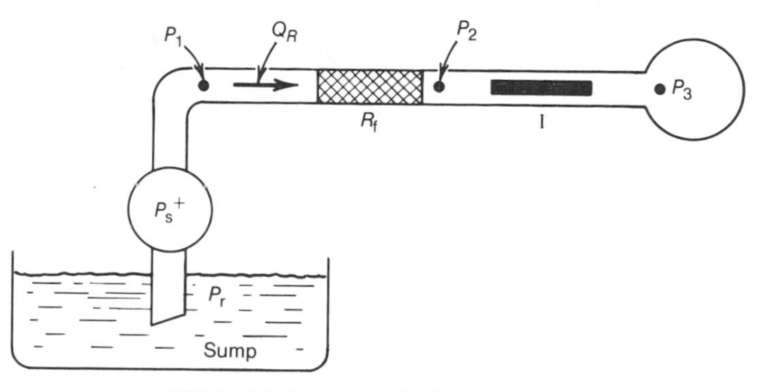

FIGURE 9.9 A simple fluid R, I, C system.

Combining Equations (9.9), (9.10) and (9.12) to eliminate P_{12} and P_{23}

$$I\frac{dQ_R}{dt} = P_s - R_f Q_R - P_{3r} \tag{9.13}$$

Rearranging (9.13) yields the first state-variable equation

$$\frac{dQ_R}{dt} = -\frac{R_f}{I}Q_R - \frac{1}{I}P_{3r} + \frac{1}{I}P_s \tag{9.14}$$

Rearranging (9.11) yields the second state-variable equation

$$\frac{dP_{3r}}{dt} = \frac{1}{C_f}Q_R \tag{9.15}$$

Combining (9.14) and (9.15) to elimiante Q_R and multiplying all terms by I yields the input-output system differential equation

$$C_f I\frac{d^2P_{3r}}{dt^2} + R_f C_f\frac{dP_{3r}}{dt} + P_{3r} = P_s \tag{9.16}$$

■

EXAMPLE 9.2

A variable-orifice nonlinear resistor is being used to modulate the flowrate Q_{NLR} and control the pressure P_{2r} in the simple fluid control system shown in Figure 9.10. The flow equation for the orifice is $Q_{\text{NLR}} = C_1A_0(P_{2r})^{.5}$. Develop the input-output differential equation relating small changes of the output pressure P_{2r} to small changes of the orifice area A_0 when the supply flow Q_s is constant.

Solution

The elemental equations for small perturbations of all variables are as follows: For the linear resistor

$$\hat{P}_{12} = R_f\hat{Q}_R \tag{9.17}$$

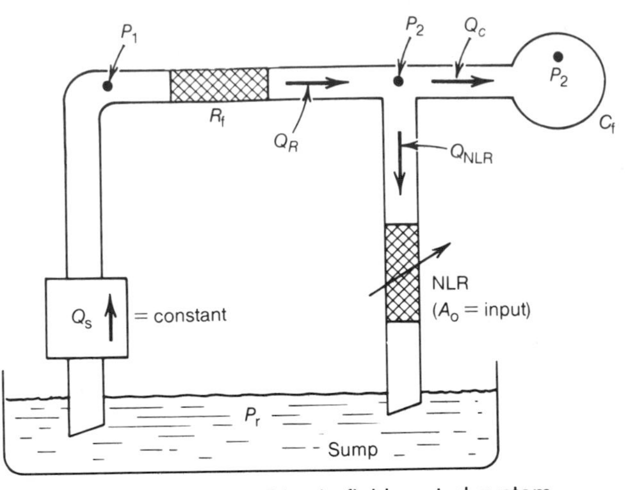

FIGURE 9.10 Simple fluid control system.

For the fluid capacitor

$$\hat{Q}_C = C_f \frac{d\hat{P}_{2r}}{dt} \tag{9.18}$$

For the time-varying nonlinear resistor

$$\hat{Q}_{\mathrm{NLR}} = \left(\frac{C_1\overline{A}_0}{2|\overline{P}_{2r}|^{.5}}\right)\hat{P}_{2r} + C_1|\overline{P}_{2r}|^{.5}\,\hat{A}_0 \tag{9.19}$$

To satisfy continuity at (1)

$$\hat{Q}_R = \hat{Q}_s \tag{9.20}$$

To satisfy continuity at (2)

$$\hat{Q}_R = \hat{Q}_C + \hat{Q}_{\mathrm{NLR}} \tag{9.21}$$

The compatability relation

$$\hat{P}_{1r} = \hat{P}_{12} + \hat{P}_{2r} \tag{9.22}$$

is not needed here because of lack of interest in finding $\hat{P}_{1r}$.

Combining Equations (9.18), (9.19), (9.20), and (9.21) to eliminate $\hat{Q}_R$, $\hat{Q}_C$, and $\hat{Q}_{\mathrm{NLR}}$ gives

$$\hat{Q}_s = C_f \frac{d\hat{P}_{2r}}{dt} + \left(\frac{C_1\overline{A}_0}{2|\overline{P}_{2r}|^{.5}}\right)\hat{P}_{2r} + C_1|\overline{P}_{2r}|^{.5}\hat{A}_0 \tag{9.23}$$

Since Q_s is constant, $\hat{Q}_s$ is zero, and the output terms on right-hand side may be rearranged to yield the system input-output differential equation for small perturbations of all variables.

$$C_f \frac{d\hat{P}_{2r}}{dt} + \left(\frac{C_1\overline{A}_0}{2|\overline{P}_{2r}|^{.5}}\right)\hat{P}_{2r} = -C_1|\overline{P}_{2r}|^{.5}\hat{A}_0 \tag{9.24}$$

■

9.4 Pneumatic Systems

Up to this point, this chapter has dealt with fluid systems in which the effects of fluid compressibility are small, but not necessarily negligible. Thus the volume occupied by a given mass of fluid has been nearly constant throughout the system in which it was being employed, regardless of the pressure changes which occurred as the fluid mass moved from one part of the system to another. For the compressibility effects in the fluid to be small enough, the fluid had to be a liquid that was not near its boiling point—in other words a hydraulic fluid such as water or oil at ambient temperature; or if gaseous—in other words a pneumatic fluid—the gross changes in pressure had to be small enough so that the density changes were only of the order of a few percent.

Thus the discussions and methods employed so far have been limited to what are commonly known as hydraulic systems. In this section we shall deal with modifications of the describing equations needed to cope with pneumatic systems in which fluid compressibility plays a much greater role than in hydraulic systems.

Of foremost concern in the analysis of pneumatic systems is the need to satisfy the continuity requirement at connecting points between elements (in other words at node points). This makes it necessary to employ mass or weight rate of flow as the T-variable instead of volume rate of flow[5]. Also it is necessary to employ different or modified elemental equations to incorporate the greater effects of compressibility on the flow relations for the pneumatic system elements.

Here we shall employ as the T-variable the weight rate of flow $Q_w = \gamma v_{ave} A$ instead of volume rate of flow $Q = v_{ave} A$, where γ is the local weight density of the fluid, and v_{ave} is the mean velocity in a flow passage of cross-sectional area A. Similar procedures may be followed if one wishes to employ mass rate of flow $Q_m = \rho v_{ave} A$ where ρ is the local mass density of the fluid, or to employ the standard volume rate of flow $Q_s = (P/T)(T_s/P_s) v_{ave} A$ where P and T are the local pressure and temperature of the fluid and P_s and T_s are the standard pressure and temperature conditions to be employed.

First the elemental equation for a fluid capacitor (9.1) is modified by multiplying both sides by the local weight density of the fluid γ.

$$Q_{wC} = C_{fw} \frac{dP_{1r}}{dt} \tag{9.25}$$

where $Q_{wC} = \gamma Q_C$ and $C_{fw} = \gamma C_f$.

The most commonly encountered form of pneumatic capacitor is a chamber or passage filled with the pneumatic fluid (usually air, but other gases such as nitrogen, oxygen, helium, etc. are used as pneumatic working fluids). Thus the expressions shown for the gaseous part of a gas-charged hydraulic accumulator in Figure 9.2 are directly available for modification.

[5]Another alternative is to employ an equivalent standard volume rate of flow in which the mass of the fluid is expressed in terms of the volume that it would occupy if it were at some standard pressure and temperature such as ambient air at sea level.

In the modifications which follow it is assumed that the gaseous fluid behaves as a perfect gas (in other words it is not so highly compressed that so as to be near its liquid state), and obeys the perfect gas law $P = \rho RT = (\gamma/g)RT$ where R is its perfect gas constant and g is the acceleration due to gravity.

Thus the expression for the modified fluid capacitance of a gas-filled chamber of constant volume is

$$C_{fw} = \begin{cases} \gamma_1 \mathcal{V}/P_1 = g\mathcal{V}/(RT_1), \text{ for very 'slow' changes of } P_1 \\ \gamma_1 \mathcal{V}/kP_1 = g\mathcal{V}/(kRT_1), \text{ for every 'fast' changes of } P_1 \end{cases} \tag{9.26}$$

where $\mathcal{V}$ is the total volume of the gas-filled chamber, P_1 is the absolute pressure of the chamber gas, the specific heat ratio $k = c_p/c_v$ for the chamber gas, and T_1 is the absolute temperature of the chamber gas. The term 'slow' here denotes changes that take place slowly enough for heat transfer from the surroundings to keep the chamber gas temperature constant—in other words the changes in the chamber are isothermal. The term 'fast' here denotes changes that occur so rapidly that heat transfer from the surroundings is negligible—in other words the changes in the chamber are adiabatic. For cases in which the rapidity of change is intermediate, a polytropic coefficient n, where $1.0 < n < 1.4$, may be used instead of k for the adiabatic case—that is the equation for 'fast' changes.

When the capacitor consists of a cylinder with a spring-restrained piston of area A_p (see Figure 9.2) or a bellows of area A_p and spring stiffness k_s enclosing a storage chamber, and the fluid is a compressible gas, the capacitance consists of two parts: (a) a part due to the compressibility of the gas in the chamber; and (b) a part due to change in volume $\mathcal{V}$ and energy storage in the spring k_s.

$$C_{fw} = \begin{cases} \dfrac{g\mathcal{V}(t)}{RT_1} + \dfrac{gP_1(t)A_p^2}{RT_1k_s}, \text{ for very 'slow' changes in } P_1 \\[2ex] \dfrac{g\mathcal{V}(t)}{kRT_1} + \dfrac{gP_1(t)A_p^2}{RT_1k_s}, \text{ for very 'fast' changes in } P_1 \end{cases} \tag{9.27}$$

Note that P_1 must be expressed as an absolute pressure, and P_1 and $\mathcal{V}$ are now functions of time. The absolute temperature T_1 also varies slightly with time, but this variation superposed on its relatively large absolute value usually represents a negligible effect. Using absolute zero pressure (a perfect vacuum) as the reference pressure (for example P_r) eliminates the problem of having some pressures being absolute pressures and some being relative (or gage) pressures.

The equation for an ideal pneumatic inertor having a uniform flow velocity profile across its cross-sectional area is obtained by modifying Equation (9.4) so that $Q_{wI} = \gamma Q_I$ and $I_w = I/\gamma = L/(gA)$

$$P_{12} = I_w \frac{dQ_{wI}}{dt} \tag{9.28}$$

where $\gamma = g(P_1 + P_2)/RT$ is the average weight density of the gas in the passage.

Note that a gas-filled passage is more likely to need to be modeled as a chamber-type pneumatic capacitor having inflow and outflow at its ends than as an inertor;

or it may even need to be modeled as a long transmission line having distributed capacitance and inertance, requiring partial differential equations, which is beyond the scope of this text[6].

The capillary or porous plug type of hydraulic resistor which was modeled earlier as a linear resistor with resistance R_f becomes nonlinear with a compressible gas flowing through it, having the nonlinear elemental equation

$$Q_{w\text{NLR}} = K_w(P_1^2 - P_2^2) \tag{9.29}$$

where K_w is a conductance factor that needs to be evaluated experimentally for best results. (This nonlinear model is limited to operation of the device with flow velocities of Mach number less than .2, because at higher Mach numbers the operation begins to resemble that of an orifice.)

For pneumatic flow through sharp-edged orifices the flow rate is related to the ratio of the downstream pressure to the upstream pressure by the classical equation for isentropic flow of a perfect gas through a converging nozzle[7], corrected for the vena contracta effect by the use of a discharge coefficient C_d.

$$\frac{Q_{w\text{NLR}}}{C_d A_0} = \begin{Bmatrix} \dfrac{C_2 P_u}{(T_u)^{.5}}, \text{ for } 0 < \dfrac{P_d}{P_u} < PR_{\text{crit}}, \text{ ``Choked flow''} \\ \dfrac{C_2 P_u}{(T_u)^{.5} C_3}\left(\dfrac{P_d}{P_u}\right)^{1/k}\left[1 - \left(\dfrac{P_d}{P_u}\right)^{k-1/k}\right]^{.5}, \text{ for} \\ PR_{\text{crit}} < \dfrac{P_d}{P_u} < 1.0, \text{ ``Unchoked flow''} \end{Bmatrix} \tag{9.30}$$

where

C_d = discharge coefficient, approximately 0.85 for air
A_0 = orifice area, in^2 [m^2]
$Q_{w\text{NLR}}$ = the weight rate of flow, lb/sec. [N/sec.]

$$C_2 = g\sqrt{\frac{k}{R}\left(\frac{k+1}{2}\right)^{k+1/k-1}} = .532 \text{ (deg R)}^{.5}/\text{sec for air} \quad [.410 \text{ (deg K)}^{.5}/\text{sec}]$$

$$C_3 = \sqrt{\frac{2\left(\dfrac{k+1}{2}\right)^{k+1/k-1}}{k-1}} = 3.872 \text{ for air}$$

[6]For discussions of transmission line models see: J. Watton, *Fluid Power Systems*, Prentice-Hall, New York, 1989, pp. 244–248. J. F. Blackburn, G. Reethof, and J. L. Shearer, *Fluid Power Control*, M.I.T. Press, Cambridge, Mass, 1960, pp. 81–88. V. L. Streeter, *Handbook of Fluid Dynamics*, McGraw-Hill Book Co., New York, 1961, Section 21, pp. 21–20 through 21–22.

[7]J. F. Blackburn, G. Reethof, and J. L. Shearer, *Fluid Power Control*, M.I.T. Press, Cambridge, Mass, 1960, pp. 214–223.

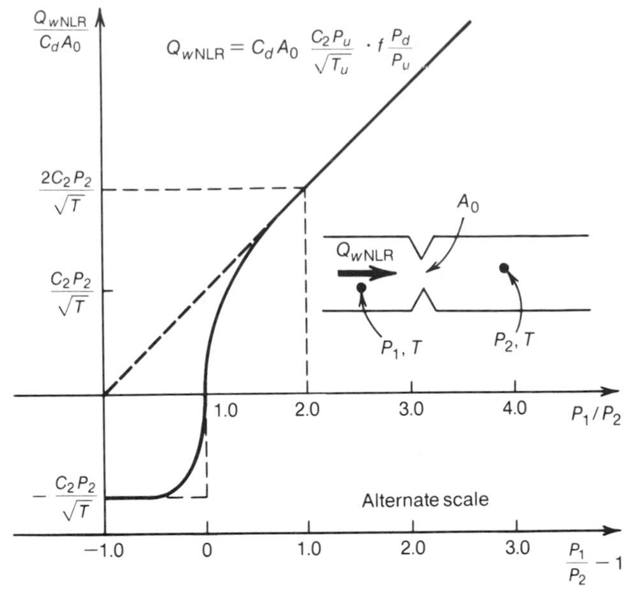

(a)

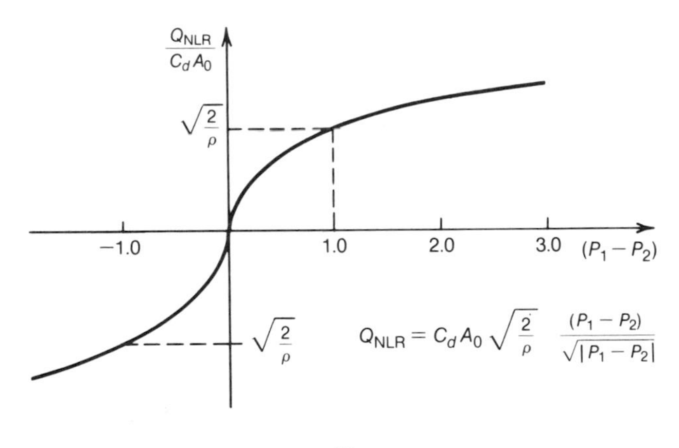

(b)

FIGURE 9.11 Weight flowrate vs. pressure graphs for flow through a sharp-edged orifice of area A_0. (a) for air; and (b) for a liquid fluid such as water or light oil.

$$PR_{crit} = \left(\frac{2}{k+1}\right)^{k/(k-1)} = .528 \text{ for air}$$

P_u = upstream pressure, lb/in.^{2} [N/m^2]
P_d = downstream pressure, lb/in.^{2} [N/m^2]
T_u = upstream temperature, deg R [deg K]
R = the perfect gas constant = 2.48×10^5in.^{2}/(sec^2 − deg R) [268 m^2/(sec^2 − deg K)]

k = the specific heat ratio $\frac{c_p}{c_v}$ = 1.4 for air

A very close approximation, requiring much less computing effort, is to use the following for gases such as air having $k = 1.4$:

$$\frac{Q_{wNLR}}{C_d A_0} = \begin{cases} \dfrac{C_2 P_u}{(T_u)^{.5}}, & \text{for } 0 < \dfrac{P_d}{P_u} < 0.5 \\ \dfrac{2C_2 P_u}{(T_u)^{.5}}\left[\dfrac{P_d}{P_u}\left(1 - \dfrac{P_d}{P_u}\right)\right]^{.5}, & \text{for } 0.5 < \dfrac{P_d}{P_u} < 1.0 \end{cases} \tag{9.31}$$

This approximation leads to values of weight flowrate for air that are consistently smaller than the values obtained by use of the ideal isentropic flow Equation (9.30), but which never depart by more than 3 percent from the ideal values.

When $P_1 > P_2$ the flow is from (1) to (2) and $P_d = P_2$ and $P_u = P_1$. However, when $P_2 > P_1$, the flow is reversed, and $P_d = P_1$ and $P_u = P_2$.

The graph for Q_{wNLR} vs. P_1/P_2 for the complete range of P_1/P_2 from zero to a very large value is plotted in Figure 9.11a. The curve shown in Figure 9.11b is the square-law graph of the volume flow rate Q vs. $(P_1 - P_2)$ for ideal hydraulic orifice flow, included for comparison purposes.

EXAMPLE 9.3

The pneumatic amplifier shown schematically in Figure 9.12 has been widely used in pneumatic instruments for measurement and automatic control in the process industries, for heating and ventilating controls, and in certain military and aerospace systems. It operates from a pressure source of clean air much as an electric circuit would operate from a battery. Usually the working fluid is air drawn continuously from the atmosphere by a pressure-controlled air compressor. After losing pressure as it flows through a fixed orifice of area A_0, the flowrate Q_{wa} approaches the branchpoint or node (2) from which part may flow as Q_{wb} to the bellows chamber and part may flow as Q_{wc} through the flapper-nozzle orifice back to the atmosphere.

The atmospheric pressure P_{atm} can serve as a secondary reference, but the primary reference is a perfect vacuum so that all pressures (except P_s) are expressed as absolute pressures.

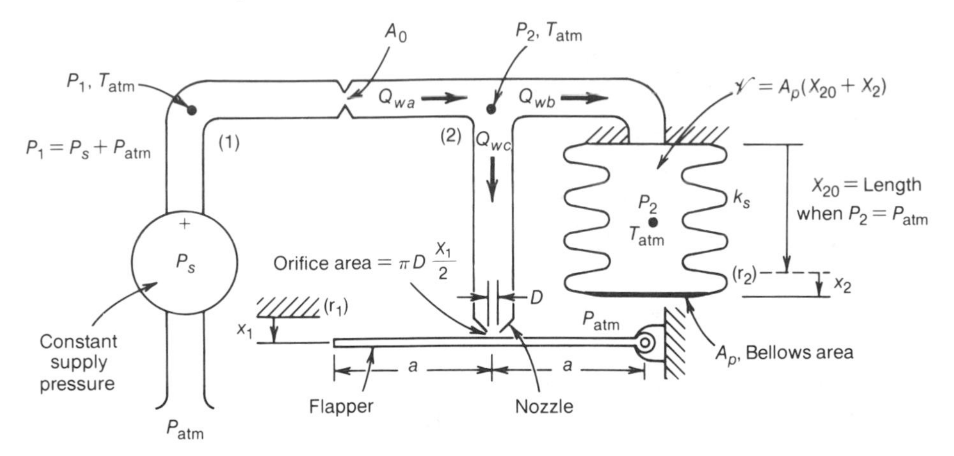

FIGURE 9.12 Schematic diagram of a pneumatic amplifier system.

This system is referred to as an amplifier because the energy or power required to actuate the input x_1 is usually only a tiny fraction of the energy or power available to produce the output x_2. As an amplifier it is then useful, when combined with other elements, for executing control functions in complete systems.

During operation at a steady-state normal operating point, the pressure P_2 and the output x_2 are constant, and the flowrate Q_{wb} is zero.

(a) Find the normal operating point values $\bar{x}_1$ and $\bar{x}_2$ when $\overline{P}_2 = 0.4P_1$, where $P_1 = 50.0$ lb/in.² abs.

(b) Write the necessary and sufficient set of describing equations for this system and develop the state variable equation for this first-order system employing P_2 as the state variable and using symbols wherever possible. Assume that changes in P_2 occur slowly (isothermal chamber).

(c) Linearize the describing equations for small perturbations and combine to form a linearized state-variable equation, using $\hat{P}_2$ as the state variable. Write the necessary output equation for $\hat{x}_2$ as the output variable.

(d) Develop the linearized system input-output differential equation relating $\hat{x}_2$ to $\hat{x}_1$, and find and sketch the response of $\hat{x}_2$ to a small step change in $\hat{x}_1$, $\hat{x}_1 = D/50$.

Solution

(a) At the normal operating point, $\overline{Q}_{wa} = \overline{Q}_{wc}$, and since $(\overline{P}_2/\overline{P}_1) < .5$, the flow through A_0 is choked, so that from Equation (9.31)

$$\overline{Q}_{wa} = \frac{C_{d1}A_0C_2\overline{P}_1}{(T_{atm})^{.5}}$$

Since $P_{atm}/\overline{P}_2 = 14.7/20.0 = 0.735$, the flow through the flapper-nozzle orifice is not choked, so that from Equation (9.31)

$$\overline{Q}_{wc} = C_{d2}\pi D\left(\frac{\bar{x}_1}{2}\right) C_2(0.4\overline{P}_1)(2.0)\left[\frac{(.735)(.265)}{T_{atm}}\right]^{.5}$$

Equating these flows then yields

$$\bar{x}_1 = \frac{A_0}{(.1764)(3.1416)(.15)} = 12.03\ A_0 \tag{9.32}$$

For $\bar{x}_2$,

$$\bar{x}_2 = \frac{(\bar{P}_1 - P_{atm})A_p}{k_s} = \frac{45.3A_p}{k_s} \tag{9.33}$$

(b) The describing equations are:
For orifice A_0, using Equation (9.31)

$$Q_{wa} = \frac{C_{d1}A_0C_2P_1}{(T_{atm})^{.5}} \tag{9.34}$$

For the flapper-nozzle orifice, using Equation (9.31)

$$Q_{wc} = C_{d2}\pi Dx_1C_2P_2\left[\frac{\left\{\left(\frac{P_{atm}}{P_2}\right)\left(1 - \frac{P_{atm}}{P_2}\right)\right\}}{T_{atm}}\right]^{.5} \tag{9.35}$$

For the bellows capacitor, using Equation (9.25)

$$Q_{wb} = C_{fw}\frac{dP_2}{dt} \tag{9.36}$$

where, using Equation (9.27)—Isothermal case,

$$C_{fw} = \frac{gA_p}{RT_{atm}}\left(x_{20} + x_2 + \frac{P_2A_p}{k_s}\right) \tag{9.37}$$

For the bellows spring,

$$x_2 = A_p\frac{P_2 - P_{atm}}{k_s} \tag{9.38}$$

And continuity at (2),

$$Q_{wa} = Q_{wb} + Q_{wc} \tag{9.39}$$

Combining (9.34) through (9.39) yields the nonlinear state-variable equation,

$$\frac{dP_2}{dt} = \left[\left(\frac{-C_{d2}C_2\pi Dx_1}{C_{fw}\sqrt{T_{atm}}}\right)\sqrt{\left(\frac{P_{atm}}{P_2}\right)\left(1 - \frac{P_{atm}}{P_2}\right)}\right]P_2 + \left[\frac{C_{d1}C_2}{C_{fw}\sqrt{T_{atm}}}\right]P_1A_0 \tag{9.40}$$

(c) Linearizing the describing equations for small perturbations, for A_0, using Equation (9.34)

$$\hat{Q}_{wa} = 0 \tag{9.41}$$

For the flapper-nozzle orifice, using Equations (9.35)

$$\hat{Q}_{wc} = k_1\hat{x}_1 + k_2\hat{P}_2 \tag{9.42}$$

where

$$k_1 = C_{d2}\pi D C_2 \left[P_{atm} \frac{\overline{P}_2 - P_{atm}}{T_{atm}} \right]^{.5} \tag{9.43}$$

$$k_2 = C_{d2}\pi D \left(\frac{\overline{x}_1}{2}\right) \frac{C_2 P_{atm}}{[T_{atm} P_{atm}(\overline{P}_2 - P_{atm}]^{.5}} \tag{9.44}$$

For the bellows capacitor, from Equations (9.36) and (9.37)

$$\hat{Q}_{wb} = \frac{gA_p\,(x_{20} + \overline{x}_2 + \overline{P}_2 A_p/k_s)}{RT_{atm}} \frac{d\hat{P}_2}{dt} \tag{9.45}$$

For the bellows spring, from (9.38)

$$\overline{x}_2 = \frac{A_p(\overline{P}_2 - P_{atm})}{k_s} \tag{9.46}$$

$$\hat{x}_2 = \left(\frac{A_p}{k_s}\right) \hat{P}_2 \tag{9.47}$$

Combining (9.45), (9.46), and (9.47)

$$\hat{Q}_{wb} = k_3 \frac{d\hat{P}_2}{dt} \tag{9.48}$$

where

$$k_3 = \frac{gA_p \left(x_{20} - \dfrac{A_p P_{atm}}{k_s} + \dfrac{2A_p \overline{P}_2}{k_s} \right)}{RT_{atm}} \tag{9.49}$$

Continuity

$$\hat{Q}_{wa} = \hat{Q}_{wb} + \hat{Q}_{wc} \tag{9.50}$$

Combining (9.41), (9.42), (9.48), and (9.50)

$$\frac{d\hat{P}_2}{dt} = \left(\frac{-k_2}{k_3}\right) \hat{P}_2 + \left(\frac{-k_1}{k_3}\right) \hat{x}_1 \tag{9.51}$$

Rearranging (9.51)

$$\frac{d\hat{P}_2}{dt} + \left(\frac{k_2}{k_3}\right) \hat{P}_2 = \left(\frac{-k_1}{k_3}\right) \hat{x}_1 \tag{9.52}$$

Solving (9.47) for $\hat{P}_2$ in terms of $\hat{x}_2$

$$\hat{P}_2 = \left(\frac{k_s}{A_p}\right) \hat{x}_2 \tag{9.53}$$

Combining (9.52) and (9.53)

$$\frac{d\hat{x}_2}{dt} + \left(\frac{k_2}{k_3}\right)\hat{x}_2 = \left(\frac{-A_p k_1}{k_s k_3}\right)\hat{x}_1 \tag{9.54}$$

The input is

$$\hat{x}_1(t) = \begin{cases} 0, \text{ for } t < 0 \\ \dfrac{D}{50}, \text{ for } t > 0 \end{cases}$$

The initial condition for the output is $\hat{x}_2(0^+) = 0$.

The system time constant $\tau = k_3/k_2$.

$$(\hat{x}_2)_{ss} = \frac{-k_1 A_p D}{(50 k_2 k_s)} \tag{9.55}$$

$$\hat{x}_2(t) = \left[\frac{-k_1 A_p D}{(50 k_2 k_s)}\right](1 - e^{-t/\tau}) \tag{9.56}$$

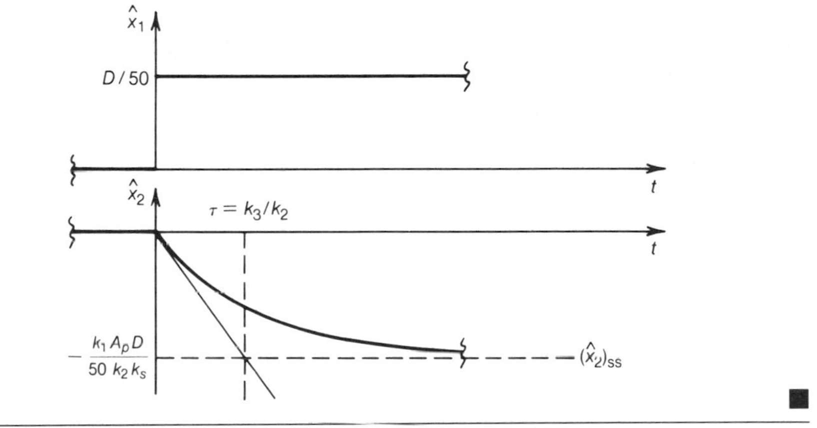

■

9.5 Synopsis

The basic fluid system elements were described and shown to be analogues of their corresponding mechanical and electrical A-, T-, and D-type elements: fluid capacitors, fluid inertors, and fluid resistors respectively. Here the A-variable is pressure and the T-variable is volume flowrate. The interconnecting laws of continuity and compatability needed for dealing with systems of fluid elements correspond to Kirchhoff's current and voltage laws used in electrical system analysis.

When the working fluid is a liquid (or sometimes a slightly compressed gas), the system usually is referred to as a hydraulic system. When the working fluid is a gas, like air, which undergoes large pressure changes and/or flows with velocities having Mach numbers greater than about 0.2, the system usually is referred to as a pneumatic system. Analysis of pneumatic systems requires modification of the describing equations so that the T-variable is weight rate of flow instead of volume rate of flow.

Both linear and nonlinear fluid resistors have been introduced, with emphasis on orifice characteristics. For pneumatic orifice flow the flow rate is seen to be a function of the ratio of upstream and downstream absolute pressures, compared to the hydraulic orifice flow relation which expresses the flow rate as a function of the difference between the upstream and downstream pressures.

Examples have been included to demonstrate the techniques of modeling and analysis for both hydraulic and pneumatic systems. Since many systems which employ working fluids also incorporate mechanical and/or electrical devices, other examples of fluid system analysis will appear in later chapters.

Fluid system components often offer significant advantages over other possible system components, such as speed of response, survivability in difficult environments, safety in hazardous environments, ease of use and/or maintenance, etc.

Problems

9.1. **(a)** Develop the system differential equation relating P_{2r} to P_s for the first-order low-pass hydraulic filter shown below in Figure P9.1a.

(b) Write the expression for the system time constant, and sketch the response, $P_{2r}(t)$ versus t for a step change ΔP_s from an initial value $P_s(0)$, which has been constant for a very long time. (Figure P9.1b)

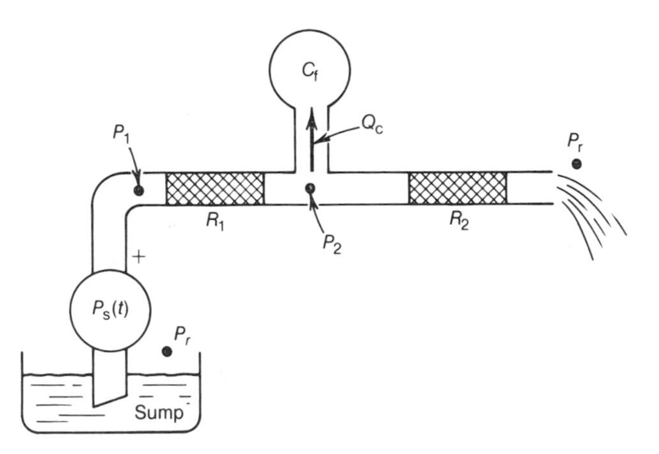

FIGURE P9.1a First-order low-pass hydraulic filter.

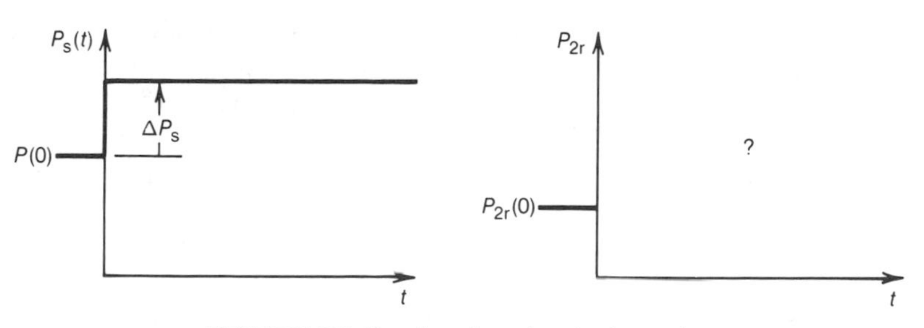

FIGURE P9.1b Input and output graphs.

9.2. **(a)** Develop the system differential equation relating P_{3r} to P_s for the second-order hydraulic low-pass filter shown in Figure P9.2a.

(b) Find expressions for the natural frequency ω_n and damping ratio ζ, and sketch the response versus time for a step change ΔP_s from an initial value $P_s(0)$, which has been constant for a very long period of time, assuming that $\zeta = 0.3$. Show clearly the period of the oscillation, the per-cycle decay ratio, and the final steady-state value of P_{3r} in Figure P9.2b.

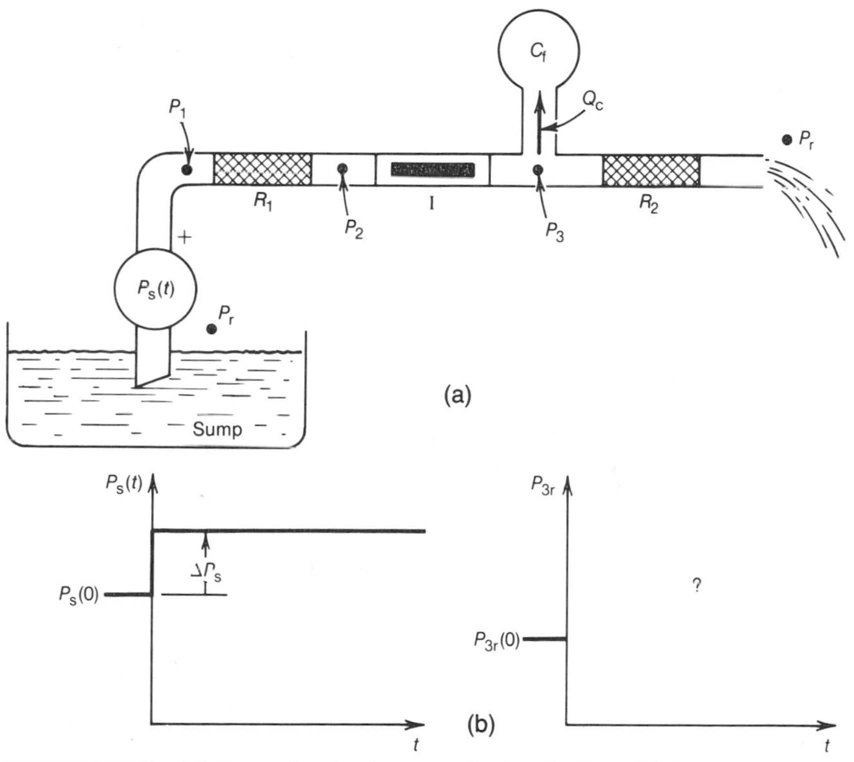

FIGURE P9.2 (a) Second-order low-pass hydraulic filter. (b) Input and output graphs.

9.3. The fluid system modeled in Figure P9.3 represents the process of filling a remote tank or reservoir in a batch process at a chemical plant.

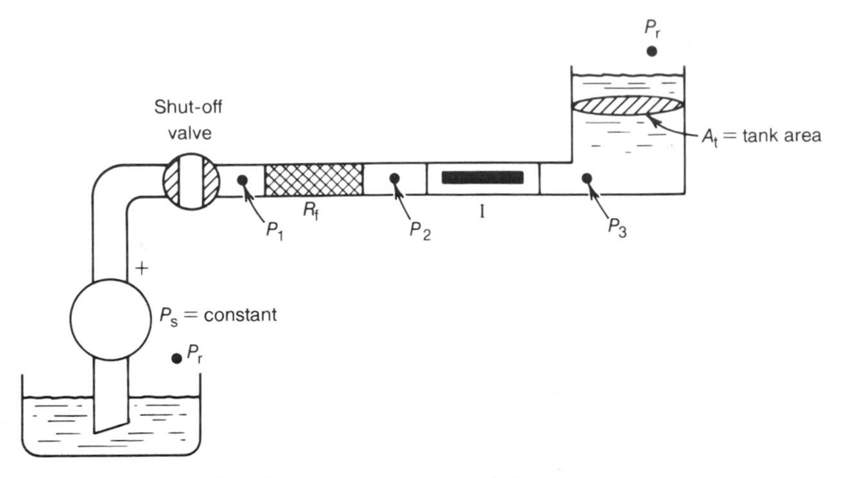

FIGURE P9.3 Reservoir filling system.

Fluid provided by an ideal pressure source P_s is suddenly turned on or off by a rotary valve that offers very little resistance to flow when it is open. The long line connecting the tank to the shut-off valve has an internal diameter of 2.0 cm and has a length of 50 m. The reservoir has an inside diameter of 25 cm.

In analyzing the response of this system to sudden opening of the valve, it has been proposed that the line be modeled as a lumped resistance in series with a lumped inertance as shown.

(a) Find the estimated lumped resistance R_f, the lumped inertance I, and the lumped capacitance C_f for the line and the capacitance C_t of the reservoir. Assume laminar flow in the line. The fluid has viscosity $\mu = 1.03 \times 10^{-2}$ N-sec/m^2, weight density $\gamma = 8.74 \times 10^{-3}$ N/m^3, and bulk modulus of elasticity $\beta = 1.38 \times 10^9$ N/m^2.

(b) Write the necessary and sufficient set of describing equations for this system based on the $R_f - I$ model of the line for the time starting with $t = 0$ when the valve is suddenly opened.

(c) Develop the system differential equation for the case when P_{3r} is the output.

(d) Calculate the natural frequency and damping ratio for this model, using pertinent values found in part (a).

(e) How do you feel about the decision to neglect the lumped capacitance, evaluated in part (a), in modeling this system?

9.4. A variable flow source $Q_s(t)$ is being used to replenish a reservoir which in turn supplies an orifice as shown in Figure P9.4a.

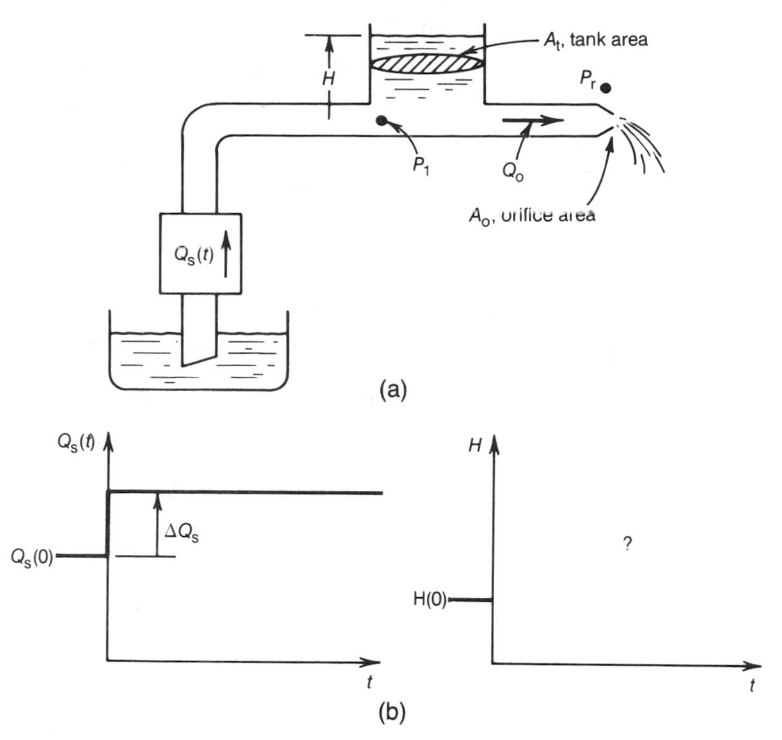

FIGURE P9.4 (a) Reservoir and discharge orifice replenished by flow source. (b) Input and output graphs.

(a) Write the necessary and sufficient set of describing equations for this system.

(b) Linearize and combine to develop the system differential equation relating $\hat{H}$ to $\hat{Q}_s(t)$.

(c) Using only symbols, solve to find and plot in Figure P9.4b the response of $\hat{H}$ to a small step change ΔQ_s from an initial value $Q_s(0)$, which has been constant for a very long time.

9.5. Two tanks are connected by a fluid line and a shut-off valve as shown in Figure P9.5.

The valve resistance is negligible when it is fully open.

(a) Write the necessary and sufficient set of describing equations for this system.

(b) Combine to form the system differential equation needed to solve for finding P_{2r} in response to suddenly opening the valve at $t = 0$ when $H_1(0)$ is greater then $H_2(0)$. Note that the input here is a sudden change in the system and that both initial conditions are needed in combining the describing equations.

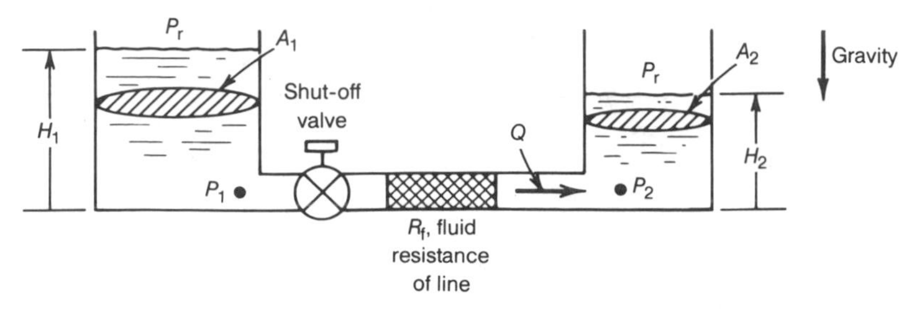

FIGURE P9.5 Two interconnected tanks.

9.6. The pneumatic system shown schematically in Figure P9.6 is a model of the air supply system for a large factory. The air compressor with its own on-off controller delivers air on a cyclic basis to the large receiving tank where the compressed air is stored at a somewhat time-varying pressure P_1. From this receiving tank, the air flows through an orifice having area A_{01} to a ballast tank which acts, together with A_{01}, as a pressure filter to provide an output pressure P_2 having much smaller cyclic variations than the fluctuations in P_1 caused by the on-off flow Q_{ws} from the compressor. The object of this problem is to develop the mathematical model needed to compare the predicted variations of P_2 with the variations in P_1.

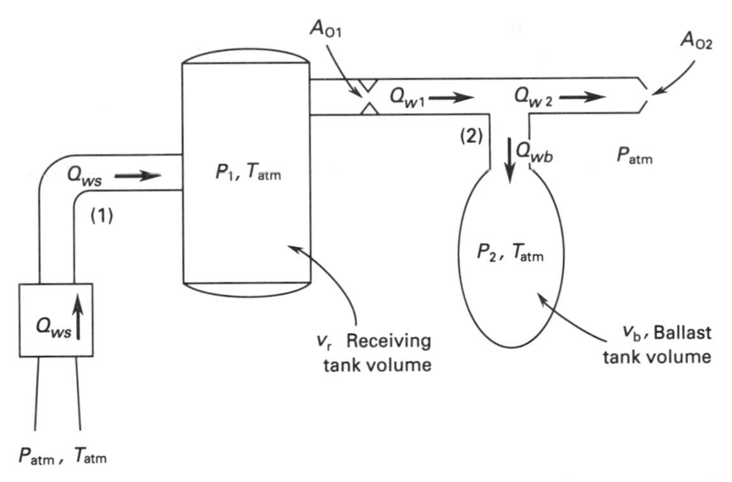

FIGURE P9.6 Schematic diagram of factory air supply system model.

The second orifice having area A_{02}, exhausting to atmosphere, is provided to simulate the load effects of all the air-consuming devices and processes in the whole factory, and the value of A_{02} is chosen to simulate the normal operating (that is, average) factory consumption rate of flow of air, $\overline{Q}_{w2} = 0.05$ N/sec (0.0051 kg/sec) when the normal operating pressure $\overline{P}_2$ is 5.0×10^5 N/m^2 absolute.

(a) Compute the value of A_{02} needed for the given desired normal operating point conditions.

(b) Write the necessary and sufficient set of describing equations for this system, considering the compressor flowrate $Q_{ws}(t)$ as the system input.

(c) Linearize the equations in (b) and form the set of state-variable equations having $\hat{P}_1$ and $\hat{P}_2$ as the system state variables.

(d) Develop the system input-output differential equation relating $\hat{P}_2(t)$ to $\hat{Q}_{ws}(t)$, and do the same for relating $\hat{P}_1(t)$ to $\hat{Q}_{ws}(t)$.

10

Mixed Systems

10.1 Introduction

In previous chapters various types of systems have been discussed, each within its own discipline: mechanical, electrical, thermal, and fluid. However, many engineering systems consist of combinations of these elementary single-discipline system elements: electromechanical, fluidomechanical, and so on. In order to combine single-discipline systems, it is necessary to employ coupling devices that convert one kind of energy or signal to another: mechanical to electrical, fluid to mechanical, and so on.

The general name transducer will be used here for ideal coupling devices. In cases where significant amounts of energy or power are involved, these coupling devices will be referred to as energy-converting transducers; when the amount of energy being transferred is minimal, they will be referred to as signal-converting transducers.

Selected nonideal energy convertors, which are modeled graphically, are also discussed in terms of typical characteristic curves that have been derived from performance tests.

10.2 Energy-Converting Transducers and Devices

The energy-converting transducers to be introduced here are ideal in that they are lossless models that contain no energy-storing or energy-dissipating elements. When energy storage or energy dissipation are present, these effects are modeled with lumped ideal elements connected at the terminals of the ideal transducer.

Mechanical Translation to Mechanical Rotation Transducers. The symbolic diagram for mechanisms that convert translational motion to rotational motion or vice versa is shown in Figure 10.1, where n is the coupling coefficient relating output motion to input motion.

The elemental equations for this transducer are

$$\Omega_{3g} = nv_{1g} \tag{10.1}$$

and

$$nT_t = F_t \tag{10.2}$$

For the device shown, the flow of power is from left to right when the product of $v_{1g}F_t$ (and the product of $\Omega_{3g}T_t$) is positive, that is, whenever the through and across variables are of the same sign.

Pulley and cable systems, lever and shaft mechanisms, and rack and gear mechanisms are examples of translation-to-rotation transducers.

Electromechanical Energy Convertors. The symbolic diagrams to be used here for ideal energy-converting transducers are shown in Figure 10.2, (a) for the case of translational mechanical motion, and (b) for the case of rotational mechanical motion. The elemental equations are also given in Figure 10.2. Since the coupling coefficient α is in many cases controllable, a feature that makes electromechanical transducers especially attractive for control systems use, it is shown as an input signal.

The translational version is an ideal solenoid. The direction of power flow for the device as shown (Figure 10.2a) is from left to right when $e_{12}i_t$ is positive.

The rotational version is an ideal electrical motor or an ideal electrical generator, depending the direction of usual power flow. The device as shown (Figure 10.2b) is operating as a motor when $e_{12}i_t$ is positive and as a generator when $e_{12}i_t$ is negative. A device that usually operates as a motor may temporarily operate as a generator when system transients occur, and vice versa.

Although these models are intended primarily for use with direct current (DC) devices, they apply reasonably well to the same devices operating with alternating current (AC) as long as root mean square (rms) values of voltage and current are

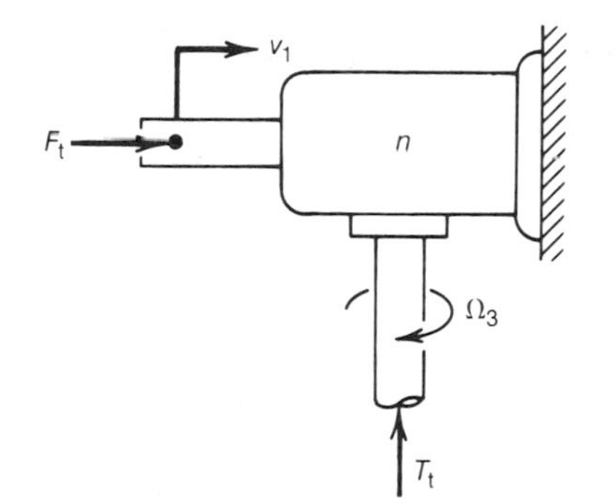

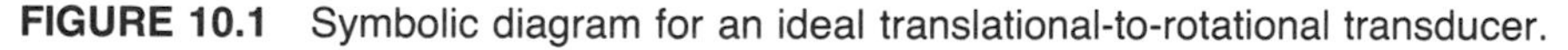
FIGURE 10.1 Symbolic diagram for an ideal translational-to-rotational transducer.

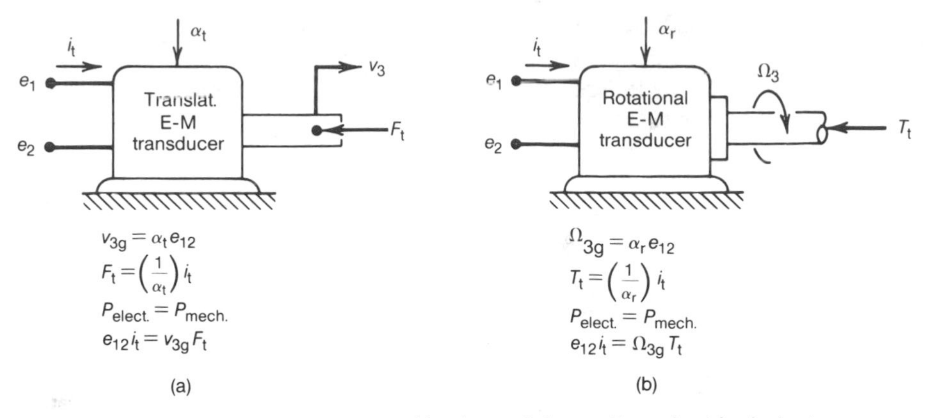

FIGURE 10.2 Symbolic diagrams with elemental equations for ideal electromechanical transducers: (a) translational mechanical and (b) rotational mechanical.

used and the dynamic response of the rest of the system is slow compared to the periodic variation of the alternating current.

AC Induction motors of the squirrel-cage type operate well only at speeds close to their no-load speed, which is synchronous with the AC frequency and usually constant. A typical torque versus speed characteristic for such a motor, operating with constant supply voltage, is shown in Figure 10.3. This characteristic includes the effects of bearing friction and windage loss in the fluid surrounding the rotating members, as well as resistance and inductance in the windings, so it is not an ideal lossless model. Thus it is necessary to know how the efficiency varies with load in order to determine the input current. When the efficiency is known, the input current, when operating as a motor, is given by

$$i_{m_{rms}} = \frac{1}{\eta_m} \cdot \frac{T\Omega_{1g}}{e_{12_{rms}}} \tag{10.3}$$

where η_m is the efficiency of the unit operating as a motor.

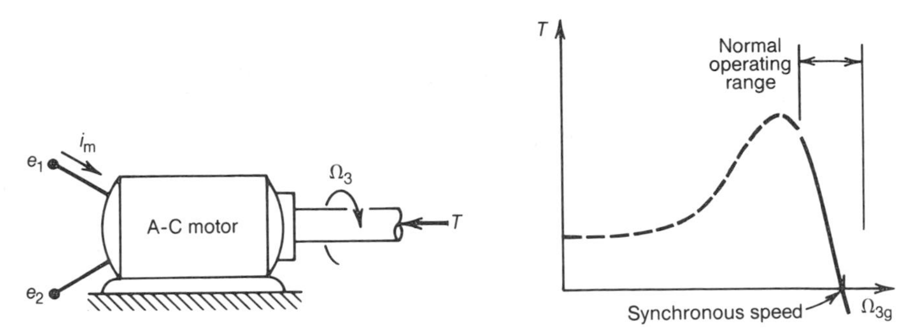

FIGURE 10.3 Steady-state torque versus speed characteristic for a squirrel-cage AC motor operating at constant supply frequency and voltage.

Fluidomechanical Energy Convertors. The symbolic diagrams to be used here for ideal energy-converting transducers are shown in Figure 10.4, (a) for the case of translational mechanical motion and (b) for the case of rotational mechanical motion. The elemental equations are also given in Figure 10.4. The coupling coefficient D is shown as an input signal for both cases, although it is not controllable for the case of translational motion.

The translational version is a fluid cylinder with reciprocating piston, which may operate in pump or motor fashion, depending on the direction of power flow. The power flow is from left to right when $P_{12}Q_t$ is positive for the device shown (Figure 10.4a).

The rotational version is an ideal positive displacement fluid motor or an ideal fluid pump, depending on the direction of usual power flow. The device as shown (Figure 10.4a) is operating as a fluid motor when $P_{12}Q_t$ is positive, and as a fluid pump when the product $P_{12}Q_t$ is negative. A device that usually operates as a pump may temporarily operate as a motor when system transients occur, and vice versa.

The coupling coefficient D is the ideal volume displaced per unit motion of the output shaft (i.e., no leakage occurring) for both the rotational and translational cases.

These models are valid only when the compressibility of the fluid employed is negligible, so that the volume rate of flow of fluid entering the transducer is the same as the volume rate of flow leaving it. Because the action of the fluid on the moving members is by means of static fluid pressure alone (i.e., momentum transfer is negligible), these devices are sometimes referred to as hydrostatic energy convertors.

Hydrokinetic energy convertors such as centrifugal pumps and turbines do involve momentum interchange between the moving fluid and moving blades and fixed walls. These devices are modeled graphically through the use of experimentally derived characteristic curves rather than by means of ideal transducers (similar to the modeling shown for squirrel-cage AC motors). Thus these models are not

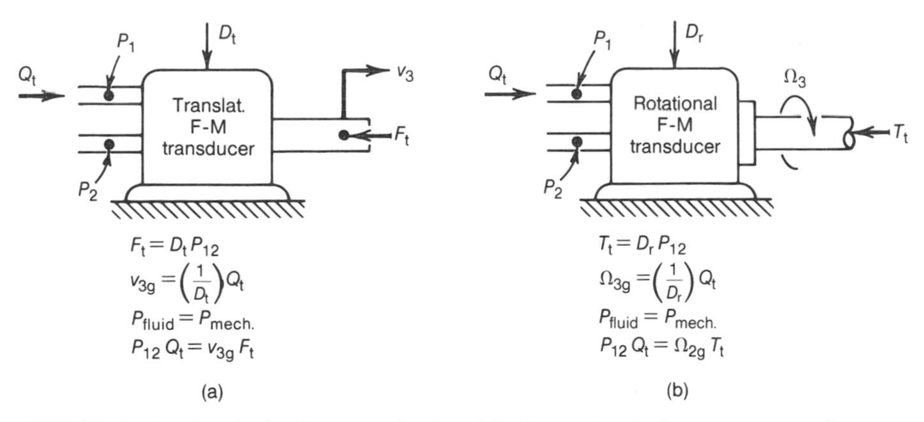

FIGURE 10.4 Symbolic diagrams for ideal fluidomechanical energy converting transducers: (a) translational and (b) rotational.

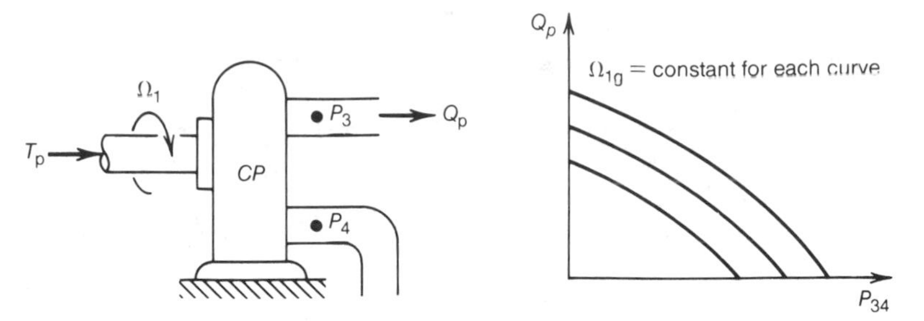

FIGURE 10.5 Symbolic diagram and characteristic curves for a centrifugal pump.

lossless, and it is necessary to know how their efficiency varies with operating conditions.

The symbolic diagram and typical characteristic curves for a centrifugal pump operating at a series of constant speeds are shown in Figure 10.5. The pump torque is given by

$$T_p = \frac{1}{\eta_p} \cdot \frac{P_{12} Q_p}{\Omega_{1g}} \tag{10.4}$$

where η_p is the pump efficiency.

The symbolic diagram and characteristic curves for operation of a hydraulic turbine at a series of constant pressure drops are shown in Figure 10.6. The turbine flowrate is given by

$$Q_t = \frac{1}{\eta_t} \cdot \frac{\Omega_{3g} T_t}{P_{12}} \tag{10.5}$$

where η_t is the turbine efficiency.

Linearization of these characteristics for small perturbations about a set of normal operating conditions is readily accomplished by employing the techniques discussed in Chapter 1.

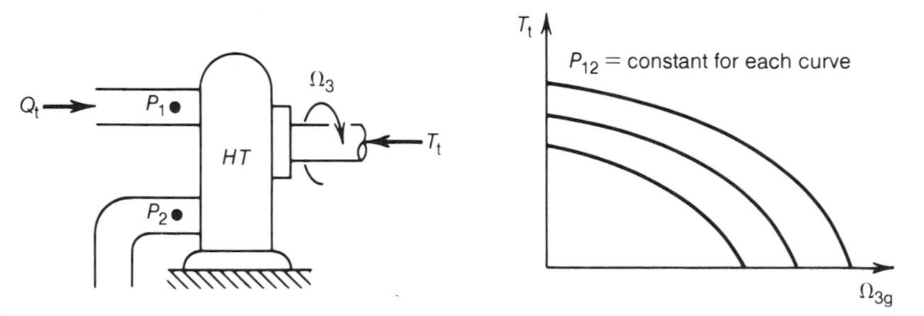

FIGURE 10.6 Symbolic diagram and typical characteristic curves for a hydraulic turbine.

10.3 Signal-Converting Transducers

Signal-converting transducers in some cases are simply energy-converting transducers that have negligible load, as for instance a tachometer generator operating into a very high-resistance load. In other cases they are specially designed devices or systems that convert one type of signal to another, as for instance a fly-weight speed sensor or a Bourdon gage pressure sensor. Still another example is the use of a resistance potentiometer, supplied with a constant voltage, to deliver an output wiper-arm signal that is a function of its wiper-arm position.

A partial list of signal-converting transducers is given below.

- Tachometer generator
- Centrifugal pump speed sensor
- Linear variable differential transformer position sensor
- Variable reluctance pressure sensor
- Flyweight speed sensor
- Linear velocity sensor
- Thermistor temperature sensor
- Variable capacitor proximity sensor
- Piezo-electric force sensor

The potential user is referred to a wide range of technical bulletins and specifications prepared by the manufacturers of such equipment. Several texts are currently in print on this topic[1].

The symbolic diagrams for ideal signal-converting transducers are controlled sources, shown in Figure 10.7 together with their describing equations: (a) shows a controlled A-variable type transducer, and (b) shows a controlled T-variable type transducer.

The input variable x denotes the variable being sensed, and e_s (or P_s or v_s or T_s) denotes the output of an A-variable type transducer. For T-variable type transducers, i_s (or Q_s or F_s or Q_{hs}) denotes the output.

[1]As a starting point one might consider *Mechanical Measurements*, by T. G. Beckwith, H. L. Buck, and R. D. Marangoni, Addison-Wesley Publishing Co., Reading, Mass., 1982, or *Measurement Systems*, by E. O. Doebelin, McGraw-Hill Publishing Co., New York, 1983.

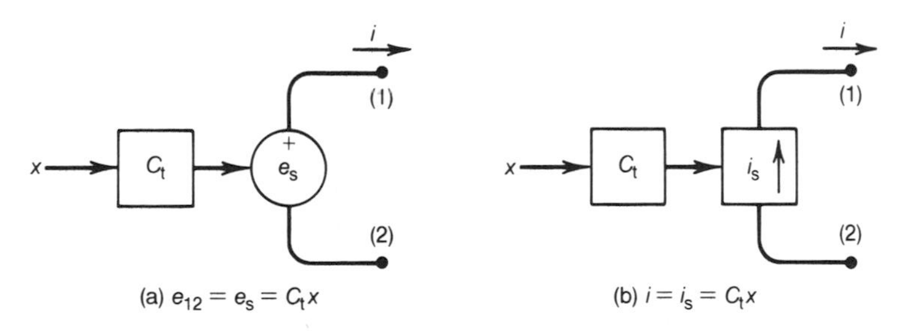

FIGURE 10.7 Symbols for signal-converting transducers.

10.4 Applications Examples

EXAMPLE 10.1

The rack-and-pinion system shown in Figure 10.8 is to be modeled as part of a large system, which is to be simulated on the digital computer. The object here is to set up the state-variable equation(s) for this subsystem, considering T_s and F_s as the system inputs and v_{3g} as the system output.

Solution

The detailed symbolic free-body diagram for this system is shown in Figure 10.9.

The elemental equations are as follows: For the pinion inertia

$$T_s - T_t = J_p \frac{d\Omega_{1g}}{dt} \tag{10.6}$$

For the transducer

$$\Omega_{1g} = nv_{3g} \tag{10.7}$$

$$T_t = (1/n)F_t \tag{10.8}$$

For the rack mass

$$F_t - F_b - F_s = m \frac{dv_{3g}}{dt} \tag{10.9}$$

where $n = 1/r$.

For the rack friction

$$F_b = bv_{3g} \tag{10.10}$$

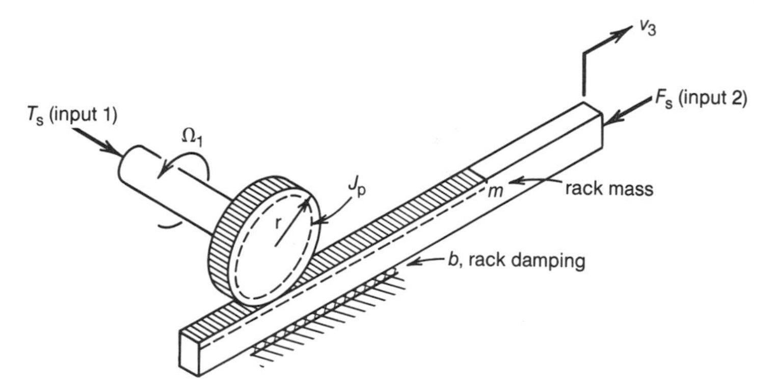

FIGURE 10.8 Rack-and-pinion system, including pinion inertia, rack mass, and friction.

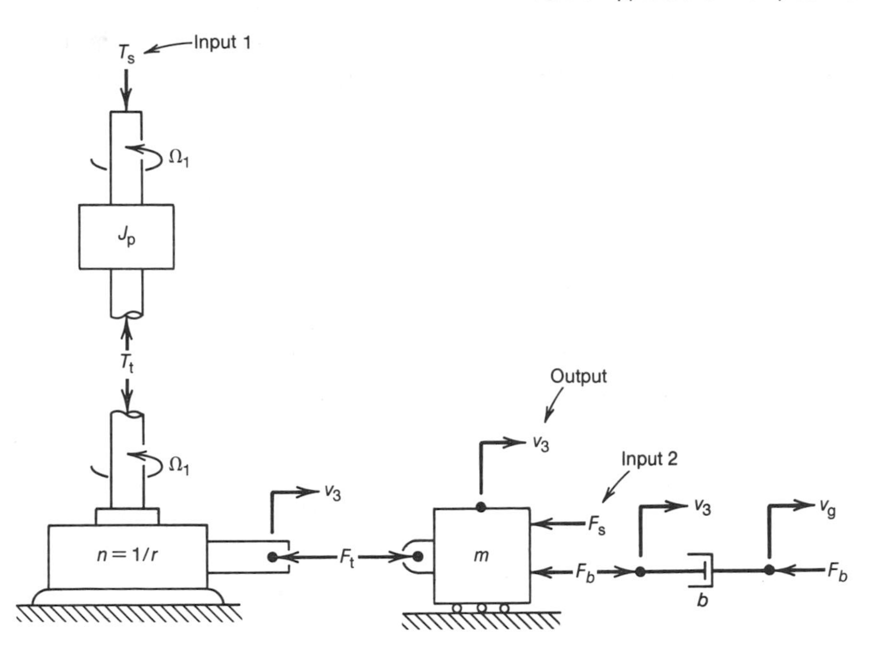

FIGURE 10.9 Detailed symbolic free-body diagram of rack-and-pinion system.

Combine Equations (10.6), (10.7), and (10.8) to eliminate T_t and Ω_{1g}.

$$T_s - \frac{1}{n} F_t = nJ_p \frac{dv_{3g}}{dt} \tag{10.11}$$

Combine (10.9) and (10.10) to eliminate F_b.

$$F_t - bv_{3g} - F_s = m \frac{dv_{3g}}{dt} \tag{10.12}$$

Combine (10.11) and (10.12) to eliminate F_t

$$T_s - \frac{1}{n}\left(m \frac{dv_{3g}}{dt} + bv_{3g} + F_s\right) = nJ_p \frac{dv_{3g}}{dt} \tag{10.13}$$

Rearranging yields a single state-variable equation.

$$\frac{dv_{3g}}{dt} = \frac{1}{n^2 J_p + m}(-bv_{3g} + nT_s - F_s) \tag{10.14}$$

Note that the inertia from the rotational part of the system becomes an equivalent mass equal to $n^2 J_p$; thus, although this system has two energy storage elements, they are coupled by the transducer so that they behave together as a single energy storage element. There is, therefore, only one state-variable equation required to describe this system. ■

The multiplication of a system parameter of one discipline by the square of the transducer constant to produce an equivalent parameter in the other discipline will be seen to occur consistently in all mixed-system analyses.

EXAMPLE 10.2

A permanent magnet DC motor has series resistance R and series inductance L in its armature winding and rotor inertia J_{m}, plus bearing and windage friction B_{m}. The motor is being used to drive a mechanical load consisting of a load inertia J_ℓ and load damping B_ℓ, as shown in Figure 10.10. Develop the system differential equation relating the output speed $\Omega_{3\mathrm{g}}$ to the input source voltage e_{s}.

Solution

The detailed symbolic circuit and free-body diagram for this system is shown in Figure 10.11.

The elemental equations are as follows: For the series resistance and inductance

$$e_{\mathrm{s}} - e_{42} = Ri_{\mathrm{t}} + L\frac{di_{\mathrm{t}}}{dt} \tag{10.15}$$

For the transducer

$$\Omega_{1\mathrm{g}} = \alpha_{\mathrm{r}} e_{42} \tag{10.16}$$

$$T_{\mathrm{t}} = \frac{1}{\alpha_r} i_{\mathrm{t}} \tag{10.17}$$

For the combined inertias and rotational damper

$$T_{\mathrm{t}} = (J_m + J_\ell)\frac{d\Omega_{1\mathrm{g}}}{dt} + (B_\ell + B_m)\Omega_{1\mathrm{g}} \tag{10.18}$$

Combining Equations (10.15), (10.16), and (10.17) gives

$$e_{\mathrm{s}} - \frac{1}{\alpha_r}\Omega_{1\mathrm{g}} = \alpha_r R T_{\mathrm{t}} + \alpha_r L\frac{dT_{\mathrm{t}}}{dt} \tag{10.19}$$

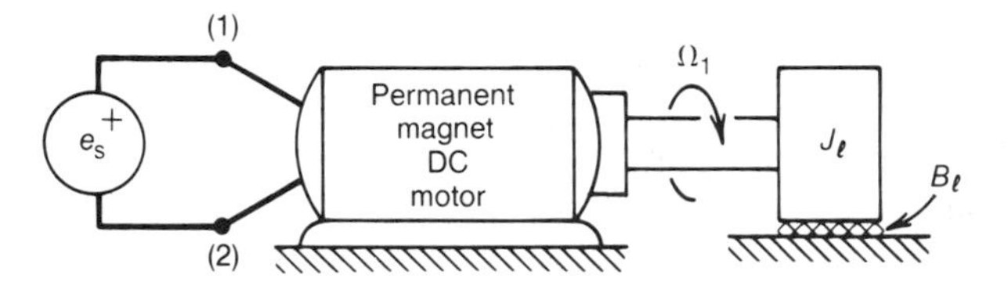

FIGURE 10.10 Schematic diagram of DC motor-driven mechanical system.

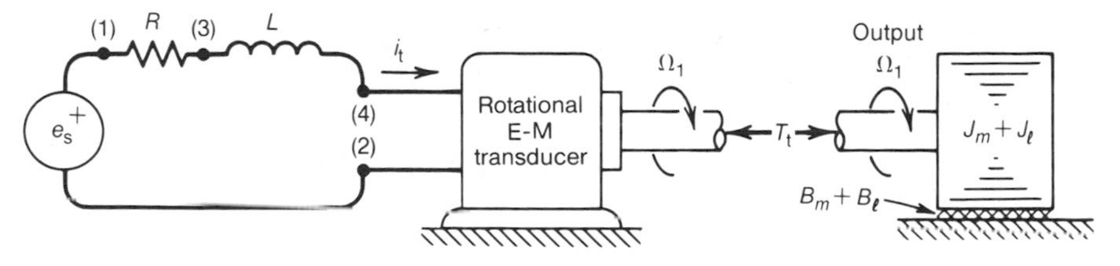

FIGURE 10.11 Detailed symbolic diagram of motor-driven system.

Now combine (10.18) and (10.19) to eliminate T_t, using the differential operator D.

$$e_s - \frac{1}{\alpha_r}\Omega_{1g} = \alpha_r(R + LD)[(J_m + J_\ell)D + (B_\ell + B_m)]\Omega_{1g} \quad (10.20)$$

Transposing and multiplying all terms by α_r

$$(J_m + J_\ell)\,\alpha_r^2 L\,\frac{d^2\Omega_{1g}}{dt^2} + [(B_m + B_\ell)\alpha_r^2 L + (J_m + J_\ell)\alpha_r^2 R]\,\frac{d\Omega_{1g}}{dt} + [(B_m + B_\ell)\alpha_r^2 R + 1]\Omega_{1g} = \alpha_r e_s \quad (10.21)$$

■

In example 10.2 the inductor, L, is transformed into an equivalent spring, $1/\alpha_r^2 L$, and the resistor, R, is transformed into an equivalent damper, $1/\alpha_r^2 R$, on the mechanical side of the system. Since the two inertias are connected to each other by a rigid shaft, they become lumped together as a single energy storage element, and the system is only a second-order system.

EXAMPLE 10.3

A variable-displacement hydraulic motor, supplied with a constant flow source is used to vary the output speed of an inertia-damper load, as shown in Figure 10.12. The displacement of the motor D_m is proportional to the stroke lever angle ψ, and the motor has leakage resistance R_f, rotor inertia J_m, and bearing and windage friction B_m. The load torque T_ℓ is a second input to the system, in addition to the motor stroke, ψ. Develop the state-variable equation(s) for this system.

Solution

The detailed symbolic diagram for this system is shown in Figure 10.13.

The elemental equations are as follows: For the leakage resistor

$$Q_R = Q_s - Q_t = \frac{1}{R_f}P_{12} \quad (10.22)$$

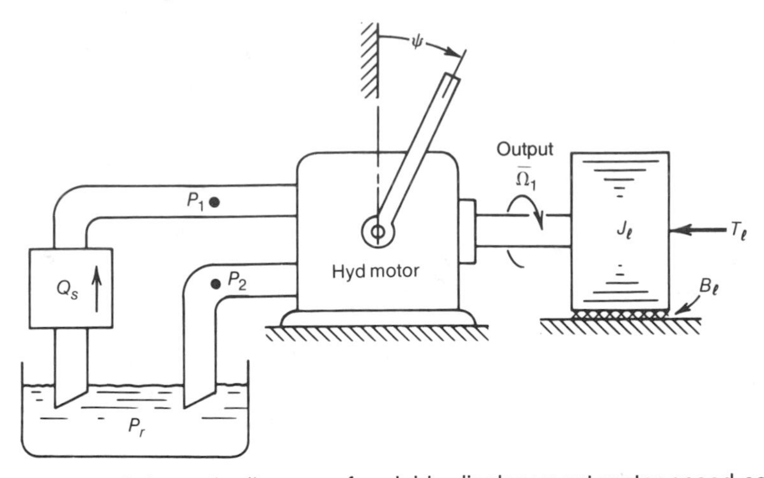

FIGURE 10.12 Schematic diagram of variable-displacement motor speed control system.

For the transducer

$$T_t = C_1\psi P_{12} \tag{10.23}$$

$$\Omega_{1g} = \frac{1}{C_1\psi} \cdot Q_t \tag{10.24}$$

For the inertias and dampers

$$T_t - T_\ell = (J_m + J_\ell)\,\frac{d\Omega_{1g}}{dt} + (B_m + B_\ell)\Omega_{1g} \tag{10.25}$$

Combining Equations (10.22), (10.23), and (10.24) and multiplying all terms by $R_f C_1$ gives

$$T_t = C_1\psi R_f Q_s - (C_1\psi)^2 R_f \Omega_{1g} \tag{10.26}$$

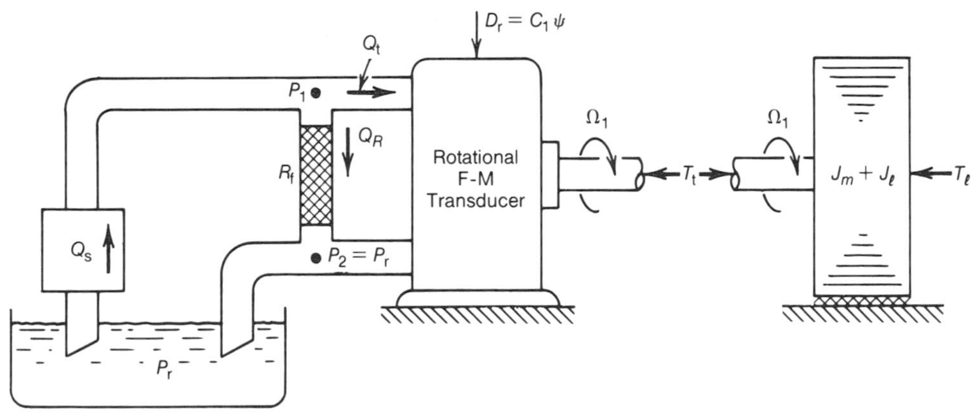

FIGURE 10.13 Detailed symbolic diagram for hydraulic motor controlled system.

Combine (10.25) and (10.26) to eliminate T_t.

$$C_1\psi R_f Q_s - (C_1\psi)^2 R_f \Omega_{1g} - T_\ell = (J_m + J_\ell)\frac{d\Omega_{1g}}{dt} + (B_m + B_\ell)\Omega_{1g} \qquad (10.27)$$

Divide all terms by $(J_m + J_\ell)$ and rearrange into state-variable format.

$$\frac{d\Omega_{1g}}{dt} = -\frac{B_m + B_\ell + (C_1\psi)^2 R_f}{J_m + J_\ell}\Omega_{1g} + \frac{C_1\psi R_f}{J_m + J_\ell}Q_s - \frac{1}{J_m + J_\ell}T_\ell \qquad (10.28)$$

■

In this example the fluid resistor R_f is transformed into an equivalent displacement-referenced damper $(C_1\psi)^2R_f$ on the mechanical side of the system.

10.5 Synopsis

In this chapter several types of ideal transducers have been employed to model the coupling devices that interconnect one type of system with another type of system resulting in what is called a mixed system. In this modeling of coupling devices nonideal characteristics are described by the careful addition of A-, T-, or D-type elements to the ideal transducer. For the cases where the use of an ideal transducer is not feasible (for instance AC machinery, hydrokinetic machinery), graphical performance data are used to describe the characteristics of the coupling device as a function of two variables employing families of curves.

Detailed examples are provided to illustrate the techniques of modeling and analysis associated with the design and development of mixed systems. The transformation of the characteristics of an element on one side of an ideal transducer to an equivalent coupling-coefficient-referenced element on the other side of the transducer is illustrated. The use of state-variable equations in system modeling is reiterated, and in each case input-output differential equations are also provided that are ready for use in analytical solution or computer simulation studies.

Problems

10.1. A rack-and-pinion mechanism has been proposed as a means of employing a DC motor to control the motion of the moving carriage of a machine tool, as shown schematically in Figure P10.1a.

The ideal current source $i_s(t)$ is capable of delivering the current i_s to the motor regardless of the voltage e_{21} required. The nonlinear friction in the rack-and-pinion mechanism has been lumped into the single nonlinear friction force F_{NLD} versus velocity v_{2g} characteristic shown in Figure P10.1b.

(a) Draw a complete free-body diagram of the system showing the ideal inertialess motor, the motor inertia J_m, the motor damper B_m, the shaft spring K, the shaft torquet T_s, the gear inertia J_g, the ideal rotational to translational transducer, the lumped bar and carriage mass $(m_b + m_c)$, and the nonlinear damper NLD. The speed at the motor end of the shaft is Ω_1.

(b) Using only symbols, write the necessary and sufficient set of describing equations for this system.

(c) Using only symbols, rearrange the equations developed in part (b) to form the set of nonlinear state-variable equations. Use Ω_{1g}, v_{2g}, and T_s as the state variables.

(d) Find the steady operating point values $\overline{\Omega}_{1g}$ and $\overline{v}_{2g}$ when $\bar{i}_s = 5.0$ amp.

(e) Using only symbols, linearize the state-variable equations and combine them to develop the system differential equation relating $\hat{v}_{2g}$ to $\hat{i}_s(t)$.

(f) Find the roots of the characteristic equation and calculate the natural frequency and damping ratio for the linearized model, using the data given in Fig. P10.1 and part (d).

(g) Write the conditional statements needed to describe the nonlinear damper NLD for all possible values of F_{NLD}.

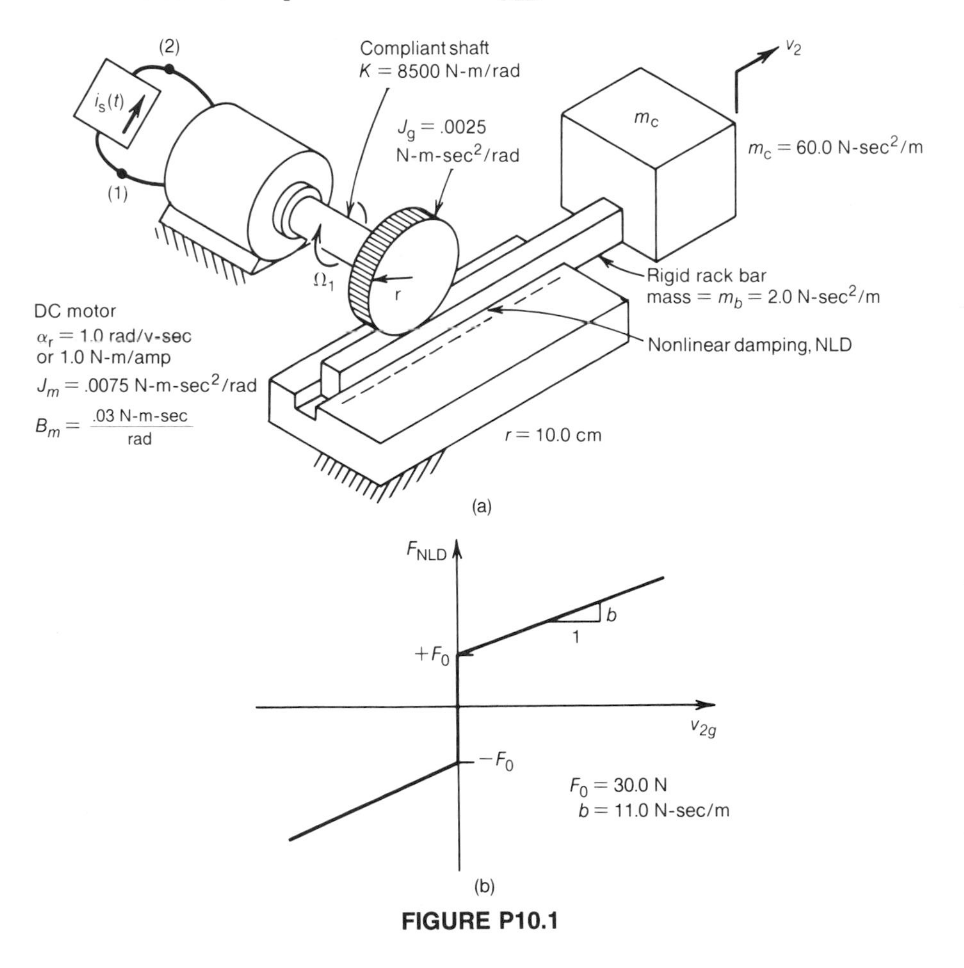

FIGURE P10.1

10.2. The schematic diagram shown in Figure P10.2a represents a model of a small hydroelectric power plant used to convert water power, diverted from a nearby stream, into DC electricity for operating a small factory.

You have been given the task of determining the dynamic response characteristics of this system. The input is the current source $i_s(t)$, supplied to the field winding of the generator which establishes the field strength, and therefore the value of the coupling coefficient α_r of the generator. The output is the voltage e_{12} with which the generator delivers power to its electrical load, modelled here as a series *R-L* circuit. (This information about the system dynamic response will be needed later for designing a voltage controller that will keep the output voltage constant under varying electrical load conditions.)

Previously one of your colleagues found an appropriate approximate analytical model for the hydraulic turbine that fitted reasonably well the torque-speed characteristics obtained by a lab technician. The technician's results are shown graphically below; your colleague's equation is

$$T_t = T_0 - \frac{T_0}{6}\left(1 - \frac{\Omega_{1g}}{\Omega_0}\right)^2 - T_0\left(1 - \frac{P_{12}}{P_0}\right)$$

Note that in this graphical presentation (Figure P10.2b) the numerical value of Ω_0 is negative.

(a) Draw a free-body diagram of the mechanical part of the system showing the inertialess turbine, the lumped turbine plus generator inertia (J_t +

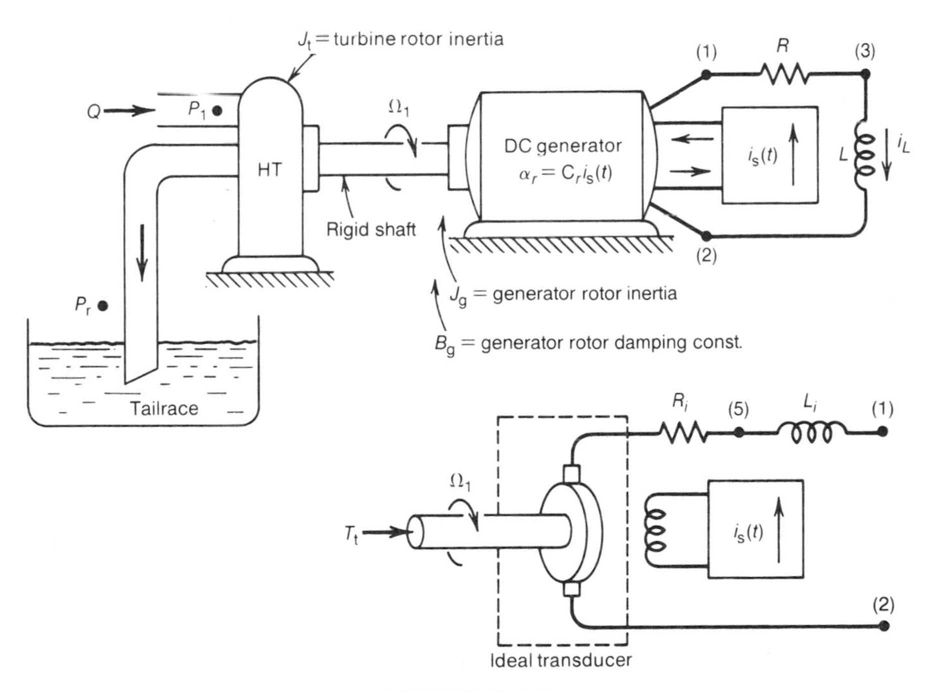

FIGURE P10.2a

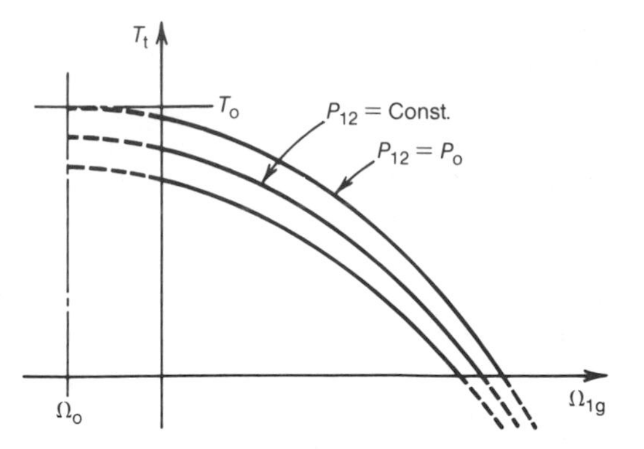

FIGURE P10.2b Steady-state turbine torque versus speed characteristics at various constant speeds.

J_g), the generator damper B_g, and the ideal inertialess generator. Also draw the complete electrical circuit including the ideal generator, the internal resistance R_i, the internal inductance L_i, and the load resistance and inductance R and L.

(b) Write the necessary and sufficient set of describing equations for this system.

(c) Rearrange to form the set of nonlinear state-variable equations employing Ω_{1g} and i_L as the state variables.

(d) Linearize the state-variable equations and provide an output equation for the system output $\hat{e}_{12}$.

(e) Combine the state-variable equations to form the system differential equation relating $\hat{e}_{12}$ to the input $\hat{\imath}_s(t)$.

10.3. A valve-controlled motor is used with a constant pressure source P_s to control the speed of an inertia J_ℓ having a load torque T_ℓ and linear damper B_ℓ acting upon it as shown in Figure P10.3.

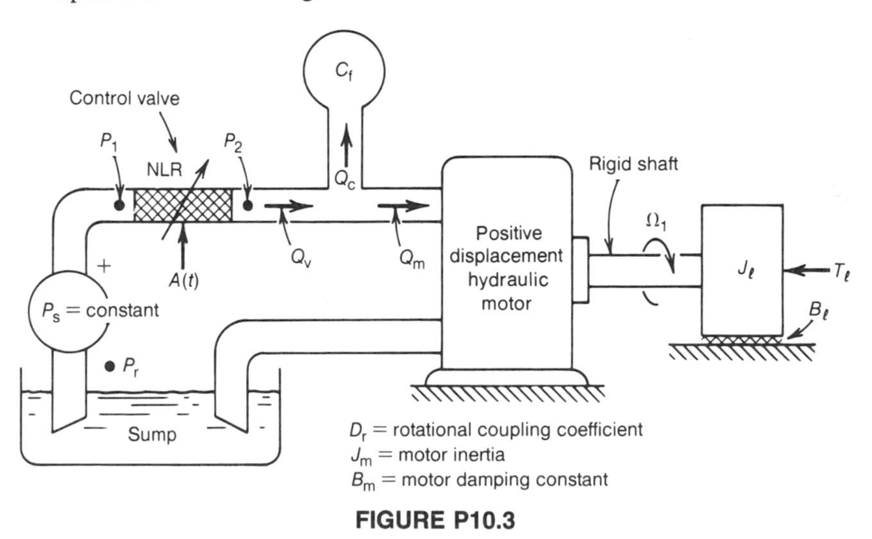

D_r = rotational coupling coefficient
J_m = motor inertia
B_m = motor damping constant

FIGURE P10.3

The lumped fluid capacitance C_f arises from the compressibility of the fluid under pressure between the control valve and the working parts (pistons, vanes, or gears) of the hydraulic motor. The valve is basically a variable nonlinear fluid resistor in the form of a variable area orifice and is described by the equation

$$Q_v = A(t)C_d\left(\frac{2}{\rho}\right)^{.5}\frac{P_{12}}{|P_{12}|^{.5}}$$

The shaft is very stiff so that the motor inertia J_m may be lumped together with the load inertia J_ℓ, and the motor damper B_m may be lumped together with the load damper B_ℓ. Leakage in the motor is negligible.

(a) Draw a free-body diagram of the mechanical part of the system showing the inertialess ideal motor shaft, the lumped inertias, the lumped dampers, and the load torque.

(b) Write the necessary and sufficient set of describing equations for this system.

(c) Rearrange the equations developed in part (b) to form the set of state-variable equations employing P_{2r} and Ω_{1g} as the state variables.

(d) Linearize and combine the state-variable equations to produce the system differential equation relating $\hat{\Omega}_{1g}$ to the input $\hat{A}(t)$.

10.4. A variable-displacement hydraulic pump, driven at constant speed Ω_1, is being used in a hydraulic power supply system for a variable resistance load R_ℓ as shown in Figure P10.4.

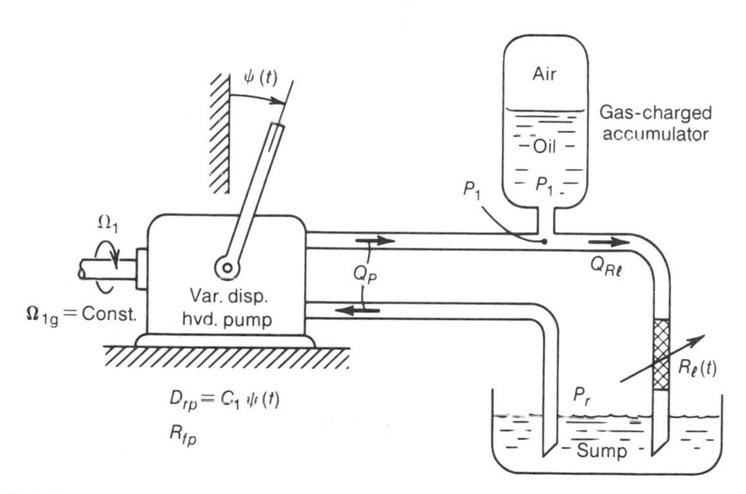

FIGURE P10.4 Schematic diagram of variable-displacement hydraulic power supply.

The displacement of the pump $D_{rp} = C_1\psi(t)$ where $\psi(t)$ is the pump stroke input to the system. The internal leakage in the pump is $Q_{\ell p} = (1/R_{fp})P_{1r}$. The air in the gas-charged accumulator occupies volume $\overline{\mathcal{V}}$ when the system is at its normal operating point, and the absolute temperature of the gas is approximately the same as the atmosphere.

(a) Develop the system differential equation relating the pressure P_{1r} to the two inputs $\psi(t)$ and $R_\ell(t)$, using only symbols.

(b) Determine the normal operating point values $\overline{\psi}$ and $\overline{Q}_p$, given the following data:

$$\begin{aligned}\overline{\Omega}_{1g} &= 180 \text{ rad/sec}\\ C_1 &= 4.4 \text{ in.}^3/\text{rad per stroke radian}\\ \overline{R}_\ell &= 40 \text{ lb-sec/in.}^5\\ \overline{P}_{1r} &= 1000 \text{ lb/in.}^2\\ R_{fp} &= 1200 \text{ lb-sec/in.}^5\end{aligned}$$

(c) Linearize the system differential equation for small perturbations of all variables about the normal operating point, and find the system time constant τ given the data which follows (assume 'fast' changes of pressure in the accumulator).

$$\begin{aligned}\overline{\mathcal{V}} &= 300 \text{ in.}^3\\ T_{atm} &= 530 \text{ deg R}\end{aligned}$$

(d) Sketch the response to a small step change $\hat{R}_\ell(t) = -2$ lb-sec/in.^{5} at $t = 0$, followed by a small step change $\hat{\psi}(t) = \overline{\psi}/20$ at $t = \tau/2$.

10.5. A variable-speed hydraulic transmission used to drive the spindle of a large turret lathe is shown schematically together with constant speed drive motor and associated output gearing in Figure P10.5a.

The variable-displacement pump has displacement $D_{rp} = C_1\psi(t)$, and the internal leakage of the pump is proportional to the pressure P_{1r}, modeled by a leakage resistance R_{fp}. The fixed-displacement motor has displacement D_{rm},

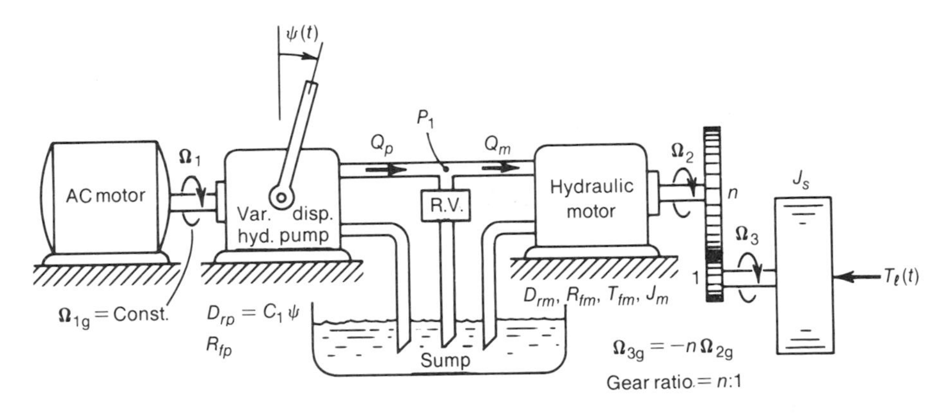

FIGURE P10.5a Schematic diagram of variable-speed drive system.

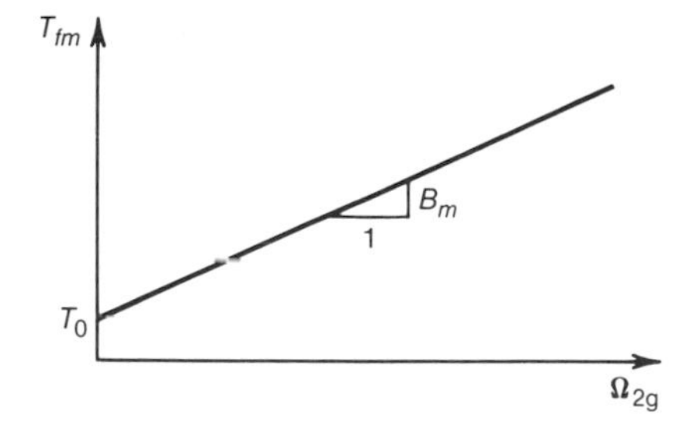

FIGURE P10.5b Graph of motor friction torque versus shaft speed.

internal leakage resistance R_{fm}, and rotational inertia J_m. The friction torque of the motor versus shaft speed Ω_{2g} is shown by the graph of Figure P10.5b

The relief valve RV, which acts only if something happens to cause P_{1r} to exceed a safe operating value, is of no concern in this problem. The combined friction of the gear and spindle bearings and the friction in the gear train are negligible compared with the friction in the motor. The gear inertias are negligible, and the combined inertia of the lathe spindle, chuck, and workpiece is J_s. The torque $T_\ell(t)$ represents the load torque exerted by the cutting tool on the workpiece being held and driven by the spindle chuck.

(a) As a starting point it will be assumed that the pressurized line between the pump and motor can be modeled as a simple fluid capacitor having volume $\mathcal{V} = AL + \mathcal{V}_{\text{int}}$ where A is the flow area of the line, L is the length of the line, and $\mathcal{V}_{\text{int}}$ is the sum of the internal volumes of the pump and the motor. In other words, $\mathcal{V}$ is the total volume between the pistons in the pump and the pistons in the motor. Based on this simplification, prepare a complete system diagram showing all variables and all the essential ideal elements in schematic-freebody diagram form (as in the solution for Example 10.3 in the text).

(b) For a desired normal operating pressure $\overline{P}_{1r}$ when the normal operating load torque is $\overline{T}_\ell$ and the normal spindle speed is $\overline{\Omega}_{3g}$, derive an expression for the motor displacement D_{rm} in terms of $\overline{P}_{1r}$, $\overline{T}_\ell$, and $\overline{\Omega}_{3g}$.

Then develop an expression for the normal operating flowrate $\overline{Q}_m = \overline{Q}_p$ in terms of $\overline{\Omega}_{3g}$, D_{rm}, and $\overline{P}_{1r}$; and also develop an expression for the normal operating value of pump displacement $\overline{D}_{rp} = C_1\overline{\psi}$ in terms of $\overline{Q}_m$, $\overline{P}_{1r}$, and $\overline{\Omega}_{1g}$.

(c) Write the necessary and sufficient set of describing equations for this system and combine to form the set of state-variable equations having P_{1r} and Ω_{2g} as state variables, and write the output equation for Ω_{3g},

(d) Linearize the state-variable equations and the output equation for small perturbations of all variables.

(e) Employing the data which follows, find the damping ratio ζ and natural frequency ω_n (or time constants) for this system using the input-output differential equation relating $\hat{\Omega}_{3g}$ to $\hat{\psi}(t)$.

$$\begin{aligned}
\overline{\Omega}_{1g} &= 190 \text{ rad/sec} \\
\overline{P}_{1r} &= 1000 \text{ lb/in.}^2 \\
R_{fp} &= 480 \text{ lb-sec/in.}^5 \\
R_{fm} &= 520 \text{ lb-sec/in.}^5 \\
\mathcal{V} &= 10 \text{ in.}^3 \\
T_0 &= 30 \text{ lb-in.} \\
B_m &= 0.65 \text{ lb-in.-sec} \\
J_m &= .11 \text{ lb-in.-sec}^2 \\
n &= 3 \\
J_s &= 6.0 \text{ lb-in.-sec}^2 \\
\overline{T}_\ell &= -1500 \text{ lb-in.} \\
\overline{\Omega}_{3g} &= -260 \text{ rad/sec}
\end{aligned}$$

Fluid Properties:

$$\begin{aligned}
\gamma &= .032 \text{ lb/in.}^3 \\
\beta &= 200{,}000 \text{ lb/in.}^2
\end{aligned}$$

(f) Compute the fluid inertance of the hydraulic line using an internal line diameter of 0.5 in. and a line length of 40 in., and compare its motor-displacement-reflected effective inertia $D_{rm}^2 I$ with the motor inertia J_m, and with the gear-ratio-reflected inertia $n^2 J_s$. How do you feel about the tentative decision made in part (a) to neglect the fluid inertance of the line?

10.6. The generator-motor arrangement shown in Figure P10.6a is one version of a Ward-Leonard variable-speed drive used to drive loads at variable speeds and loads, using power from a constant speed source Ω_1. This system is similar in many respects to the hydraulic variable speed drive studied in Problem 10.5.

The coupling coefficient for the generator is $\alpha_{rg} = C_1 i_s(t)$, where $i_s(t)$ is a system input, and the generator armature winding has resistance R_g and inductance L_g.

The coupling coefficient for the motor is $\alpha_{rm} = C_2 i_f$, which is constant. The motor armature winding has resistance R_m and inductance L_m. The motor rotor has inertia J_m, and the motor friction torque as a function of speed is

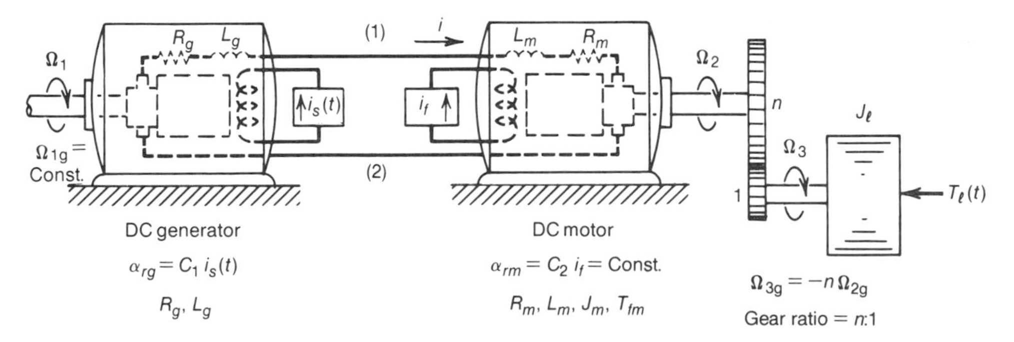

FIGURE P10.6a Schematic diagram of Ward-Leonard variable speed drive system.

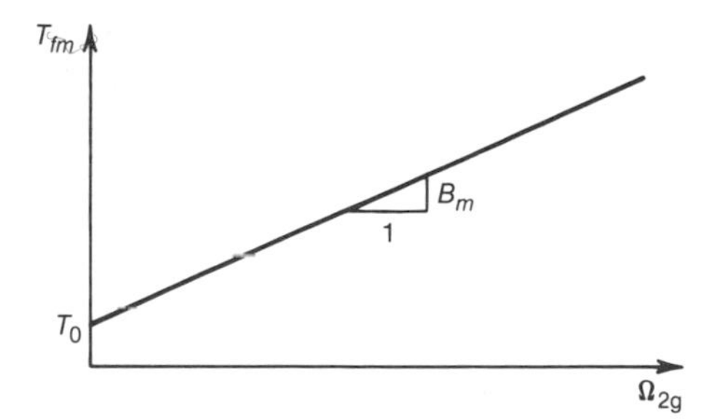

FIGURE P10.6b Graph of motor friction torque versus shaft speed.

shown in the graph of friction torque versus speed in Figure P10.6b. The combined friction of gear and load inertia bearings and the friction in the gear train are negligible. The gear inertias are also negligible. The load inertia is J_ℓ. The torque $T_\ell(t)$ represents the external load which is a second input to the system.

(a) It will be assumed that the conductors between the generator and the motor have negligible resistance, inductance, and capacitance. Prepare a complete system diagram showing all variables and all the essential ideal elements in schematic circuit-freebody diagram form (as in Example 10.2 of the text).

(b) For a desired normal operating current $\bar{i}$ when the normal operating load is $\overline{T}_\ell$ and the normal operating speed is $\overline{\Omega}_{3g}$, derive an expression for the motor coupling coefficient α_{rm} in terms of $\bar{i}$, $\overline{T}_\ell$, and $\overline{\Omega}_{3g}$. Then develop an expression for $\bar{e}_{12}$ in terms of $\overline{\Omega}_{3g}$, α_{rm}, and $\bar{i}$. Also develop an expression for the normal operating value $\bar{\alpha}_{rg} = C_1\bar{i}_s$ in terms of $\bar{e}_{12}$, $\bar{i}$, and $\overline{\Omega}_{1g}$.

(c) Write the necessary and sufficient set of describing equations for this system and combine to form the state-variable equations having i and Ω_{2g} as the state-variables. Also write the output equation for Ω_{3g}.

(d) Linearize the state-variable equations and the output equation for small perturbations about the normal operating point.

(e) Employing the data which follows, find the damping ratio ζ and the natural frequency ω_n (or time constants) for this system, using the system input-output differential equation relating $\hat{\Omega}_{3g}$ to $\hat{i}_s(t)$.

$$
\begin{aligned}
\overline{\Omega}_{1g} &= 190 \text{ rad/sec} \\
\bar{i} &= 30 \text{ amp} \\
R_g &= .5 \text{ ohm} \\
R_m &= .5 \text{ ohm} \\
L_g &= 5 \text{ h} \\
L_m &= 5 \text{ h} \\
T_0 &= 10.0 \text{ lb-in.} \\
B_m &= .4 \text{ lb-in.-sec} \\
J_m &= 30 \text{ lb-in.-sec}^2 \\
J_\ell &= 6 \text{ lb-in.-sec}^2 \\
\overline{T}_\ell &= -1500 \text{ lb-in.} \\
\overline{\Omega}_{3g} &= -260 \text{ rad/sec} \\
n &= 3
\end{aligned}
$$

10.7. Find expressions for the steady-state response of x_2 to a small step change in the input $x_1(t)$, and the damping ratio for the valve-controlled pneumatic system shown in Figure P10.7, assuming k_s is negligible, and F_ℓ is constant.

Assume that $\overline{P}_2 < .45\overline{P}_1$ and employ linearized equations for your analysis.

Using parameters supplied by you or your instructor, calculate the system natural frequency ω_n (or time constants) of this system. Note that $A_{02} = \pi D x_1/2$ should be less than $\pi D^2/8$, in other words x_1 should be less than $D/4$, for the flapper-nozzle valve to function effectively. Furthermore A_{02}/A_{01} must satisfy the desired $\overline{P}_2/\overline{P}_1$ ratio in order to proceed with the calculation.

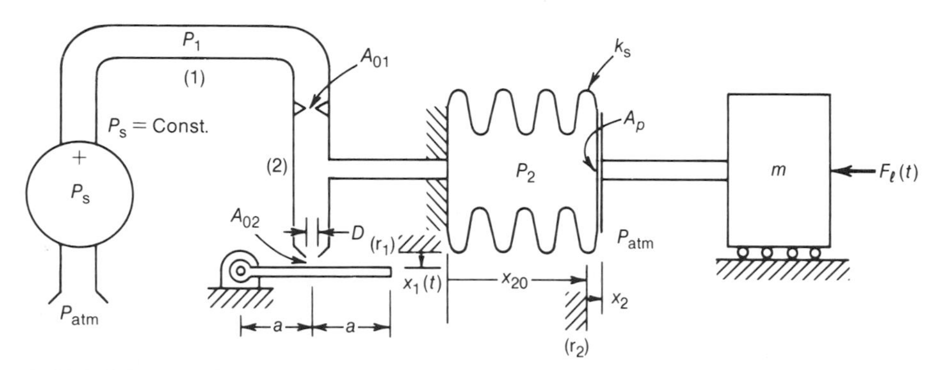

FIGURE P10.7 Schematic diagram of pneumatically-driven mass and force load.

10.8. Find expressions for the steady-state response of Ω_{1g} to a small step change in e_s, and the damping ratio for the amplifier-driven DC motor and load system shown in Figure P10.8.

Assume that the gain k_a of the power amplifier is negative and very large in magnitude, at least 10^5, and that negligible current is required at its input terminal.

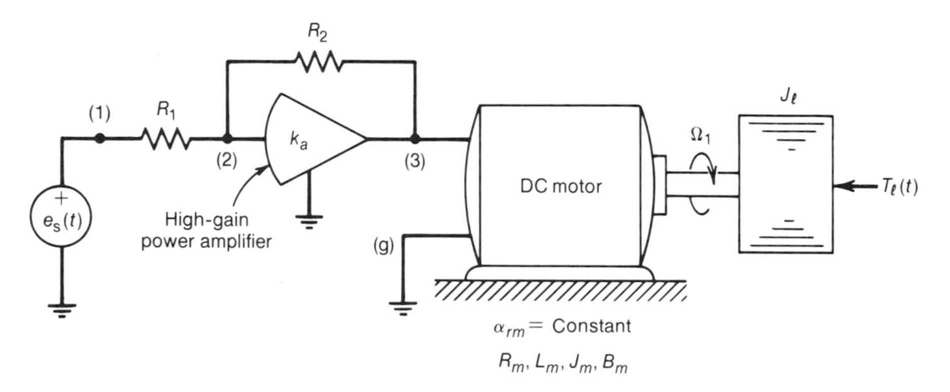

FIGURE P10.8 Schematic diagram of amplifier-driven motor and load system.

11

System Transfer Functions

11.1 Introduction

In dealing with linear systems, a significant body of theory has been developed to deal with single-input single-output systems without having to go through the classical methods of solving the input-output differential equation for the system. This body of theory involves the use of the complex variable $s = \sigma + j\omega$, sometimes known as a complex frequency variable. This variable is essentially the same as the Laplace transformation variable s in many respects, but its use does not need to involve the complex (in both senses of the word) transformation problems of assuring convergence of integrals having limits approaching infinity, nor does it involve the need to carry out the tedious process of inverse transformation by means of partial fraction expansion and the use of a table of transformation pairs. Here the variable s is simply considered to be the coefficient in the exponential input function e^{st}, and the transfer function emerges from solving for the particular or forced part of the response to this input. The notion of transfer function is then combined with the use of simple input-output block diagrams to express symbolically the input-output characteristics of the dynamical behavior of the system.

The systems described and analyzed in previous chapters have nearly all incorporated naturally occurring feedback effects. The presence of these feedback effects is readily observed when a simulation block diagram has been prepared for the system. In addition to the natural feedback effects, we will need to deal with intentional, man-made feedback of the kind employed in feedback amplifiers and automatic control systems. The use of transfer functions and input-output block diagrams described in this chapter will prove to be of great value in the analysis of all types of feedback, or closed-loop, systems, beginning with the next chapter on frequency response analysis.

11.2 Examples of Exponential Inputs

The exponential input function e^{st} exhibits different aspects, depending on whether $s = \sigma + j\omega$ is real, imaginary, or complex, as shown in Figures 11.1 and 11.2.

(a) When both σ and ω are zero, $e^{st} = 1$, representing a constant input with time, as for instance the value of a unit step input for $t > 0$.
(b) When only ω is zero, the input is a growing or decaying real exponential, depending on whether σ is positive or negative, as shown in Figure 11.1.
(c) When only σ is zero, the input is complex, having sinusoidal real and imaginary parts and it is represented by a unit vector rotating at speed ωt, as shown in Figure 11.2.

Examination of Figure 11.2 readily reveals the basis for the Euler identity

$$e^{j\omega t} = \cos \omega t + j \sin \omega t \tag{11.1}$$

Here it is seen that the input $e^{j\omega t}$ can be used to represent either a sine wave or a cosine wave, or even both, as the occasion demands. As such $e^{j\omega t}$ in Equation (11.1) will be used to represent sinusoidal inputs in general, and $e^{j\omega t}$ forms the basis for the frequency response transfer function that will be developed in Chapter 12.

When neither σ nor ω are zero, the input is represented on the complex plane by a rotating vector, which is initially unity at $t = 0$, and then grows or decays exponentially with time, as shown in Figure 11.3. This input may be used to represent growing or decaying sinusoids, the latter often occurring naturally as the output of a preceding underdamped system responding to its own step input.

11.3 System Response to an Exponential Input

Here the homogeneous part of the response of a stable system to an input will be considered of minor interest (it usually dies away soon and, in a linear system, in no way affects the particular, or forced, part). The forced part of the response is used here as the basis for developing the notion of a transfer function that relates the output of a linear system to its input.

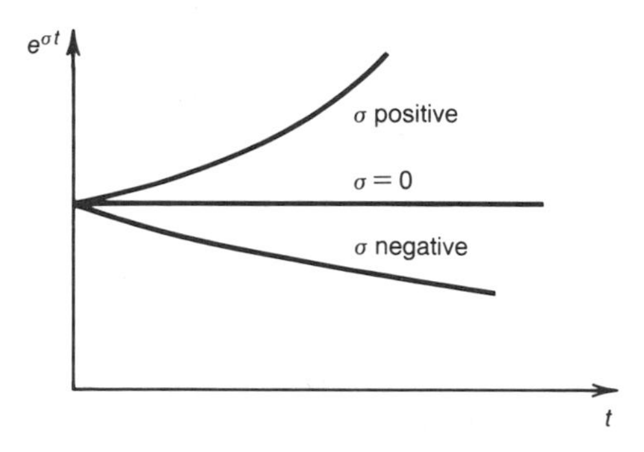

FIGURE 11.1 Variation of $e^{\sigma t}$ with time for different values of σ.

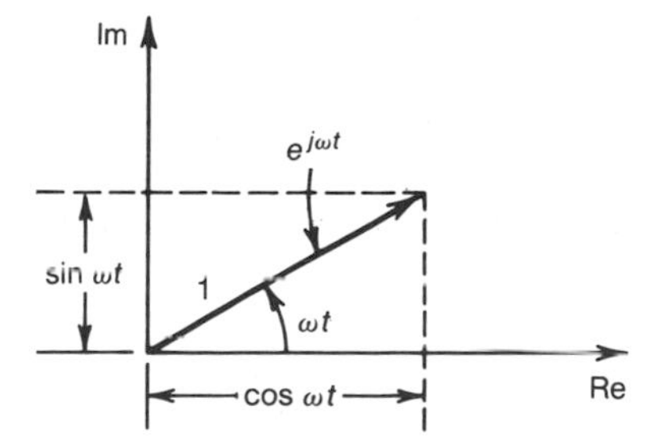

FIGURE 11.2 Complex plane representation of $e^{j\omega t}$.

Recalling that the particular or forced part of the response is of the same form as the input and/or its derivatives[1], we see that the response to an exponential input must also be an exponential of the form $\mathbf{C}e^{st}$, where $\mathbf{C}$ is an undetermined complex coefficient[2]. As an illustration, consider a third-order system having the following input-output differential equation.

$$a_3 \frac{d^3y}{dt^3} + a_2 \frac{d^2y}{dt^2} + a_1 \frac{dy}{dt} + a_0 y = b_1 \frac{du}{dt} + b_0 u \tag{11.2}$$

where the input u and the output y are functions of time. Now with $u = \mathbf{U}e^{st}$, the forced part of the output is taken to be $y = \mathbf{Y}e^{st}$, which is then substituted in Equation (11.2) to yield

$$(a_3 s^3 + a_2 s^2 + a_1 s + a_0)\mathbf{Y}e^{st} = (b_1 s + b_0)\mathbf{U}e^{st} \tag{11.3}$$

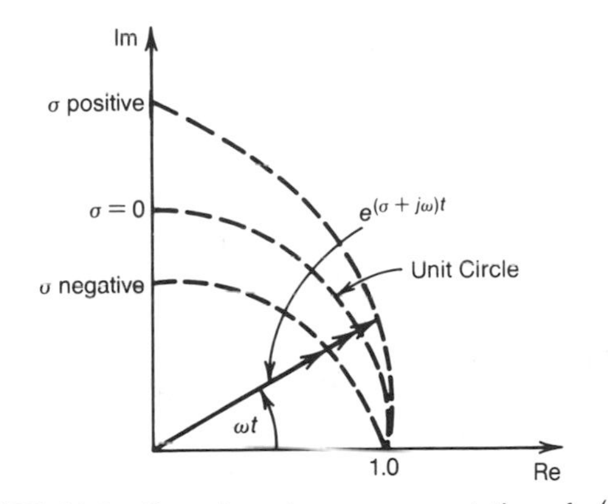

FIGURE 11.3 Complex plane representation of $e^{(\sigma + j\omega)t}$.

[1]The additional exponential term Atest is part of the particular integral for the case when s is precisely equal to one of the system poles. However this additional term, which represents the mechanism by which the particular solution may grow with time, is of no direct interest here.

[2]J. L. Shearer, A. T. Murphy, and H. H. Richardson, *Introduction to System Dynamics*, Addison-Wesley Publishing Co., Reading, Mass., 1967, pp. 324–325.

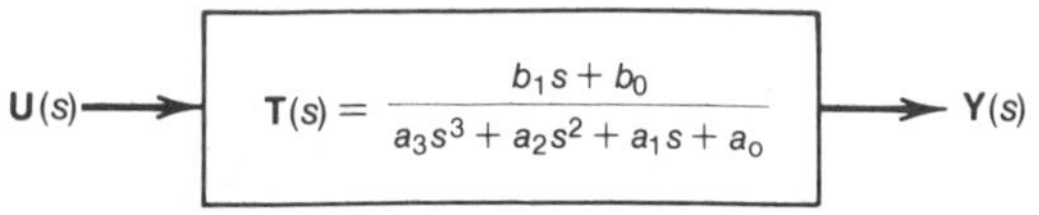

$$\mathbf{U}(s) \longrightarrow \boxed{\mathbf{T}(s) = \frac{b_1 s + b_0}{a_3 s^3 + a_2 s^2 + a_1 s + a_0}} \longrightarrow \mathbf{Y}(s)$$

FIGURE 11.4 Block diagram representation of the transfer function relating **U** and **Y**.

Solving for **Y** yields

$$\mathbf{Y} = \frac{(b_1 s + b_0)\mathbf{U}}{a_3 s^3 + a_2 s^2 + a_1 s + a_0} \tag{11.4}$$

or

$$\mathbf{Y} = \mathbf{T}(s)\mathbf{U} \tag{11.5}$$

where

$$\mathbf{T}(s) = \frac{b_0 s + b_1}{a_3 s^3 + a_2 s^2 + a_1 s + a_0} \tag{11.6}$$

The transfer function $\mathbf{T}(s)$ obtained in this simple fashion happens to be the same as the Laplace transfer function for the system having all initial conditions equal to zero, obtained through the use of the complex Laplace integral (see Appendix 2). The, until now, undetermined complex output amplitude coefficient **Y** is seen to be a complex coefficient that is readily obtained by multiplying the input amplitude coefficient **U** by the transfer function $\mathbf{T}(s)$.

With Equation (11.4) in mind, the system block diagram shown in Figure 11.4 can be used to express symbolically the relationship between the output **Y** and the input **U**. The similarity to the Laplace transform representation for the system with all initial conditions equal to zero is obvious[3]. However, the Laplace interpretation is not necessary here. The use of Laplace transform methods to obtain system responses is unnecessarily tedious and time consuming, given the ready availability of high-speed digital computers with programs that provide solutions in a few seconds of computing time.

It is important to note that the use of transfer functions to model nonlinear systems is not valid unless the system has been linearized and the limitations imposed by the linearization are thoroughly understood.

11.4 Use of Block Diagrams in System Modeling

11.4.1 Elementary Block Diagrams

The development of the concept of a system transfer function operating within a single block to represent an input-output relationship for a given linear system makes it possible to use block diagrams as "building blocks." The building blocks

[3]J. L. Shearer, A. T. Murphy, and H. H. Richardson, *Introduction to System Dynamics*, Addison-Wesley Publishing Co., Reading, Mass., 1967, pp. 395–396.

may then be used to assemble a complete system from a number of individual elementary parts in a manner similar to that employed for the time domain in Chapter 3, except that now the domain is that of the complex exponential coefficient s, also known as the complex frequency variable s.

The four basic block diagrams employed in linear system modeling in the s domain correspond to the time domain processes discussed earlier of summing, multiplying by a constant coefficient, integrating with respect to time, and time delay, as shown in Figure 11.5.

11.4.2 Combining Elementary "Building Block" Diagrams

There are four basic interconnections used to combine the elementary block diagrams: (a) sequential-cascaded, (b) sequential-noncascaded, (c) parallel, and (d) closed loop, as shown in Figure 11.6. Each combination denotes a corresponding algebraic expression involving the elemental transfer functions for the blocks involved.

11.4.3 Illustrative Example

The simple combinations illustrated in Figure 11.6 are then used in various ways to model complete systems, and the corresponding algebraic expressions are developed to give the desired overall input-output transfer function for a given system. As an example, consider the modeling in the s domain of a simple mass, spring, dashpot system being acted upon by an input force applied to the mass, as shown in Figure 11.7. The goal is to obtain the input-output transfer function relating the output $\mathbf{X}$ to the input $\mathbf{F}$ through the use of "block diagram algebra."

The complete detailed block diagram with outlines showing the key equations developed in Figure 11.7 is shown in Figure 11.8.

Using block diagram algebra with the equations given in Figure 11.7, it is seen that

$$\mathbf{V} = \frac{1}{ms}\left(\mathbf{F} - \frac{k}{s}\mathbf{V} - b\mathbf{V}\right) \tag{11.7}$$

FIGURE 11.5 Four basic building blocks.

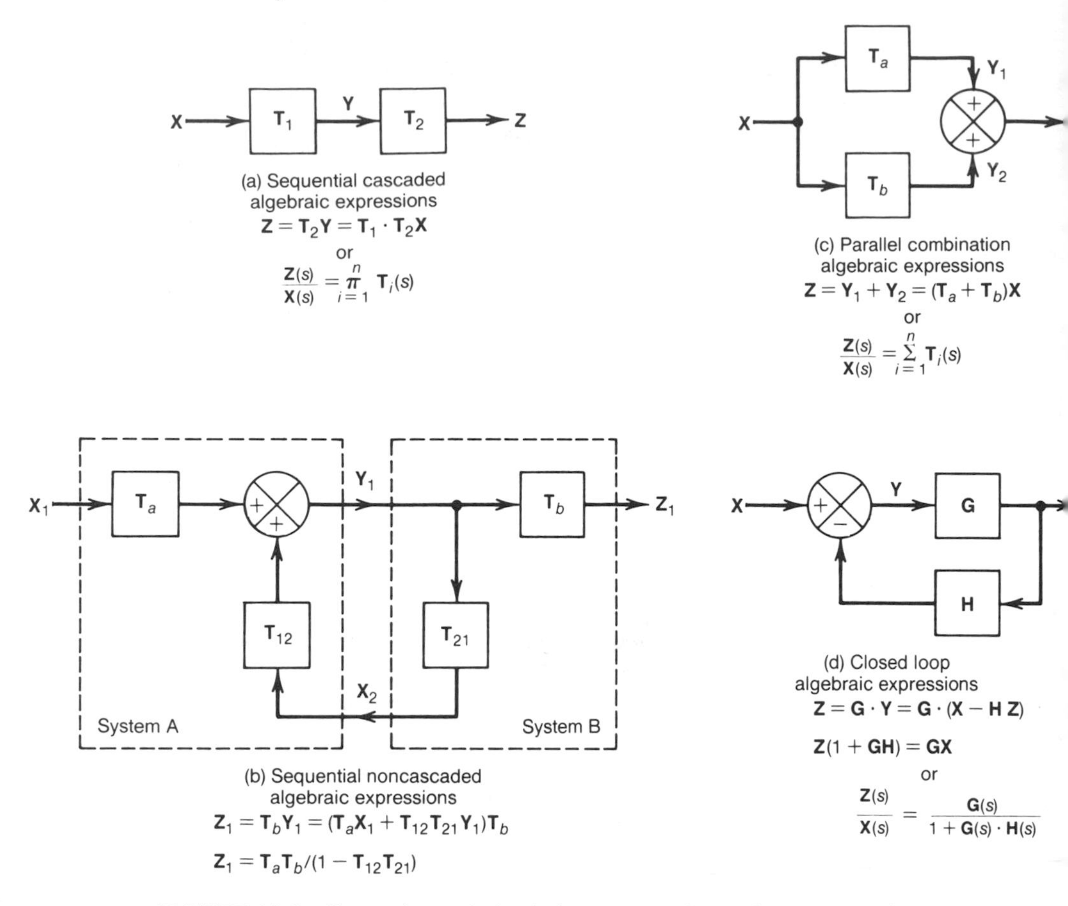

FIGURE 11.6 Illustrations of simple interconnections of elementary blocks.

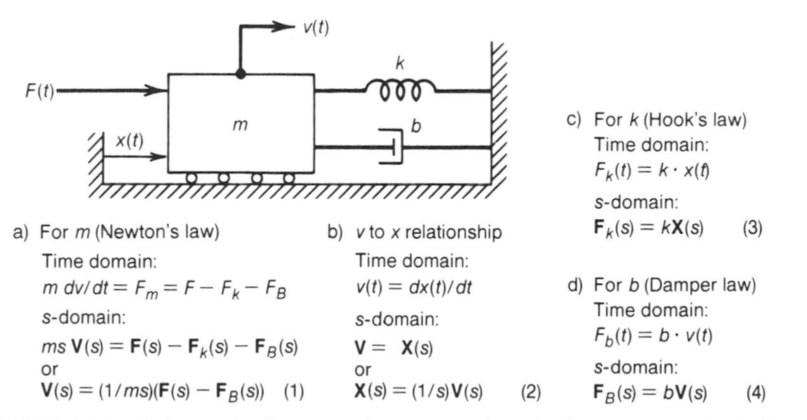

FIGURE 11.7 Schematic diagram of mass, spring, dashpot system including elemental and interconnection equations.

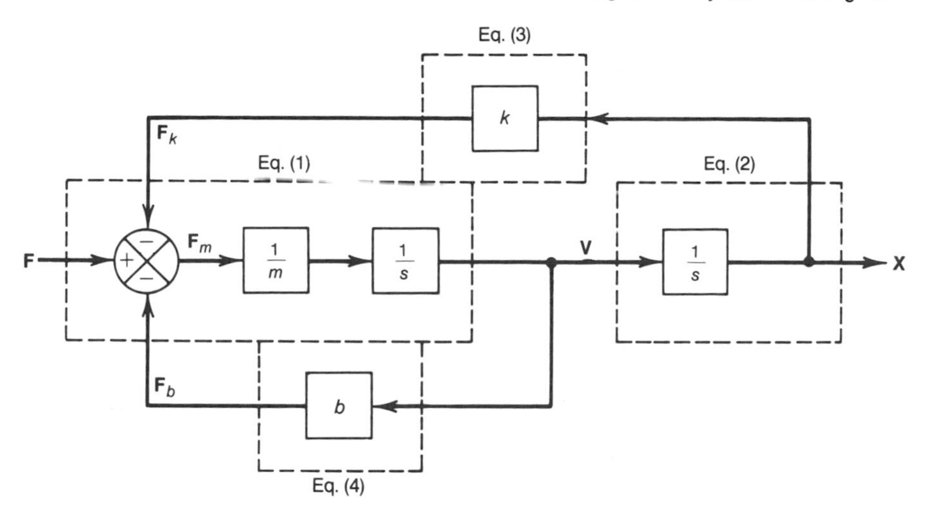

FIGURE 11.8 Complete detailed block diagram of the mass, spring, dashpot system of Figure 11.7

Rearranging yields

$$\left(1 + \frac{k}{ms^2} + \frac{b}{ms}\right) \mathbf{V} = \frac{1}{ms} \mathbf{F} \tag{11.8}$$

Solving for **V** then gives

$$\mathbf{V} = \frac{\dfrac{1}{ms}\mathbf{F}}{1 + \dfrac{k}{ms^2} + \dfrac{b}{ms}} = \frac{s\mathbf{F}}{ms^2 + bs + k} \tag{11.9}$$

Since $\mathbf{X} = (1/s)\mathbf{V}$, the desired transfer function format emerges from

$$\mathbf{X} = \frac{\mathbf{F}}{ms^2 + bs + k}$$

so that we may write,

$$\mathbf{T}(s) = \frac{\mathbf{X}(s)}{\mathbf{F}(s)} = \frac{1}{ms^2 + bs + k} \tag{11.10}$$

And the overall transfer function block diagram is seen in Figure 11.9

Note that the corresponding differential equation relating the output x to the input F in the time domain is

$$m\frac{d^2x}{dt^2} + b\frac{dx}{dt} + kx = F \tag{11.11}$$

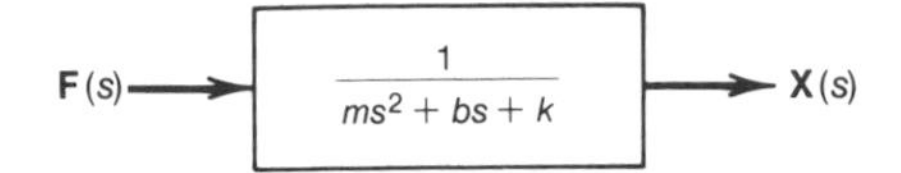

FIGURE 11.9 Overall transfer function block diagram relating **X** to **F**.

where the term involving the ith derivative of x in Equation (11.11) corresponds to the term in the denominator of $\mathbf{T}(s)$ involving the ith power of s.

Note also that the denominator of Equation (11.10), when set equal to zero, is another form of the system characteristic equation

$$ms^2 + bs + k = 0 \tag{11.12}$$

Here the roots (poles) are values of s which satisfy Equation (11.12)

$$m(s - p_1)(s - p_2) = ms^2 + bs + k = 0$$

where

$$p_1, p_2 = [-b \pm \sqrt{b^2 - 4mk}]/2m \tag{11.13}$$

Using the damping ratio $\zeta = b/(2\sqrt{mk})$, when $\zeta \geq 1.0$, the poles correspond to inverse time constants

$$p_1, p_2 = -1/\tau_1, -1/\tau_2 \tag{11.14}$$

and when $0 < \zeta < 1.0$, the poles correspond to the conjugate exponential coefficients for a damped sinusoid

$$p_1, p_2 = -\zeta\omega_n \pm j\omega_n\sqrt{1 - \zeta^2} \tag{11.15}$$

where

$$\omega_n = \sqrt{k/m}$$

11.5 Transfer Functions and Performance Characteristics

The overall system transfer function, for example Equation 11.10 contains all the information needed to predict system response, and to write the system differential equation.

- First, the final value of the steady-state response to a unit step input is the same as the value of the system transfer function when $s \to 0$. Thus, for example, the final steady-state displacement of the mass x is $1/k$ times the new steady value of the suddenly applied force F.
- Second, the change in value of the output response at $t = 0^+$ from its value at $t = 0^-$, when a unit step is applied at $t = 0$, is equal to the limiting value of the transfer function as $s \to \infty$. For example the value of x at $t = 0^+$ is the same as its value at $t = 0^-$.

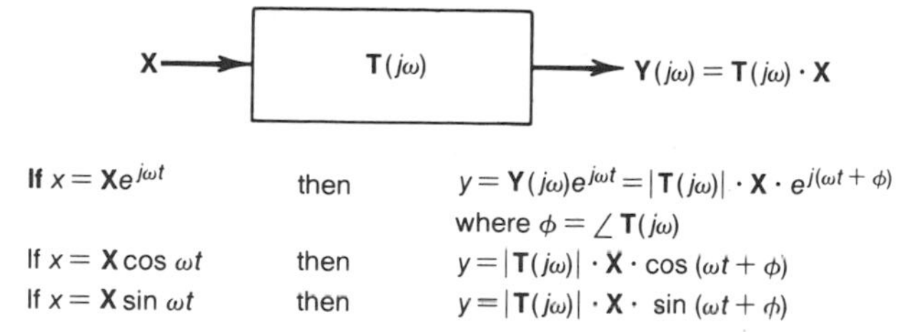

FIGURE 11.10 Summary of the use of $e^{j\omega t}$ as the input for the case of sinusoidal forcing, so that $s = j\omega$.

The two conditions just described correspond to the results obtained by using the initial and final value theorems of Laplace transform theory.

- Third, the amplitude of the steady sinusoidal response to a sinusoidal input of unit amplitude is the magnitude of the transfer function when $s = j\omega$ and the phase angle of the output sinusoid relative to the input sinusoid is the angle of the same transfer function.[4]

It is important to note that using $s = j\omega$ for the case when the input is a cosine wave implies using the real part of $e^{j\omega t}$ for the input; this means that the real part of the response $\mathbf{T}(j\omega)e^{j\omega t}$ is to be taken as the output. Similarly, when the input is a sine wave, the imaginary part of the output $\mathbf{T}(j\omega)e^{j\omega t}$ is to be taken as the output. The case of sinusoidal excitation is summarized in Figure 11.10.

- Fourth, the roots of the denominator of the overall system transfer function, called the system poles, are the roots of the system characteristic equation, each real root corresponding to a system time constant and each complex conjugate pair of roots corresponding to a natural frequency and a damping ratio of the system.

Note finally that the term containing the ith power of s in the denominator of the transfer function corresponds to the term in the system differential equation involving the ith derivative of the output, whereas the term containing the jth power of s in the numerator of the transfer function corresponds to the term in the system differential equation involving the jth derivative of the input. Thus the system differential equation is readily recovered, by simple observation, from inspecting the numerator and denominator of the system transfer function.

11.6 Synopsis

The concept of a linear system transfer function has been developed here for the case when the input to the system is an exponential of the form $u(t) = \mathbf{U}e^{st}$ where $\mathbf{U}$ is a constant that may be complex or real, s is the complex variable $\sigma + j\omega$, and the particular solution is $y(t) = \mathbf{Y}e^{st}$ where $\mathbf{Y}$ is a complex coefficient.

[4] J. L. Shearer, A. T. Murphy, and H. H. Richardson, *Introduction to System Dynamics*, Addison-Wesley Publishing Co., Reading, Mass., 1967, pp. 328–329.

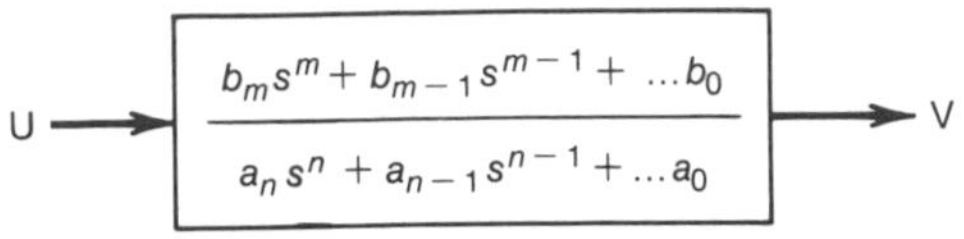

$$U \longrightarrow \boxed{\frac{b_m s^m + b_{m-1} s^{m-1} + \ldots b_0}{a_n s^n + a_{n-1} s^{n-1} + \ldots a_0}} \longrightarrow V$$

FIGURE 11.11 General form of ratio-of-polynomial type system transfer function where $m \le n$.

For a linear system having an input-output differential equation of the form:

$$a_n \frac{d^n y}{dt^n} + a_{n-1} \frac{d^{n-1} y}{dt^{n-1}} + \cdots + a_0 y = b_m \frac{d^m u}{dt^m} + b_{m-1} \frac{d^{m-1} u}{dt^{m-1}} + \cdots b_0 u \tag{11.6}$$

the system transfer function $\mathbf{T}(s)$ is expressed in block diagram form in Figure 11.11:

The denominator of $\mathbf{T}(s)$, when set equal to zero is a form of the system characteristic equation

$$a_n s^n + a_{n-1} s^{n-1} + \cdots a_0 = 0 \tag{11.17}$$

which may also be expressed in terms of its roots (poles)

$$a_n(s - p_1)(s - p_2) \cdots (s - p_n) = 0 \tag{11.18}$$

The real poles p_i correspond to the inverse system time constants

$$p_i = -1/\tau_i \tag{11.19}$$

And each pair of conjugate complex roots p_k, p_{k+1} correspond to the exponential coefficients of a damped sinusoid

$$p_k, p_{k+1} = \sigma_k \pm j\omega_{dk} = \zeta\omega_{nk} \pm j\omega_{nk}\sqrt{1 - \zeta_k^2} \tag{11.20}$$

where

$$\zeta_k = \sigma_k / \sqrt{\sigma_k^2 + \omega_{dk}^2}$$

$$\omega_{nk} = \sqrt{\sigma_k^2 + \omega_{dk}^2}$$

Using $s = j\omega$ results in a transfer function $\mathbf{T}(j\omega)$ which will be used in the next chapter for dealing with the steady response of a system to steady sinusoidal inputs.

Problems

11.1. **(a)** Prepare a detailed transfer function block diagram for the linearized equations of the rotational system of Example 2.7 in Chapter 2.

(b) Express the overall input-output transfer functions with block diagrams for the linearized system relating (i) $\mathbf{T}_K$ to the input torque $\mathbf{T}_e$, with $\mathbf{T}_w = 0$, and (ii) $\mathbf{T}_K$ to the other input torque $\mathbf{T}_w$, with $\mathbf{T}_e = 0$.

11.2. **(a)** Prepare detailed transfer function block diagrams for each of the two subsystems of Example 2.6 in Chapter 2 [i.e., for the subsystems described by Equations (2.41) and (2.42)], after having linearized the nonlinear damper to obtain an incremental damping coefficient $b_{\text{inc}} = \left.\dfrac{df_{\text{NL}}}{dv_{3g}}\right|_{\bar{v}_{3g}}$.

(b) Combine the diagrams prepared in part (a) into a single transfer function block diagram relating $\mathbf{X}_3$ to the input $\mathbf{X}_1$.

(c) Combine the diagrams prepared in part (a) into a single transfer function block diagram relating $\mathbf{X}_2$ to the input $\mathbf{X}_1$.

11.3. **(a)** Prepare a detailed transfer function block diagram for the electric circuit of Example 7.1 in Chapter 7.

(b) Express the overall transfer function with a block diagram relating $\mathbf{E}_0$ to $\mathbf{E}_s$.

11.4. **(a)** Prepare a detailed transfer function block diagram for the linearized state-variable equations of Example 7.3 in Chapter 7.

(b) Express the overall transfer function with a block diagram relating $\mathbf{E}_{2g}$ to $\mathbf{E}_s$.

11.5. **(a)** For the case when $T_w = C_2|\Omega_{2g}|\Omega_{2g}$, prepare a detailed transfer function block diagram for the linearized rotational system of Example 2.7 in Chapter 2.

(b) Combine the linearized equations to develop the input-output system differential equation relating $\hat{\Omega}_{2g}$ to $\hat{T}_e$.

(c) Express the overall transfer function with a block diagram relating $\boldsymbol{\Omega}_{2g}$ to $\mathbf{T}_e$.

11.6. **(a)** Prepare a detailed transfer function block diagram for the motor-driven inertia system of Example 10.2 in Chapter 10.

(b) Express the overall transfer function with a block diagram relating $\boldsymbol{\Omega}_{1g}$ to $\mathbf{E}_s$.

11.7. **(a)** Prepare a detailed transfer function block diagram for the linearized fluid control system of Example 9.2 in Chapter 9.

(b) Express the overall transfer function with a block diagram relating $\mathbf{P}_{2r}$ to $\mathbf{A}_0$.

11.8. **(a)** Prepare a detailed transfer function block diagram for the third-order system having the following three state-variable equations.

$$\frac{dx_1}{dt} = a_{11}x_1 + a_{12}x_2 + b_{11}u_1$$

$$\frac{dx_2}{dt} = a_{21}x_1 + a_{23}x_3$$

$$\frac{dx_3}{dt} = a_{31}x_1 + a_{32}x_2 + a_{33}x_1$$

(b) Use Cramer's rule with determinants to find the transfer function relating $\mathbf{X}_3$ to $\mathbf{U}_1$, and express the overall transfer function with a block diagram for this input-output combination.

(c) For an output $y_1 = c_{12}x_2 + c_{13}x_3$, find the transfer function relating $\mathbf{Y}_1$ to $\mathbf{U}_1$, and express the overall transfer function with a block diagram for this input-output combination.

12
Frequency Analysis

12.1 Introduction

The response of linear systems to sinusoidal inputs forms the basis of an extensive body of theory dealing with the modeling and analysis of dynamic systems. This theory was developed first in the field of communications (telephone and radio) and later became extended and then widely used in the design and development of automatic control systems. The theory is useful not only in determining the sinusoidal response of a system but also in specifying performance requirements (as in a high fidelity sound system or in a radar tracking system), in finding the response to other periodic inputs (square wave, sawtooth wave, etc.)[1], and for predicting the stability of feedback control systems (amplifiers, regulators, and automatic controllers). A graphical portrayal of the input and output sinusoids of a typical linear system is provided in Figure 12.1.

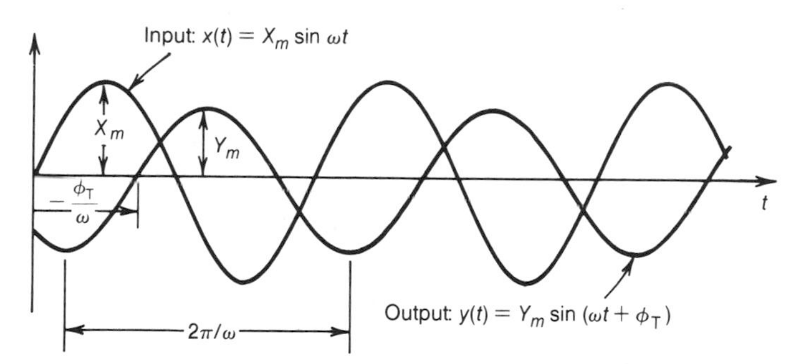

FIGURE 12.1 Sinusoidal input and output waveforms for a linear dynamic system.

[1]See Appendix 1 for a review of the Fourier series and its use in describing a periodic time function in terms of its harmonic components.

12.2 Frequency Response Transfer Functions

It was shown in Chapter 11 how the exponential e^{st} can be used to describe a sinusoid by setting $s = j\omega$; then the concept of a transfer function based on the complex variable s was carried over, for the special case with $s = j\omega$, to yield the frequency response transfer function $\mathbf{T}(j\omega)$, relating the complex amplitude coefficient $\mathbf{Y}$ of the output to the complex amplitude coefficient $\mathbf{X}$ of the input, as shown in Figure 12.2. Thus the transfer function $\mathbf{T}(j\omega)$ provides all the information needed to describe the sinusoidal response of a single-input single-output linear system.

Very often the input amplitude coefficient $\mathbf{X} = Xe^{j\phi_x}$ is real so that $\phi_x = 0$, and $\mathbf{X} = X$, and in the discussion that follows, this will be assumed. As developed in Chapter 11, the use of the block diagram shown in Figure 12.2 of course implies that when $x(t) = Xe^{j\omega t}$, the output is given by

$$y(t) = \mathbf{T}(j\omega)Xe^{j\omega t} = \mathbf{Y}(j\omega)e^{j\omega t}, \quad \text{where } \mathbf{Y}(j\omega) = \mathbf{T}(j\omega)X$$

Since $\mathbf{T}(j\omega) = T(\omega)e^{j\phi_T}$, where $T(\omega) = |T(j\omega)|$ and $\phi_T(\omega) = \underline{/\mathbf{T}(j\omega)}$, then $y(t) = T(\omega)Xe^{j(\omega t+\phi_T)}$. Also implied is that when $x(t) = X \sin \omega t$, $y(t) = T(\omega)X \sin(\omega t + \phi_T)$ and when $x(t) = X \cos \omega t$, $y(t) = T(\omega)\, X \cos(\omega t + \phi_T)$, as shown in Figure 11.10 (repeated here for convenience).

Thus the magnitude of the output sinusoid is equal to the magnitude of the input sinusoid times the magnitude of the complex transfer function $\mathbf{T}(j\omega)$, and the phase of the output sinusoid relative to the input sinusoid is the angle of the transfer function. In the discussion that follows, the salient features of $\mathbf{T}(j\omega)$ as described by its magnitude $T(\omega)$ and its phase $\phi_T(\omega)$ will be discussed in detail with emphasis on how they vary with frequency for various types of systems.

12.3 Bode Diagrams

Because of the pioneering work by H. W. Bode on feedback amplifier design[2], the use of logarithmic charts to portray the magnitude and phase characteristics has led to the development of Bode diagrams. These diagrams are now widely used in describing the dynamic performance of linear systems.

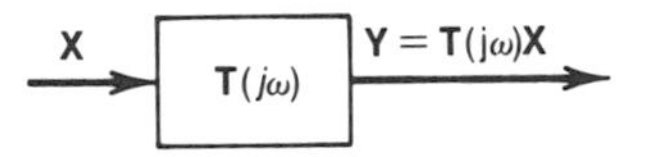

FIGURE 12.2 Block diagram representation of frequency response transfer function

[2]H. W. Bode, *Network Analysis and Feedback Amplifier Design*, Van Nostrand, Princeton, N. J., 1945.

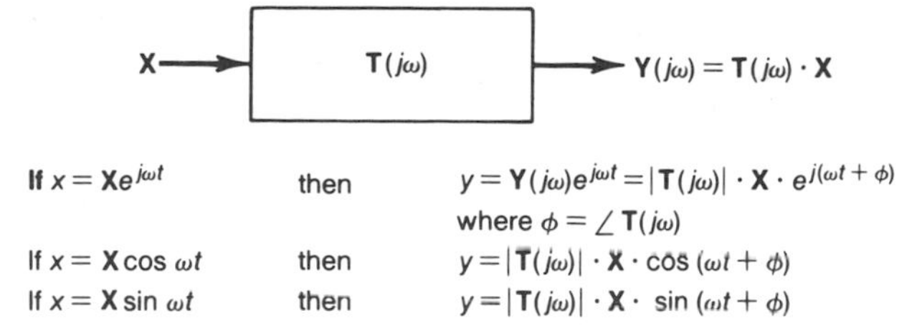

FIGURE 11.10 Summary of the use of e^{st} as the input for the case of sinusoidal forcing (repeated from Chapter 11)

Since the transfer function $\mathbf{T}(j\omega)$ is complex, it may be expressed in complex exponential form.

$$\mathbf{T}(j\omega) = Te^{j\phi_T} \tag{12.1}$$

where T and ϕ_T, both functions of frequency, are often expressed as follows:

$$T(\omega) = |\mathbf{T}(j\omega)| \tag{12.2}$$

and

$$\phi_T(\omega) = \underline{/\mathbf{T}(j\omega)} \tag{12.3}$$

or, using Cartesian coordinates, the transfer function may be expressed in terms of its real and imaginary parts.

$$\mathbf{T}(j\omega) = \text{Re}\,[\mathbf{T}(j\omega)] + j\,\text{Im}\,[\mathbf{T}(j\omega)] = T\cos\phi_T + jT\sin\phi_T \tag{12.4}$$

Further discussion of coordinates to be used for complex plane plots will be deferred to the next section.

With the Bode Diagrams, the magnitude $T(\omega)$ versus frequency ω and the phase angle $\phi_T(\omega)$ versus frequency ω characteristics are drawn on separate plots that share a logarithmic frequency axis. The amplitude axis is either logarithmic (log T) or quasilogarithmic (decibels), whereas the phase axis is linear.

The emphasis here will be on the use of Bode diagrams for analytical studies. However, these diagrams are also used in portraying the results of frequency response measurements that are obtained experimentally by applying an input sinusoid through a range of frequencies and measuring the resultant magnitude and phase of the output signal at each of the applied input frequencies. Each measurement is made after a sinusoidal steady state has been achieved.

The procedure for preparing a set of Bode diagram curves analytically is as follows:

Step 1. Determine the system transfer function $\mathbf{T}(s)$.
Step 2. Convert to sinusoidal transfer function $\mathbf{T}(j\omega)$ by letting $s = j\omega$.
Step 3. Develop expressions for the magnitude, $T(\omega) = |\mathbf{T}(j\omega)|$, and phase angle, $\phi_T(\omega) = \underline{/\mathbf{T}(j\omega)}$ of $\mathbf{T}(j\omega)$.

Step 4. Plot $T(\omega)$ versus ω on log-log paper (which would be log T versus log ω on plain linear graph paper), and plot ϕ_T versus ω on semi-log paper (which would be ϕ_T versus log ω on plain linear paper). For plotting the magnitude in decibels, increase the log scale by a factor of 20 and plot on a linear scale. (When $T(\omega) = 10$, $\log T(\omega) = 1.0$ and the magnitude in decibels is 20). Because converting back to actual magnitude from decibels is messy, using log-log paper to plot $T(\omega)$ versus ω is worth the trouble of providing log paper for the magnitude curve[3].

This procedure is illustrated by Example 12.1.

EXAMPLE 12.1

Prepare Bode diagram curves for the first-order system described by the following differential equation.

$$\frac{dy}{dt} + \frac{y}{\tau} = \frac{kx}{\tau} \tag{12.5}$$

where $x(t)$ and $y(t)$ are the input and output, respectively.

Solution

Step 1. Transforming to the s-domain yields

$$s\mathbf{Y}(s) + \frac{\mathbf{Y}(s)}{\tau} = \frac{k\mathbf{X}(s)}{\tau} \tag{12.6}$$

The system transfer function is

$$\mathbf{T}(s) = \frac{\mathbf{Y}(s)}{\mathbf{X}(s)} = \frac{k}{\tau s + 1} \tag{12.7}$$

Step 2. The sinusoidal transfer function is

$$\mathbf{T}(j\omega) = \mathbf{T}(s)\big|_{s=j\omega} = \frac{k}{j\omega\tau + 1} \tag{12.8}$$

Step 3. To simplify obtaining expressions for $T(\omega)$ and $\phi_T(\omega)$, note that $\mathbf{T}(j\omega)$ can be presented as a ratio of complex functions $\mathbf{N}(j\omega)$ and $\mathbf{D}(j\omega)$.

$$\mathbf{T}(j\omega) = \frac{\mathbf{N}(j\omega)}{\mathbf{D}(j\omega)} = \frac{N(\omega)e^{j\phi_N(\omega)}}{D(\omega)e^{j\phi_D(\omega)}} \tag{12.9}$$

[3]Plotting the decibel magnitude curve on linear graph paper produces a linear interpolation between decade values (20, 40, 60, etc.); it is thus necessary to divide the decibel reading by 20 and then find the inverse log to determine the actual magnitude $T(\omega)$ from a decibel plot. Conversion charts and tables to expedite this process are available, as discussed in the *Handbook of Automation, Computation, and Control*, Vol. I, by E. M. Grabbe, S. Ramo, and D. E. Wooldridge, John Wiley and Sons, New York, 1958, Sect. 21, pp. 43–46.

or

$$\mathbf{T}(j\omega) = \frac{N(\omega)}{D(\omega)} e^{j[\phi_N(\omega) - \phi_D(\omega)]} \tag{12.10}$$

so that the magnitude $T(\omega)$ is given by

$$T(\omega) = \frac{N(\omega)}{D(\omega)} \tag{12.11}$$

In this case $N(\omega) = k$ and $D(\omega) = |j\omega\tau + 1|$, so that $T(\omega)$ is

$$T(\omega) = \frac{k}{|j\omega\tau + 1|} = \frac{k}{\sqrt{1 + \omega^2\tau^2}} \tag{12.12}$$

and the phase angle ϕ_T is given by

$$\phi_T(\omega) = \phi_N(\omega) - \phi_D(\omega) \tag{12.13}$$

where

$$\phi_N(\omega) = \tan^{-1}(0/k) = 0 \tag{12.14}$$

and

$$\phi_D(\omega) = \tan^{-1}(\omega\tau/1) = \tan^{-1}(\omega\tau) \tag{12.15}$$

Hence

$$\phi_T(\omega) = 0 - \tan^{-1}(\omega\tau) = -\tan^{-1}(\omega\tau) \tag{12.16}$$

Step 4. The Bode diagrams for amplitude and phase can now be plotted using calculations based on the expressions for $T(\omega)$ and $\phi_T(\omega)$ developed in Step 3. The resulting magnitude and phase curves for this first order system are shown in Figure 12.3. ■

In a preliminary system analysis it is often sufficient to approximate the amplitude and phase curves by the use of simple straight-line asymptotes, which are very easily sketched by hand. These straight-line asymptotes are included in Figure 12.3. The first asymptote for the magnitude curve corresponds to the case when ω is very small, approaching zero. Using Equation (12.12) for the magnitude as $\omega \to 0$, yields

$$\lim_{\omega \to 0} (\log T(\omega)) = \lim_{\omega \to 0} \left(\log \frac{k}{\sqrt{1 + \omega^2\tau^2}} \right) = \log k \tag{12.17}$$

The second asymptote for the amplitude curve is based on the case when ω is very large. Using Equation (12.12) with ω approaching infinity yields

$$\lim_{\omega \to \infty} (\log T(\omega)) = \lim_{\omega \to \infty} \left(\log \frac{k}{\sqrt{1 + \omega^2\tau^2}} \right) = \log k - \log \omega\tau \tag{12.18}$$

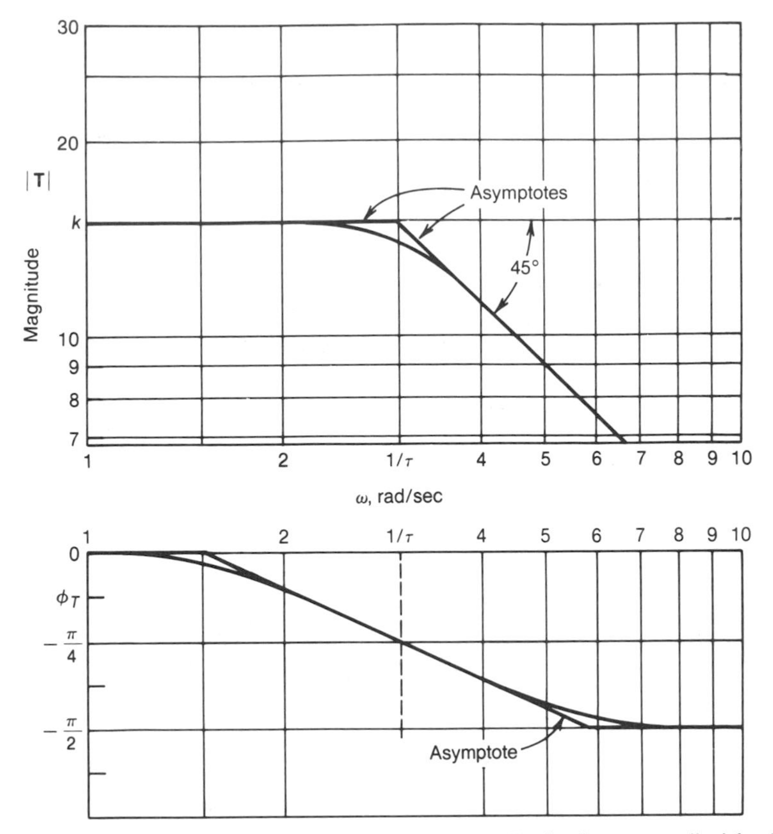

FIGURE 12.3 Amplitude and phase curves for the Bode diagram called for in Example 12.2

Hence at high frequencies

$$\log (T(\omega)) = \log k - \log \tau - \log \omega \tag{12.19}$$

Note that the slope of the second asymptote on log-log paper is minus 45 deg, which corresponds to -6 decibels per octave on the decibel plot.

The two asymptotes described by Equations (12.17) and (12.19) thus constitute the asymptotic magnitude versus frequency characteristic. The maximum error incurred by using the asymptotic approximation occurs at the "break frequency" where $\omega = 1/\tau$, i.e., where the two asymptotes intersect, also sometimes referred to as the "corner frequency."

The asymptotic approximation for the phase angle curve consists of three straight lines. The first, for very low frequency is given by

$$\lim_{\omega \to 0} (-\tan^{-1} (\omega\tau)) = -\tan^{-1} (0) = 0 \tag{12.20}$$

The second is a straight line tangent to the phase angle curve at its corner frequency, having a slope given by

$$\frac{d\phi_T(\omega)}{d(\log \omega)} = \frac{d\phi_T(\omega)}{d\omega} \frac{d\omega}{d(\log \omega)} = -\frac{\omega\tau}{1 + \omega^2\tau^2} \tag{12.21}$$

Thus at the corner frequency $\omega = 1/\tau$, the phase angle is -45 deg, and the slope is given by

$$\left.\frac{d\phi_T(\omega)}{d(\log \omega)}\right|_{\omega = 1/\tau} = -1/2 \tag{12.22}$$

The third asymptotic approximation occurs at very large values of ω where

$$\lim_{\omega \to \infty} (-\tan^{-1}(\omega\tau)) = -\tan^{-1}(\infty) = -\pi/2 \tag{12.23}$$

It can be seen from Figure 12.3 that the system behaves like a low-pass filter, having an output that drops off increasingly as the corner frequency is exceeded. Thus the system is unable to respond at all as the frequency approaches infinity. The range of frequencies over which this system can deliver an effective output is referred to as its bandwidth; the bandwidth of a low pass filter extends from very low frequency up to its corner frequency, where the output has dropped to about 0.7 of its low frequency value. Thus bandwidth is determined by the value of $1/\tau$; the smaller the time constant τ, the larger the bandwidth.

This means that the bandwidth needs to be large enough (i.e., the time constant small enough) to enable the output to follow the input without excessive attenuation. However, providing extended bandwidth is not only costly but also makes it possible for the system to follow undesirable high-frequency noise in its input. These conflicting requirements require either some compromise with the use of this system or finding a better system (probably higher order) that will follow the desired range of input frequencies but filter out the undesired higher frequency noise.

Example 12.2 is provided to reinforce the use of the required analytical procedures and to demonstrate how a higher-order system might be devised to help meet these conflicting design requirements.

EXAMPLE 12.2

Prepare Bode diagrams for a second-order system described by Equation (12.24).

$$\frac{d^2y}{dt^2} + 2\zeta\omega_n \frac{dy}{dt} + \omega_n^2 y = \omega_n^2 x \tag{12.24}$$

Step 1. The system transfer function is

$$\mathbf{T}(s) = \frac{\mathbf{Y}(s)}{\mathbf{X}(s)} = \frac{\omega_n^2}{s^2 + 2\zeta\omega_n s + \omega_n^2} \tag{12.25}$$

Step 2. The sinusoidal transfer function is

$$\mathbf{T}(j\omega) = \frac{\omega_n^2}{(j\omega)^2 + 2\zeta\omega_n j\omega + \omega_n^2} = \frac{1}{1 - (\omega/\omega_n)^2 + j2\zeta(\omega/\omega_n)} \tag{12.26}$$

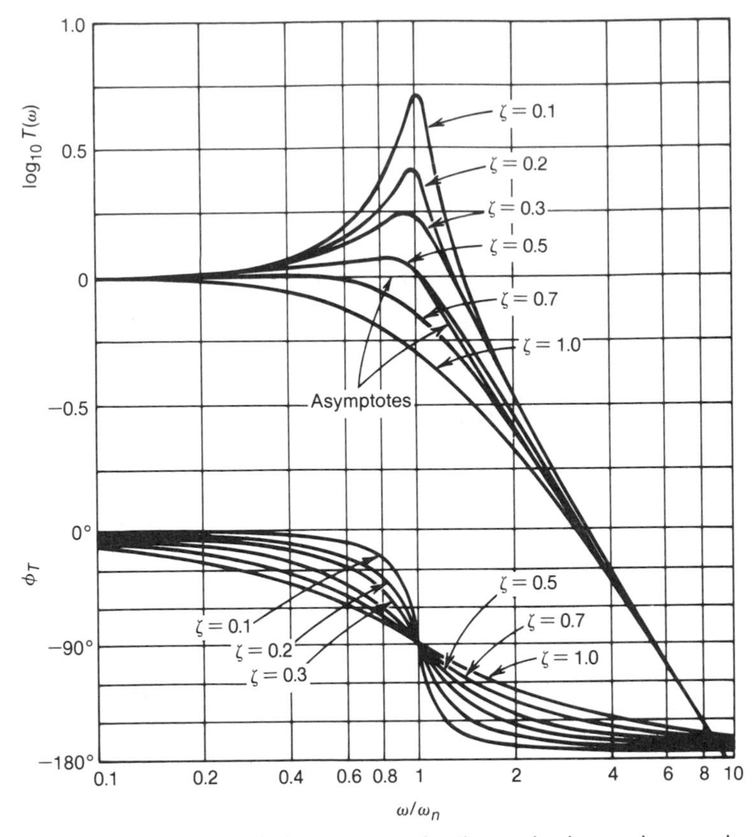

FIGURE 12.4 Magnitude and phase curves for the underdamped second-order system of Example 12.2

Step 3. Develop expressions for $T(\omega)$ and $\phi_T(\omega)$. First the magnitude

$$T(\omega) = \frac{1}{\sqrt{[1 - (\omega/\omega_n)^2]^2 + 4\zeta^2(\omega/\omega_n)^2}} \tag{12.27}$$

Then the phase angle is

$$\phi_T(\omega) = -\tan^{-1}\frac{2\zeta(\omega/\omega_n)}{1 - (\omega/\omega_n)^2} \tag{12.28}$$

Step 4. Prepare the Bode diagrams as shown in Figure 12.4. ■

In Example 12.2 a normalized frequency scale based on (ω/ω_n) has been employed so that the frequency response of underdamped ($\zeta < 1$) linear systems can be emphasized. For overdamped second-order systems, the characteristic equation has real roots and two first-order terms are used. The procedure is given in Example 12.3. Note that the slope of the high-frequency straight-line approximation is now

-2 on the log-log plot (-12 db per octave on the decibel plot), and that the high-frequency phase angle is -180 deg and the corner frequency phase angle is -90 deg.

EXAMPLE 12.3

Prepare asymptotic magnitude and phase characteristics for the system having the following transfer function:

$$\mathbf{T}(s) = \frac{b_1 s + b_0}{s(a_2 s^2 + a_1 s + a_0)} \tag{12.29}$$

Solution

The numerator has a single root $r = -(b_0/b_1)$ so that its time constant is

$$\tau_1 = -(1/r) = (b_1/b_0)$$

The denominator has three roots, or poles, p_1, p_2, and p_3, with $p_1 = 0$. Assuming that the second-order factor in the denominator is overdamped (in contrast to the underdamped case used in Example 12.2) the damping ratio $\zeta = a_1/2\sqrt{a_0 a_2}$ is greater than one.

Thus the poles associated with this second-order factor are real, and it may be expressed in terms of its time constants τ_2 and τ_3.

$$a_2 s^2 + a_1 s + a_0 = a_0(\tau_2 s + 1)(\tau_3 s + 1)$$

where

$$\tau_2 = -\frac{1}{p_2} = \frac{a_1}{2a_0}\left(1 + \sqrt{1 - \frac{4a_0 a_2}{a_1^2}}\right)$$

and

$$\tau_3 = -\frac{1}{p_3} = \frac{a_1}{2a_0}\left(1 - \sqrt{1 - \frac{4a_0 a_2}{a_1^2}}\right)$$

Then the transfer function may be written as follows

$$\mathbf{T}(s) = \frac{b_0}{a_0}\frac{(\tau_1 s + 1)}{s(\tau_2 s + 1)(\tau_3 s + 1)} \tag{12.30}$$

And the sinusoidal transfer function is

$$\mathbf{T}(j\omega) = \frac{b_0}{a_0}\frac{(j\omega\tau_1 + 1)}{j\omega(j\omega\tau_2 + 1)(j\omega\tau_3 + 1)} \tag{12.31}$$

which can be expressed in terms of five individual transfer functions:

$$\mathbf{T}(j\omega) = \frac{\mathbf{N}_1(j\omega)\mathbf{N}_2(j\omega)}{\mathbf{D}_1(j\omega)\mathbf{D}_2(j\omega)\mathbf{D}_3(j\omega)} \tag{12.32}$$

where $\mathbf{N}_1 = b_0/a_0$; $\mathbf{N}_2 = j\omega\tau_1 + 1$; $\mathbf{D}_1 = j\omega$; $\mathbf{D}_2 = j\omega\tau_2 + 1$; and $\mathbf{D}_3 = j\omega\tau_3 + 1$.

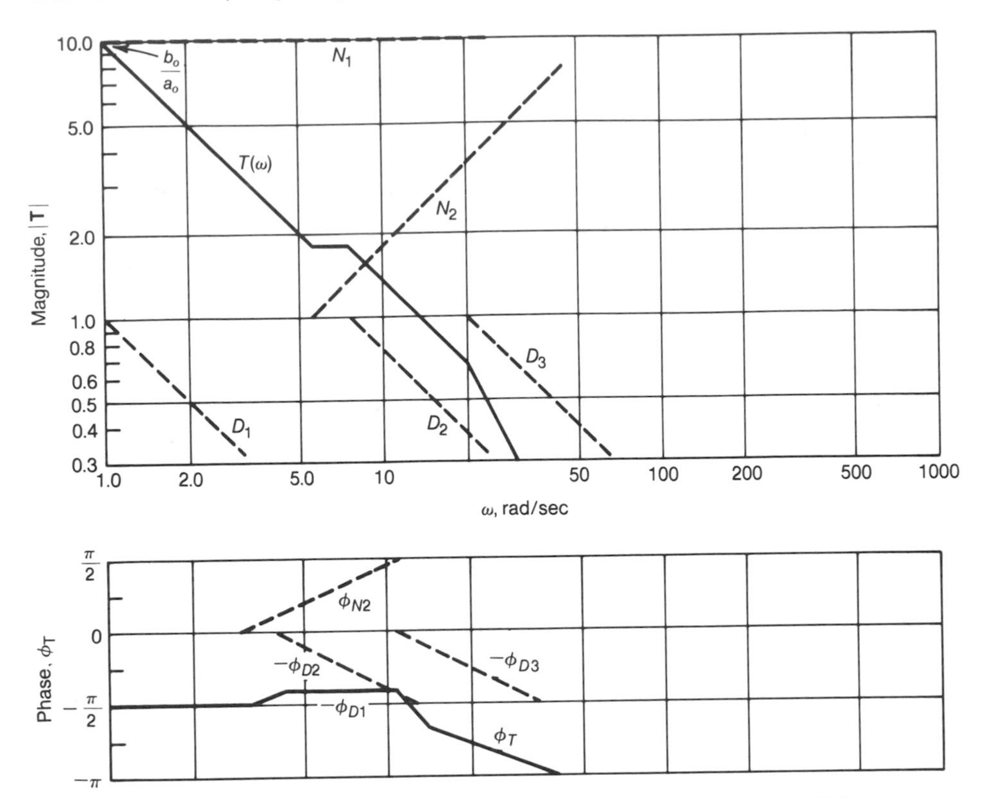

FIGURE 12.5 Bode diagram characteristics for magnitude and phase of the system in Example 12.3.

Now the individual magnitudes and phase angles may be employed to give

$$\mathbf{T}(j\omega) = \frac{N_1(\omega)e^{j\phi_{N_1}}N_2(\omega)e^{j\phi_{N_2}}}{D_1(\omega)e^{j\phi_{D_1}}D_2(\omega)e^{j\phi_{D_2}}D_3(\omega)e^{j\phi_{D_3}}}$$

or

$$\mathbf{T}(j\omega) = \frac{N_1N_2}{D_1D_2D_3}\, e^{j(\phi_{N_1} + \phi_{N_2} - \phi_{D_1} - \phi_{D_2} - \phi_{D_3})} \tag{12.33}$$

The magnitude of the overall transfer function is

$$T(\omega) = \frac{N_1(\omega)N_2(\omega)}{D_1(\omega)D_2(\omega)D_3(\omega)} \tag{12.34}$$

where $N_1 = b_0/a_0$; $N_2 = \sqrt{\omega^2\tau_1^2 + 1}$; $D_1 = \omega$; $D_2 = \sqrt{\omega^2\tau_2^2 + 1}$; and $D_3 = \sqrt{\omega^2\tau_3^2 + 1}$, and the phase angle of the overall transfer function is

$$\phi_T = \phi_{N_1} + \phi_{N_2} - \phi_{D_1} - \phi_{D_2} - \phi_{D_3} \tag{12.35}$$

where $\phi_{N_1} = 0$; $\phi_{N_2} = \tan^{-1}(\omega\tau_1)$; $\phi_{D_1} = 90$ deg.; $\phi_{D_2} = \tan^{-1}(\omega\tau_2)$; and $\phi_{D_3} = \tan^{-1}(\omega\tau_3)$.

The magnitude and phase curves for the overall transfer function are then prepared by summing the ordinates of the individual transfer functions, using the straight-line asymptotic approximations developed earlier. Usually the small departures at the corner frequencies are not of interest, so they can be ignored. Also it is very easy to implement these procedures on a digital computer and use an *x-y* plotter to by-pass a lot of tedious calculation and plotting by hand, in which case the results will be accurate at all frequencies.

The magnitude and phase curves using straight-line asymptotic approximations are presented in Figure 12.5, where $\tau_1 > \tau_2 > \tau_3$. ■

12.4 Polar Plot Diagrams

In Section 12.3 the magnitude $T(\omega)$ and phase angle $\phi_T(\omega)$ characteristics of the frequency response transfer function $\mathbf{T}(j\omega)$ were depicted by separate curves on a Bode diagram. Since they are each unique functions of frequency ω, they are also directly related to each other. This relationship was expressed earlier by Equation (12.4) which shows how the transfer function $\mathbf{T}(j\omega)$ may be represented on the complex plane, either in cartesian or polar coordinates.

$$\mathbf{T}(j\omega) = \text{Re}\,[\mathbf{T}(j\omega)] + j\,\text{Im}\,[\mathbf{T}(j\omega)] = T\cos\phi_T + jT\sin\phi_T \quad (12.4)$$

For many applications the use of polar plots on polar-coordinate graph paper is preferred, but for actual plotting on ordinary graph paper and for computer-generated plots on an x-y plotter, cartesian coordinates are preferred. Figure 12.6 illustrates how $\mathbf{T}(j\omega)$ at a given frequency ω appears on the complex plane. Note that this form of representing $\mathbf{T}(j\omega)$ does not show the frequency explicitly.

The end-points of the vectors $\mathbf{T}(j\omega)$ plotted for successive values of ω from zero to infinity form the basis of a characteristic curve called the polar plot. This form of representing $\mathbf{T}(j\omega)$ has proved to be very useful in the development of system stability theory and in the experimental evaluation and design of closed-loop control systems.

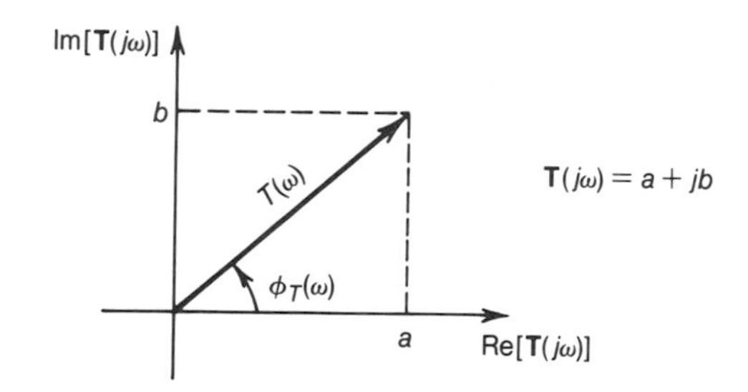

FIGURE 12.6 Complex plane representation of $\mathbf{T}(j\omega)$.

EXAMPLE 12.4

Prepare a polar plot for a first-order system having the system transfer function given in Equation (12.36).

$$\mathbf{T}(s) = \frac{10}{2s + 1} \tag{12.36}$$

The frequency response transfer function is

$$\mathbf{T}(j\omega) = \frac{10}{2j\omega + 1} \tag{12.37}$$

On the one hand the polar plot represents graphically the relationship between the magnitude $T(\omega)$ and the phase angle $\phi_T(\omega)$; on the other hand it shows the relationship between the real part $T \cos \phi_T$ and the imaginary part $T \sin \phi_T$ of the transfer function $\mathbf{T}(j\omega)$. Choosing the latter, the expressions for the real and imaginary parts of $\mathbf{T}(j\omega)$ are found to be

$$\text{Re}\,[\mathbf{T}(j\omega)] = \frac{10}{4\omega^2 + 1} \tag{12.38}$$

and

$$\text{Im}\,[\mathbf{T}(j\omega)] = -\frac{20\omega}{4\omega^2 + 1} \tag{12.39}$$

The numerical results for computations carried out at several frequencies are given in Table 12.1, which also includes corresponding values for $T(\omega)$ and $\phi_T(\omega)$. The polar plot prepared from Table 12.1 is shown in Figure 12.7. It can be shown analytically that this curve is a semi-circle with its center at the 5.0 point on the real axis.

Note that the sign convention for phase angle is counter-clockwise positive. Thus the phase angle $\phi_T(\omega)$ for this system is always negative. Although each computed point for this polar plot corresponds to a certain frequency, the frequency, ω, does not appear explicitly as an independent variable (unless successive tick-marks are labeled for the curve, as was done in Figure 12.7). ■

TABLE 12.1 Numerical data for plot of $\mathbf{T}(j\omega) = 10/(2j\omega + 1)$.

ω, rad/sec	**0.0**	**0.1**	**0.25**	**0.5**	**1.0**	**5.0**	**10.0**	. . .	∞
Re [**T**(*j*ω)]	10	9.62	8.0	5.0	2.0	0.10	0.025	. . .	0
Im [**T**(*j*ω)]	0	−1.92	−4.0	−5.0	−4.0	−1.0	−0.5	. . .	0
$T(\omega)$	10	9.8	8.94	7.07	4.47	1.0	0.5	. . .	0
$\phi_T(\omega)$, deg	0	−11.3	−26.6	−45.	−63.4	−84.3	−87.1	. . .	−90

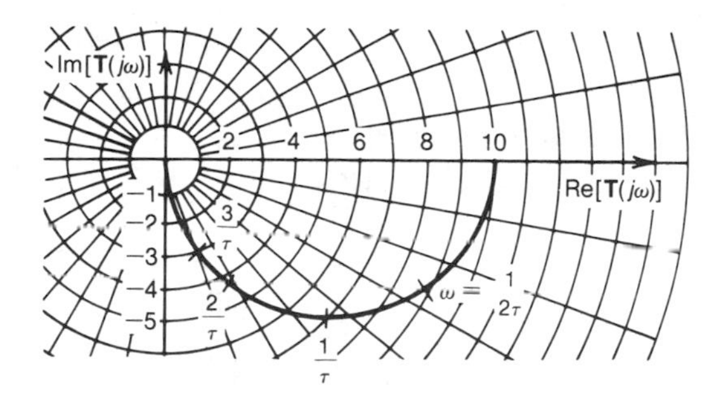

FIGURE 12.7 Polar plot for $\mathbf{T}(j\omega) = 10/(2j\omega + 1)$.

For complicated transfer functions the process for computing the values for the data points is straight-forward but becomes tedious. However it is an easy job for the digital computer, especially when a completely debugged program (there are many on the market) is available.

Polar plot diagrams display most of the same information as Bode diagrams, especially if the frequency tick-marks are included, but in a more compact form. Although they are not as easy to sketch by hand as the straight-line asymptotes of the Bode diagrams, they are especially useful in determining the stability of feedback control systems, discussed in Chapter 13.

A few comments and hints are in order here to aid in the preparation and/or verification (especially with computer-generated data) of polar plots. First, the points for zero frequency and infinite frequency are readily determined. The point for zero frequency is obtained by using $\omega = 0$ in $\mathbf{T}(j\omega)$. The point for infinite frequency is always at the origin because of the inability of real physical systems to have any response to very high frequencies. Stated mathematically

$$\lim_{\omega\to\infty} \mathbf{T}(j\omega) = 0 \tag{12.40}$$

Second, as the frequency approaches infinity, the terminal phase angle is $(\pi/2)(m - n)$ where m is the order of the numerator and n is the order of the denominator (assuming that the transfer function is a ratio of polynomials). For example if

$$\mathbf{T}(s) = \frac{b_0 + b_1 s + \cdots + b_m s^m}{a_0 + a_1 s + \cdots + a_n s^n} \qquad m \leq n$$

then the limit of the phase angle for frequency approaching infinity is given by

$$\lim_{\omega\to\infty} \phi_T(\omega) = (m - n)\,\frac{\pi}{2} \tag{12.41}$$

Suggestion: Use this rule to check the validity of the curve shown in Figure 12.7.

12.5 Synopsis

The concept of using a transfer function to describe a linear system has been applied in this chapter to systems subjected to steady sinusoidal inputs through the use of $s = j\omega$. The steady state output $\mathbf{Y}e^{j\omega t}$ of a linear system is the product of the input expressed by $\mathbf{U}e^{j\omega t}$ multiplied by the sinusoidal transfer function $\mathbf{T}(j\omega)$. Since the phase of the input $\angle \mathbf{U}$ is usually zero, $\mathbf{U} = U$, and then the phase angle of the output $\angle \mathbf{Y}$ is equal to the phase angle of the transfer function $\angle \mathbf{T}(j\omega)$. Thus when the input is a sine wave, $u(t) = \text{Im}\,[Ue^{j\omega t}] = U \sin \omega t$, the output is a phase shifted sine wave of the same frequency, $y(t) = \text{Im}\,[Ye^{j(\omega t + \phi)}] = Y \sin(\omega t + \phi)$ where $Y = U|\mathbf{T}(j\omega)|$ and $\phi = \angle \mathbf{Y} = \angle \mathbf{T}(j\omega)$.

Two techniques for graphing the variation of the transfer function $\mathbf{T}(j\omega)$ with varying frequency ω were introduced: (a) Bode diagrams showing amplitude and phase versus frequency on separate plots and (b) polar plots showing amplitude and phase in a single plot. Several rules have been provided to assist in the routine preparation of frequency response characteristics and examples have been given to illustrate the techniques involved.

The frequency response methods presented in this chapter are very useful in studying stability and dynamic response characteristics of automatic control systems. They are also commonly used in marketing and specifying products for use in industry.

Problems

12.1. Sketch asymptotic Bode diagrams for the following transfer functions:

(a) $\mathbf{T}(s) = k/s$
(b) $\mathbf{T}(s) = k/s(\tau s + 1)$
(c) $\mathbf{T}(s) = ks/(\tau s + 1)$
(d) $\mathbf{T}(s) = k/(\tau s + 1)^2$
(e) $\mathbf{T}(s) = k(\tau_1 s + 1)/[s(\tau_2 s + 1)]$
(f) $\mathbf{T}(s) = 10s(0.1s + 1)/(s + 1)^2$

12.2. Sketch polar plots for the following transfer functions:

(a) $\mathbf{T}(s) = k/s$
(b) $\mathbf{T}(s) = k/s(\tau s + 1)$
(c) $\mathbf{T}(s) = \omega_n^2/(s^2 + 2\zeta\omega_n s + \omega_n^2)$
(d) $\mathbf{T}(s) = ke^{-s\tau_0}$
(e) $\mathbf{T}(s) = (s + 1)/(s + 2)(s + 5)$
(f) $\mathbf{T}(s) = 100/[s(s + 5)(s + 10)]$

12.3. The Bode diagrams and the polar plot for the same system are shown in Figures P12.3a and b.

(a) Use the Bode diagrams to find the values of the frequencies ω_ϕ and ω_g marked on the polar plot.
(b) Find the value of the static gain, k.

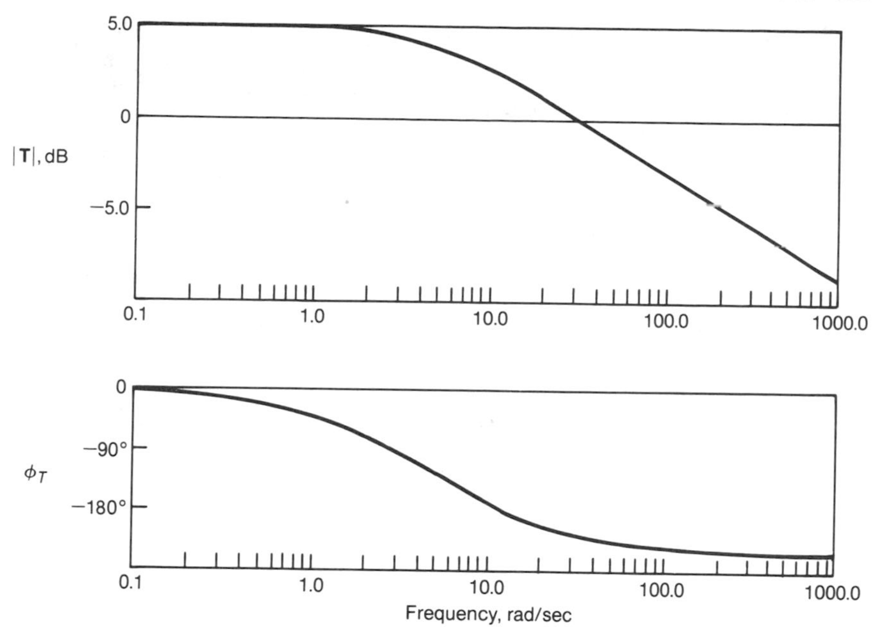

(a)

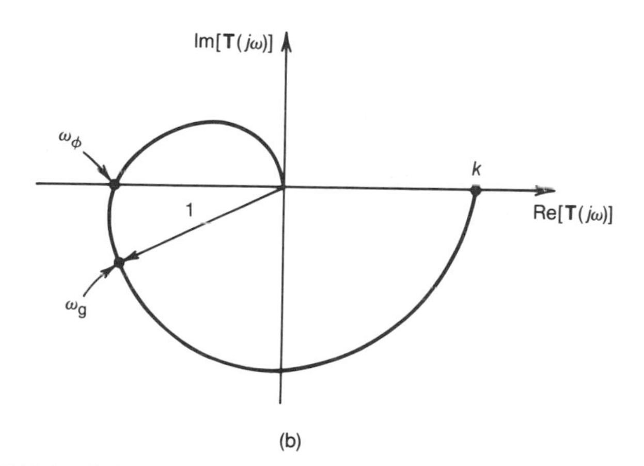

(b)

FIGURE P12.3 Polar plot and Bode diagrams for Problem 12.3.

12.4. A given system input-output equation is

$$a_2 \frac{d^2y}{dt^2} + a_1 \frac{dy}{dt} + a_0 y = b_2 \frac{d^2u}{dt^2} + b_1 \frac{du}{dt} + b_0 u$$

Find the expression for the output $y(t)$ when the input is a sinusoidal function of time, $u(t) = U \sin \omega_f t$ with $\omega_f = (a_0/a_2)^{0.5}$.

12.5. The detailed transfer function block diagram of a certain second-order system is shown in Figure P12.5. The values of the system parameters are

$$a_{11} = -0.1 \text{ sec}^{-1} \qquad a_{12} = -4.0 \text{ N/m}^3$$
$$a_{21} = 2.0 \text{ m}^3\text{/N-sec}^2 \qquad a_{22} = 0$$
$$b_{11} = 35.0 \text{ N/m}^2\text{-sec}$$

(a) Find the system transfer function $\mathbf{T}(s) = \mathbf{Y}(s)/\mathbf{U}(s)$.

(b) Find the amplitude A_0 and phase ϕ of the output $y(t)$ when the input is $u(t) = 0.001 \sin 1.5t$.

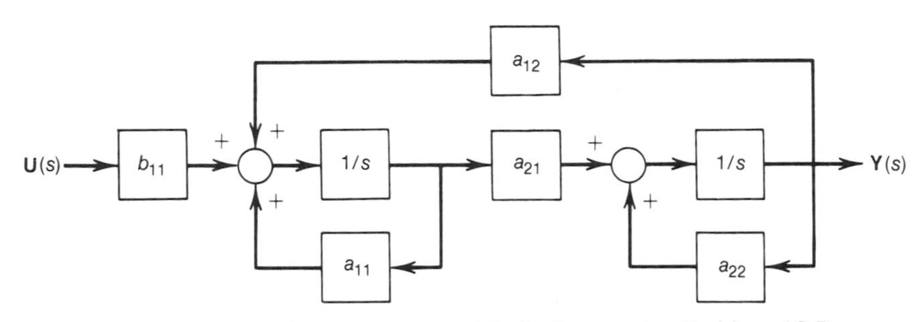

FIGURE P12.5 Transfer function block diagram for Problem 12.5.

12.6. In a buffer tank, shown in Figure P12.6, the input flowrate $Q_i(t)$ has a constant component equal to 0.5 m³/sec, an incremental sinusoidal component of amplitude 0.1 m³/sec, and frequency 0.4 rad/sec. The density of the liquid is $\rho = 1000$ kg/m³. The output flow conditions are assumed to be linear with the hydraulic resistance, $R_L = 39{,}240$ N-sec/m⁵.

(a) Write the state-variable equation for this system using a liquid height in the tank of $h(t)$ as the state variable.

(b) Find the average liquid height in the tank, $\bar{h}$.

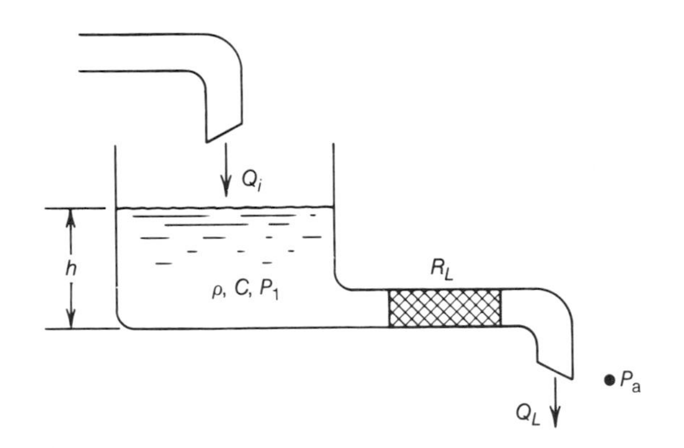

FIGURE P12.6 Buffer tank considered in Problem 12.6.

(c) Determine the condition for the cross-sectional area of the tank necessary to limit the amplitude of liquid height oscillations, due to sinusoidal oscillations in the input flowrate, to less than 0.2 m.

(d) What will be the time delay between the peaks of the sinusoidal input flowrate and the corresponding peaks of the liquid height if the cross-sectional area of the tank is $A = 1\ \text{m}^2$.

(e) Find the expression for the output flowrate $Q_L(t)$ for $A = 1\ \text{m}^2$.

12.7. A closed-loop system consisting of a process of transfer function $\mathbf{T}_\text{P}(s)$ and a controller of transfer function $\mathbf{T}_\text{C}(s)$ has been modeled as shown by the transfer function block diagram shown in Figure P12.7, where

$$\mathbf{T}_\text{P}(s) = \frac{5}{5s + 1} \quad \text{and} \quad \mathbf{T}_\text{C}(s) = \frac{2(s + 1)}{(2s + 1)}$$

(a) Find the system closed-loop transfer function $\mathbf{T}_\text{CL}(s)$.

(b) Find the amplitude and the phase angle of the output signal of the closed-loop system, $y(t)$, when $u(t) = 0.2 \sin 3t$.

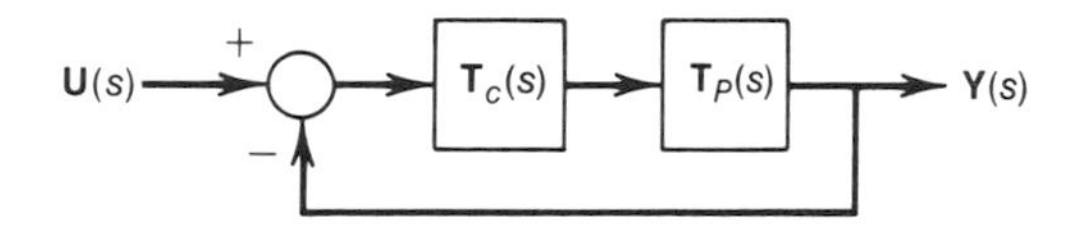

FIGURE P12.7 Block diagram of the system considered in Problem 12.7.

12.8. A hot water storage tank has been modeled by the equation

$$8000 \frac{dT_\text{w}}{dt} = 3T_\text{a} - 3T_\text{w}$$

where T_w is the temperature of water and T_a is the temperature of ambient air. For several days the ambient air temperature has been varying in a sinusoidal fashion with the maximum of 10°C at noon and the minimum of −10°C at midnight of each day. Determine the maximum and minimum temperature of water in the storage tank during those days and find at what times of the day the maximum and minimum temperatures ocurred.

13

Closed-Loop Systems and System Stability

13.1 Introduction

Up to this point the modeling and analysis in this text has dealt mainly with systems resulting from the straightforward interconnection of A-type, T-type, and D-type elements together with energy-converting transducers. The detailed reticulation of these system models by the use of simulation block diagrams has revealed the widespread natural occurrance of closed loops containing one or more integrators, each loop of which involves feedback to a summing point. The techniques of analysis employed so far with these passive systems have led to describing their dynamic characteristics by means of sets of state-variable equations and/or input-output differential equations and transfer functions. These systems are considered passive because no attempts have been made intentionally to close additional loops with signal amplifying or signal modifying devices. They are simply collections of naturally-occurring phenomena which, to be naturally-occurring, must be inherently stable in order to survive.

A system is considered stable if the following circumstances apply:

(a) The system remains in equilibrium at a steady normal operating point when left undisturbed, in other words after all transients due to previous inputs have died out.

(b) It responds with finite variations of all of its state variables when forced with a finite disturbance.

(c) It regains equilibrium at a steady operating point after the transient response to a step or pulse input has decayed to zero; or its state variables vary cyclically about steady operating point values when the input varies cyclically about a steady normal operating point value (as for instance, a system responding to a sinusoidal input variation about a constant value).

The use of intentional feedback in industrial systems seems to have started in the latter part of the Eighteenth century when James Watt devised a flyweight governor to sense the speed of a steam engine and use it as a negative feedback to control the flow of steam to the engine, thereby controlling the speed of the engine. It was almost a hundred years later that Clerk Maxwell modeled and analyzed such a system in a celebrated paper presented to the Royal Society in 1868! Since that time, inventors, engineers, and scientists have increasingly employed sensors and negative feedback[1] to improve system performance, to improve the control of quality of products, and to improve rates of production in industry. And in many instances, the hardware preceded the modeling and analysis! Thus the use of ingenuity in design and the use of physical reasoning have been very important factors in the development of the field of automatic control. These factors when combined with the techniques of modeling and analysis have made it possible to propose new concepts and to evaluate them before trying to build and test them, thereby saving time, effort, and money.

The use of negative feedback to control a passive system results in an active system having dynamic characteristics where maintaining stability may be a serious problem. The improved system traits resulting from employing intentional negative feedback control such as faster response and decreased sensitivity to loading effects will be discussed in Chapter 14.

The following example of a very commonly encountered feedback control system, the biomechanical system consisting of an automobile and driver, may be helpful in gaining an understanding of the way in which negative feedback can affect system stability.

Consider the process of steering an automobile along a straight stretch of open road with strong gusts of wind blowing normal to the direction of the roadway. The driver observes the deviation of the heading of the automobile relative to the roadway as the wind gusts deflect the heading and the driver then manipulates the steering wheel in such a way as to reduce the deviation, keeping the automobile on the road. A simplified block diagram model of this system is shown in Figure 13.1.

The desired input signal represents the direction of the roadway, and the output signal represents the heading of the automobile as it proceeds along the highway. Both signals are processed as the human visual system and brain seem to process them, that is they are compared with the effects of the heading signal acting negatively with respect to the road direction signal. This perceived deviation is acted upon by the brain and nervous system so as to generate a motion by the hands on the steering wheel that causes the automobile to change its heading, thereby reducing the deviation of the automobile heading from the road direction.

Barring extreme weather conditions and excessively high speed of the vehicle, this negative feedback by a human controller works well in the hands of moderately skilled but calm drivers. However if a driver becomes nervous and over-reacts with too much zeal to the wind gusts, the deviations of the vehicle heading from the

[1]Positive feedback is seldom used because in most cases its use leads to degraded system performance.

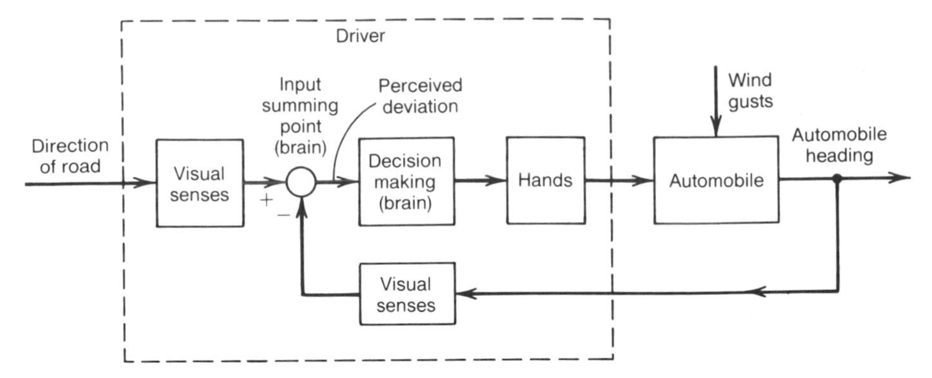

FIGURE 13.1 Simplified block diagram of the driver-automobile feedback system.

road direction may become excessive, possibly causing the vehicle to leave the road, or even worse, to collide with another vehicle.

This example illustrates the common everyday use of negative feedback based on a simplified model of the sensory and control capabilities of a human operator, and illustrates how inappropriate feedback can lead to system instability. (A more complex model of the human operator has shown that an experienced driver does more than simply observe the heading of the vehicle relative to the road, and that his control capability is tempered by experience and the ability to make rapid adjustments to unusual situations, for instance to sense the incipient effects of wind gusts before the vehicle responds to them.)

There is a persistent problem with stability that is likely to arise when intentional feedback is employed to produce closed-loop or automatic control, a problem that seldom occurs with passive, open-loop systems. Designing a system that will remain stable while achieving desired speed of response and reduced errors constitutes a problem of major importance in the development of automatic control systems.

This chapter presents basic methods for the analysis of stability of linear dynamic systems. Analysis of the stability of nonlinear systems is beyond the scope of this text. In Section 13.2 basic definitions related to stability analysis are introduced and general conditions for stability of linear systems are formulated. The next two sections are devoted to analytical methods for determining the stability of dynamic systems having zero inputs. In Section 13.3 algebraic stability criteria are presented, including the Hurwitz and Routh methods. The algebraic stability criteria are very simple and easy to use, but their applicability is limited to systems whose mathematical models are known. The Nyquist criterion described in Section 13.4 can be applied to closed-loop control systems whose open-loop frequency characteristics are known, either in an analytical form or in the form of sets of experimentally acquired frequency response data. In Section 13.5 gain and phase margins for stability are introduced. Finally, a brief outline of the root locus method, a very powerful tool in analysis and design of feedback systems, is given in Section 13.6.

13.2 Basic Definitions and Terminology

A linear system is commonly considered to be stable if its response meets the conditions listed in Section 13.1. Although this definition is intuitively correct, a much more precise definition of stability was proposed by Lyapunov. In order to explain Lyapunov's definition of stability, consider a dynamic system described by the following state equation

$$\dot{\mathbf{q}} = \mathbb{A}\mathbf{q} + \mathbb{B}\mathbf{u} \tag{13.1}$$

where $\mathbf{q}$ is a state vector, $\mathbb{A}$ is a system matrix, $\mathbf{u}$ is an input vector, and $\mathbb{B}$ is an input matrix. The system described by vector Equation (13.1) is stable in the sense of Lyapunov under the following conditions: for a given initial state $\mathbf{q}_0$ inside a hypersphere δ there exists another hypersphere ϵ such that, if the input vector is zero, $\mathbf{u} = 0$, the system will always remain in the hypersphere ϵ. Figure 13.2 illustrates the Lyapunov definition of stability for a two-dimensional system. In this case a dynamic system is considered stable if, for initial conditions inside the area δ, there exists an area ϵ such that the system will never move outside this area as long as the input signal is zero.

A stronger term than stability is *asymptotic stability*. A dynamic system described by Equation (13.1) is asymptotically stable in the sense of Lyapunov if, for any initial state $\mathbf{q}_0$ inside δ and with zero input, $\mathbf{u} = 0$, the system state vector will approach zero as time is approaching infinity

$$\lim_{t\to\infty} \|\mathbf{q}\| = 0 \quad \text{for } \mathbf{u} = 0 \tag{13.2}$$

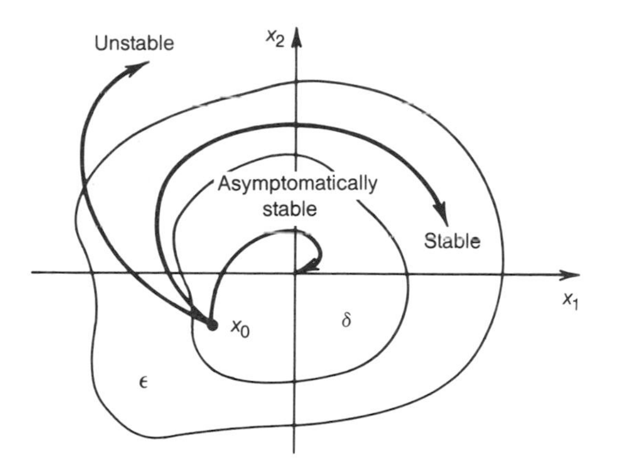

FIGURE 13.2 Illustration of stability in the sense of Lyapunov.

where $\|\mathbf{q}\|$ denotes the norm of the state vector $\mathbf{q}$.[2] If a system is assumed to be linear, its dynamics can be represented by a linear nth order differential equation

$$a_n \frac{d^n y}{dt^n} + \cdots + a_1 \frac{dy}{dt} + a_0 y = b_m \frac{d^m u}{dt^m} + \cdots + b_1 \frac{du}{dt} + b_0 u \quad (13.3)$$

where $m \leq n$. A linear system described by Equation (13.3) is asymptotically stable if, whenever $u(t) = 0$, the limit of $y(t)$ for time approaching infinity is zero for any initial conditions; that is,

$$\lim_{t \to \infty} y(t) = 0 \quad \text{if } u(t) = 0 \quad (13.4)$$

If $u(t)$ is equal to zero, the system input-output equation becomes homogenous.

$$a_n \frac{d^n y}{dt^n} + \cdots + a_1 \frac{dy}{dt} + a_0 y = 0 \quad (13.5)$$

Assuming for simplicity that all roots of the characteristic equation, $r_1, r_2, \ldots, r_n$, are distinct, which does not limit the generality of further considerations, the solution of Equation (13.5) is

$$y(t) = \sum_{i=1}^{n} K_i e^{r_i t} \quad (13.6)$$

The output signal given by Equation (13.6) will satisfy condition (13.4) if and only if all roots of the characteristic equation $r_1, r_2, \ldots, r_n$, are real and negative or are complex and have negative real parts. This observation leads to a theorem defining conditions for stability of linear systems. According to this theorem, a linear system described by Equation (13.3) is asymptotically stable if and only if all roots of the characteristic equation lie strictly in the left half of the complex plane, that is if

$$\text{Re}\,[r_i] < 0 \quad \text{for } i = 1, 2, \ldots, n \quad (13.7)$$

It can thus be concluded that stability of linear systems depends only on the location of the characteristic roots in the complex plane. Note also that stability of linear systems does not depend on the input signals. In other words, if a linear system is stable, it will remain stable regardless of the type and magnitude of input signals applied to the system. It should be stressed that this is true only for linear systems. Stability of nonlinear systems, which is beyond the scope of this book, is affected not only by the specific system properties but also by the input signals.

[2]The norm of the state vector $\mathbf{q}$ in Euclidean space is defined by

$$\|\mathbf{q}\| = \sqrt{\sum_{i=1}^{n} q_i^2}$$

that is the length of $\mathbf{q}$.

13.3. Algebraic Stability Criteria

The transfer function of the linear system described by the input-output Equation (13.3) is a ratio of polynomials

$$\mathbf{T}(s) = \frac{\mathbf{B}(s)}{\mathbf{A}(s)} \tag{13.8}$$

where the polynomials $\mathbf{A}(s)$ and $\mathbf{B}(s)$ are

$$\mathbf{A}(s) = a_n s^n + \cdots + a_1 s + a_0 \tag{13.9}$$

$$\mathbf{B}(s) = b_m s^m + \cdots + b_1 s + b_0 \tag{13.10}$$

The roots of the system characteristic equation are, of course, the same as the roots of the equation $\mathbf{A}(s) = 0$, which are the poles of the system transfer function. The transfer function of the feedback system shown in Figure 13.3 is

$$\mathbf{T}_{\mathrm{CL}}(s) = \frac{\mathbf{G}(s)}{1 + \mathbf{G}(s)\mathbf{H}(s)} \tag{13.11}$$

where $\mathbf{G}(s) = \mathbf{N}_g(s)/\mathbf{D}_g(s)$
$\mathbf{H}(s) = \mathbf{N}_h(s)/\mathbf{D}_h(s)$

The system characteristic equation is thus

$$1 + \mathbf{G}(s)\mathbf{H}(s) = 0 \quad \text{or } \mathbf{N}_g(s)\mathbf{N}_h(s) + \mathbf{D}_g(s)\mathbf{D}_h(s) = 0 \tag{13.12}$$

All roots of Equation (13.12) must be located in the left half of the complex plane for asymptotic stability of the system.

An obvious way to determine whether the conditions of asymptotic stability are met would be to find the roots of Equation (13.12) and check if all have negative real parts. This direct procedure may be cumbersome, and it usually requires a computer if a high-order system model is involved. Several stability criteria have been developed that allow for verification of the stability conditions for a linear system without calculating the roots of the characteristic equation. In this section two such methods, Hurwitz and Routh stability criteria, will be introduced. Both methods are classified as algebraic stability criteria since they both require some

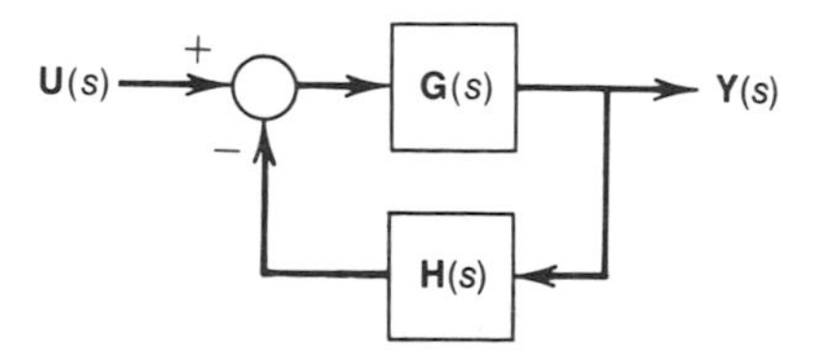

FIGURE 13.3 A linear feedback system.

algebraic operations to be performed on the coefficients of the system characteristic equation. The algebraic methods are widely used because of their simplicity, although their applicability is limited to problems in which an analytical form of the system characteristic equation is known.

In both the Hurwitz and the Routh methods the same necessary conditon for stability is formulated. Consider an nth order characteristic equation

$$a_n s^n + \cdots + a_1 s + a_0 = 0 \tag{13.13}$$

A necessary (but not sufficient) condition for all the roots of Equation (13.13) to have negative real parts is that all the coefficients a_i, $i = 0, 1, 2, \ldots, n$, have the same sign and that none of the coefficients vanishes. This necessary condition for stability can therefore be verified by inspection of the characteristic equation and this inspection serves as an initial means of screening for instability. (or as a means of detecting that an algebraic sign error has been made in the analysis of an inherently stable passive system.)

Hurwitz Criterion. The Hurwitz necessary and sufficient set of conditions for stability are based on the set of determinants which are formed as follows.

$$D_j = \begin{vmatrix} \overbrace{a_{n-1}}^{j=1} & a_{n-3} & a_{n-5} & \cdots & a_{n-(2j-1)} & \cdots & 0 \\ a_n & a_{n-2} & a_{n-4} & \cdots & a_{n-(2j-2)} & \cdots & 0 \\ 0 & a_{n-1} & a_{n-3} & \cdots & a_{n-(2j-3)} & \cdots & 0 \\ 0 & a_n & a_{n-2} & \cdots & a_{n-(2j-4)} & \cdots & 0 \\ \vdots & \vdots & \vdots & & \vdots & & \vdots \\ 0 & \cdots & \cdots & \cdots & a_{n-j} & \cdots & a_{n-n} = a_0 \end{vmatrix}$$

(Braces over the columns indicate the first $j = 1$, $j = 2$, $j = 3$, j, and n columns.)

The coefficients not present in the characteristic equation are replaced by zeros in the Hurwitz determinants.[3]

The necessary and sufficient set of conditions for all the roots of Equation (13.13) to have negative real parts is that all the Hurwitz determinants, D_j, $j = 1, 2, \ldots, n$, must be positive.

[3]See A.F.D: Sousa, *Design of Control Systems,* Prentice-Hall, Inc., Englewood Cliffs, NJ, 1988, pp. 194–196.

It can be shown that if the necessary condition for stability is satisfied, that is if all the coefficients a_i ($i = 0, 1, \ldots, n$) have the same sign and none of the coefficients vanishes, then the first and the nth Hurwitz determinants, D_1 and D_n, are positive. Thus, if the necessary condition is satisfied, it is sufficient to check if only $D_2, D_3, \ldots, D_{n-1}$ are positive.

EXAMPLE 13.1

Determine the stability of systems whose characteristic equations are

$$\text{(a) } 7s^4 + 5s^3 - 12s^2 + 6s - 1 = 0$$

The coefficients in the characteristic equation have different signs and thus the set of necessary conditions for stability is not met. The system is unstable, and thus there is no need to look at the Hurwitz determinants.

$$\text{(b) } s^3 + 6s^2 + 11s + 6 = 0$$

The necessary condition is satisfied. Next, the Hurwitz determinants must be examined (a more stringent test). Since the necessary condition is satisfied, only D_2 needs to be checked in this case. It is

$$D_2 = \begin{vmatrix} 6 & 6 \\ 1 & 11 \end{vmatrix} = 66 - 6 = 60 > 0$$

Thus all determinants are positive and the system is stable.

$$\text{(c) } 2s^4 + s^3 + 3s^2 + 5s + 10 = 0$$

The necessary condition is satisfied. Checking the determinants, D_2 and D_3

$$D_2 = -7; \quad D_3 = -45$$

The set of necessary and sufficient conditions is not satisfied because both D_2 and D_3 are negative; the system is unstable. ■

EXAMPLE 13.2

Determine the range of values for the gain K for which the closed-loop system shown in Figure 13.4 is stable.

Solution

First, find the system transfer function.

$$\mathbf{T}_{\mathrm{CL}}(s) = \frac{\dfrac{K(s + 40)}{s(s + 10)}}{1 + \dfrac{K(s + 40)}{s(s + 10)(s + 20)}}$$

Hence the system characteristic equation is

$$s(s + 10)(s + 20) + K(s + 40) = 0$$

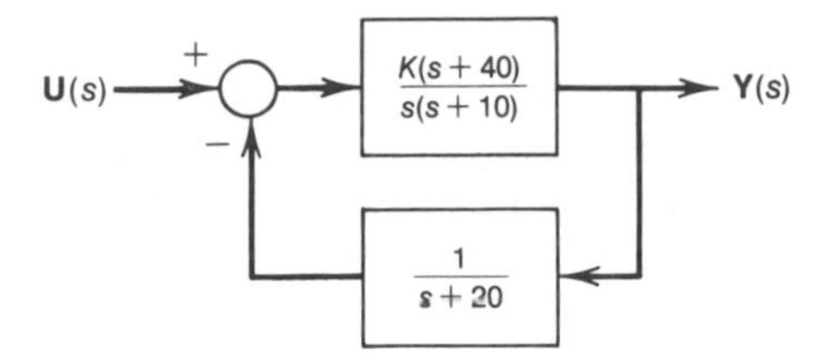

FIGURE 13.4 Closed-loop system considered in Example 13.2.

which, after multiplying, gives

$$s^3 + 30s^2 + (200 + K)s + 40K = 0$$

The necessary condition for stability is

$$200 + K > 0 \quad \text{and} \quad 40K > 0$$

Both inequalities are satisfied for $K > 0$, which satisfies the necessary conditions for system stability. But this is not enough! The Hurwitz necessary and sufficient conditions for stability of this third-order system are $D_1 > 0$ and $D_2 > 0$. The first inequality is satisfied because $D_1 = a_2 = 30$. The second Hurwitz determinant D_2 is

$$D_2 = \begin{vmatrix} a_2 & a_0 \\ a_3 & a_1 \end{vmatrix} = \begin{vmatrix} 30 & 40K \\ 1 & (200 + K) \end{vmatrix} = (6000 - 10K)$$

Combining the necessary and sufficient set of conditions yields the range of values of K for which the system is stable

$$0 < K < 600$$

■

Routh Criterion. The Hurwitz criterion becomes very laborious for higher-order systems for which large determinants must be evaluated. The Routh criterion offers an alternative method for checking the sufficient conditions of stability.

The Routh method involves a different set of necessary and sufficient stability conditions based on an array, which is formed as follows.

$$\begin{array}{l|lllll}
s^n & a_n & a_{n-2} & a_{n-4} & a_{n-6} & \cdots \\
s^{n-1} & a_{n-1} & a_{n-3} & a_{n-5} & a_{n-7} & \cdots \\
s^{n-2} & b_1 & b_2 & b_3 & b_4 & \cdots \\
s^{n-3} & c_1 & c_2 & c_3 & c_4 & \cdots \\
\vdots & & & & & \\
s^1 & \cdots & & & & \\
s^0 & \cdots & & & &
\end{array}$$

where a's are the coefficients in the characteristic Equation (13.13), and the other coefficients, b, c, . . . , are calculated as follows

$$b_1 = (a_{n-1}a_{n-2} - a_n a_{n-3})/a_{n-1}$$
$$b_2 = (a_{n-1}a_{n-4} - a_n a_{n-5})/a_{n-1}$$
$$\vdots$$
$$c_1 = (b_1 a_{n-3} - a_{n-1}b_2)/b_1$$
$$c_2 = (b_1 a_{n-5} - a_{n-1}b_3)/b_1$$
$$\vdots$$

The necessary and sufficient set of conditions for stability is that all elements in the first column of the Routh array, a_n, a_{n-1}, b_1, c_1, . . . , must have the same sign. If this set of conditions is not satisfied, the system is unstable and the number of sign changes in the first column of the Routh array is equal to the number of roots of the characteristic equation located in the right half of the complex plane. If an element in the first column is zero, the system is unstable, but additional coefficients for column 1 may be found by substituting a small number for the zero at this location and proceeding. For a more detailed discussion relating to the occurrance of zero coefficients see D'Souza.[4]

EXAMPLE 13.3

Determine the stability of a system that has the following characteristic equation

$$s^3 + 2s^2 + 4s + 9 = 0$$

Solution

The Routh array is formed as follows.

s^3	1	4
s^2	2	9
s^1	-0.5	
s^0	9	

The system is unstable and there are two roots in the right half of the complex plane, as indicated by two sign changes in the first column of the Routh array, from 2 to -0.5, and from -0.5 to 9. ■

[4]A. F. D'Souza, *Design of Control Systems*, Prentice-Hall, Inc., Englewood Cliffs, N.J., 1988, pp. 199–200.

EXAMPLE 13.4

Determine the range of values of K for which the system having the characteristic equation given below is stable.

$$s^3 + s^2 + (5 + K)s + 3K = 0$$

Solution

The Routh array is formed as follows.

$$\begin{array}{c|cc} s^3 & 1 & 5 + K \\ s^2 & 1 & 3K \\ s^1 & 5 - 2K & \\ s^0 & 3K & \end{array}$$

There will be no sign changes in the first column of the Routh array if

$$5 - 2K > 0 \quad \text{and} \quad 3K > 0$$

which gives the following range of K for stability.

$$0 < K < 2.5$$

■

13.4 Nyquist Stability Criterion

The method proposed by Nyquist[5,6] many years ago allows for determination of the stability of a closed-loop system on the basis of the frequency response of the open-loop system. The Nyquist criterion is particularly attractive because it does not require knowledge of the mathematical model of the system. Thus it can be applied also in all those cases when a system transfer function is not available in an analytical form but the system open-loop frequency response may be obtained experimentally.

To introduce the Nyquist criterion, the polar plots that, so far, have been drawn for positive frequency only, $0 < \omega < \infty$, must be extended to include negative frequencies, $-\infty < \omega < +\infty$. It can be shown that the polar plot of $\mathbf{T}(j\omega)$ for negative frequencies is symmetrical about the real axis with the polar plot of $\mathbf{T}(j\omega)$ for positive frequencies. An example of a polar plot for ω varying from $-\infty$ to $+\infty$ is shown in Figure 13.5. Notice that in order to close the contour, a curve is

[5] H. Nyquist, "Regeneration Theory," *Bell Systems Technical Journal*, **11**: 126–147, (1932).

[6] D. M. Auslander, Y. Takahashi, M. J. Rabins, *Introducing Systems and Control,* McGraw-Hill Book Co., 1974, pp. 371–375.

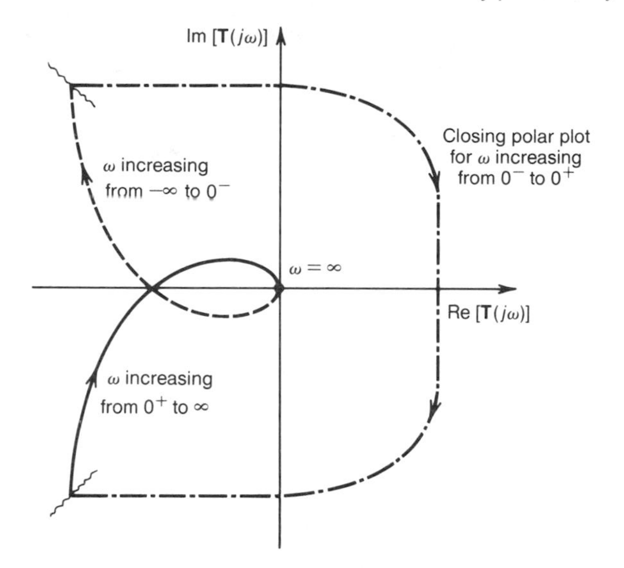

FIGURE 13.5 Example of polar plot of $\mathbb{T}(j\omega)$ for $-\infty < \omega < \infty$

drawn from the point corresponding to $\omega = 0^-$ to the point $\omega = 0^+$ in a clockwise direction, retaining the direction of increasing frequency.

Consider the same closed-loop sytstem shown in Figure 13.3. The Nyquist criterion can be stated as follows: If an open-loop transfer function $\mathbf{G}(s)\mathbf{H}(s)$ has k poles in the right half of the complex plane, then for stability of the closed-loop system the polar plot of the open-loop transfer function must encircle the point $(-1, j0)$ k times in the clockwise direction. An important implication of the Nyquist criterion is that if the open-loop system is stable and thus does not have any poles in the right half of the complex plane, $k = 0$, then the closed-loop system is stable if $\mathbf{G}(j\omega)\mathbf{H}(j\omega)$ does not encircle the point $(-1, j0)$.

If an open-loop system is stable, one can measure the system response to a sinusoidal input over a wide enough range of frequency and then plot the experimentally obtained polar plot to determine whether the curve encircles the critical point $(-1, j0)$. This experimental procedure can only be applied if the open-loop system is stable; otherwise no measurements can be taken from the system.

As one gains experience, it becomes possible to use the Nyquist criterion by inspecting the polar plot of the open-loop transfer function for positive frequency only. The general rule is that for a system to be stable, the critical point $(-1, j0)$ must be to the left of an observer following the polar plot from $\omega = 0^+$ to $+\infty$.

Figure 13.6 shows polar plots for the system having the open-loop transfer function $\mathbf{T}_{OL} = \mathbf{G}(s)\mathbf{H}(s) = K/(s^3 + s^2 + s + 1)$ for three different values of gain K. According to the Nyquist criterion, the closed-loop system is asymptotically stable for $K = K_1$; it is unstable for $K = K_3$. For $K = K_2$ the closed-loop system is said to be marginally stable.

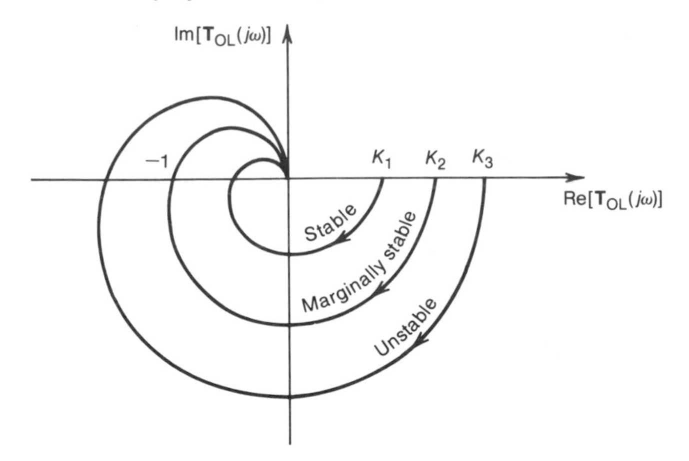

FIGURE 13.6 Polar plots for stable, marginally stable, and unstable systems.

EXAMPLE 13.5

Determine stability of the closed-loop system that has the open-loop transfer function $\mathbf{T}_{\mathrm{OL}}(s) = K/s(s + 1)(2s + 1)$.

Solution

The sinusoidal transfer function of the open-loop system, $\mathbf{T}_{\mathrm{OL}}(j\omega)$, is

$$\mathbf{T}_{\mathrm{OL}}(j\omega) = \frac{K}{j\omega(j\omega + 1)(2j\omega + 1)} \tag{13.14}$$

The real and imaginary parts of $\mathbf{T}_{\mathrm{OL}}(j\omega)$ are

$$\mathrm{Re}\,[\mathbf{T}_{\mathrm{OL}}(j\omega)] = -\frac{3K}{9\omega^2 + (2\omega^2 - 1)^2} \tag{13.15}$$

$$\mathrm{Im}\,[\mathbf{T}_{\mathrm{OL}}(j\omega)] = \frac{K(2\omega^2 - 1)}{9\omega^3 + \omega(2\omega^2 - 1)^2} \tag{13.16}$$

According to the Nyquist criterion, the closed-loop system will be stable if

$$|\mathbf{T}_{\mathrm{OL}}(j\omega_p)| < 1$$

where ω_p is the frequency at the point of intersection of the polar plot with the negative real axis. The imaginary part of $\mathbf{T}_{\mathrm{OL}}(j\omega)$ at this point is zero, so the value of ω_p can be found by solving the equation

$$\mathrm{Im}\,[\mathbf{T}_{\mathrm{OL}}(j\omega)] = 0 \tag{13.17}$$

Substituting into (13.17) the expression for Im $[\mathbf{T}_{\mathrm{OL}}(j\omega)]$, (13.16), gives

$$2\omega_p^2 - 1 = 0$$

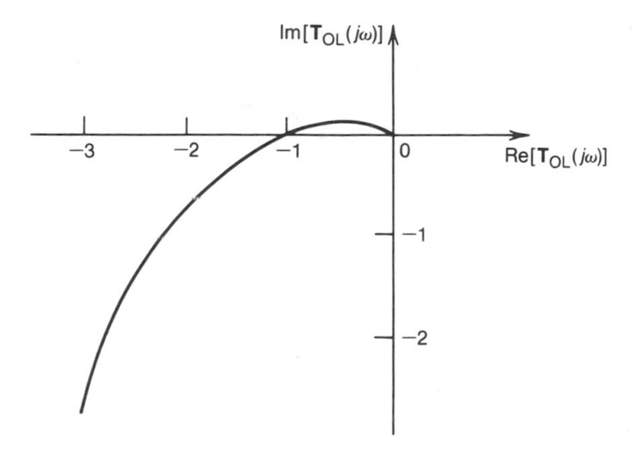

FIGURE 13.7 Polar plot for the system considered in Example 13.5.

and hence $\omega_p = 1/\sqrt{2}$ rad/sec. To assure stability of the closed-loop system the real part of $\mathbf{T}_{OL}(j\omega)$ must be less than 1 at $\omega = \omega_p$, i.e.

$$|\text{Re}\ [\mathbf{T}(j\omega_p)]| < 1 \tag{13.18}$$

Now substitute into (13.18) the expression for the real part, (13.15), to obtain

$$\frac{3K}{9\omega_p^2 + (2\omega_p^2 - 1)^2} < 1$$

and solve for K.

$$0 < K < 1.5$$

The closed-loop system will be stable for K less than 1.5. The polar plot of the open-loop system transfer function for $K = 1.5$ is shown in Figure 13.7. ■

13.5 Quantitative Measures of Stability

As stated earlier in this chapter, an unstable control system is useless. An asymptotic stability constitutes one of the basic requirements in design of control systems. However, the mere fact of meeting the stability conditions does not guarantee a satisfactory performance of the system. A system designed to barely meet the stability condition is very likely to become unstable because its actual parameter values may be slightly different from the values used in the design. In addition, very often the actual system parameters change in time due to aging, which again may push the system over the stability limit if that limit is too close. It is therefore necessary in the design of control systems to determine not only whether the system is stable but also how far the system is from instability. To answer this question, some quantitative measure of stability is needed. Two such measures, a gain margin and a phase margin, are derived from the Nyquist stability criterion.

If the $\mathbf{G}(s)\mathbf{H}(s)$ curve passes through the point $(-1, j0)$, which represents critical conditions for stability in the Nyquist criterion, the open-loop system gain is equal to one and the phase angle is -180 deg, which can be written mathematically as

$$|\mathbf{G}(j\omega)\mathbf{H}(j\omega)| = 1 \tag{13.19}$$

and

$$\angle \mathbf{G}(j\omega)\mathbf{H}(j\omega) = -180 \text{ deg} \tag{13.20}$$

If both the magnitude condition (13.19) and the phase condition (13.20) are satisfied, the closed-loop system is said to be marginally stable. The use of stability margins provides the means to indicate how far the system is from one of the critical conditions, (13.19) or (13.20), while the other condition is met.

The gain margin k_g is defined as

$$k_g = \frac{1}{|\text{Re}\,[\mathbf{G}(j\omega_p)\mathbf{H}(j\omega_p)]|} \tag{13.21}$$

where ω_p is the frequency for which the phase angle of the open-loop transfer function is -180 deg. When the gain margin has been expressed in decibels it is calculated as

$$k_{\text{gdb}} = 20 \log \left(\frac{1}{|\text{Re}\,[\mathbf{G}(j\omega_p)\mathbf{H}(j\omega_p)]|} \right) \tag{13.22}$$

The phase margin γ is defined as

$$\gamma = 180 \text{ deg} + \angle \mathbf{G}(j\omega_g)\mathbf{H}(j\omega_g) \tag{13.23}$$

where ω_g is the frequency for which the magnitude of the open-loop transfer function is unity.

$$|\mathbf{G}(j\omega_g)\mathbf{H}(j\omega_g)| = 1 \tag{13.24}$$

Figure 13.8 presents a graphical interpretation of the gain and phase stability margins.

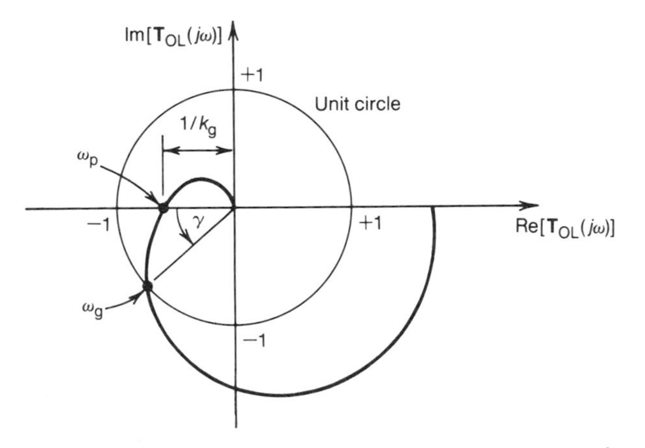

FIGURE 13.8 Graphical interpretation of stability margins.

In typical control system applications the lowest acceptable gain margin is usually between 1.2 and 1.5 (or 1.6 and 3.5 db). The typical range for the acceptable phase margin is from 30 deg to 45 deg.

Example 13.6

Determine the open-loop gain necessary for the closed-loop system shown in Figure 13.9 to be stable with the gain margin, $k_g \geq 1.2$, and the phase margin, $\gamma \geq 45$ deg.

Solution

The open-loop system transfer function is

$$\mathbf{T}_{OL}(s) = \frac{K}{6s^3 + 11s^2 + 6s + 1} \tag{13.25}$$

When $s = j\omega$, the sinusoidal transfer function is found to be

$$\mathbf{T}_{OL}(j\omega) = \frac{K}{(1 - 11\omega^2) + j6\omega(1 - \omega^2)} \tag{13.26}$$

or, equally,

$$\mathbf{T}_{OL}(j\omega) = \frac{K[(1 - 11\omega^2) - j6\omega(1 - \omega^2)]}{[(1 - 11\omega^2)^2 + 36\omega^2(1 - \omega^2)^2]} \tag{13.27}$$

The frequency ω_p, for which the phase angle is -180 deg, is found by solving the equation

$$\text{Im}\,[\mathbf{T}_{OL}(j\omega)] = 0 \tag{13.28}$$

Substituting into (13.28) the expression for the imaginary part of the open-loop transfer function from (13.27) gives

$$\omega_p = 1 \text{ rad/sec} \tag{13.29}$$

The real part of the open-loop transfer function for this frequency is

$$\text{Re}\,[\mathbf{T}_{OL}(j\omega_p)] = -\frac{K}{10} \tag{13.30}$$

Substituting (13.30) into the condition for the gain margin, $k_g \geq 1.2$, gives

$$\frac{10}{K} \geq 1.2 \qquad K \leq 8.3333$$

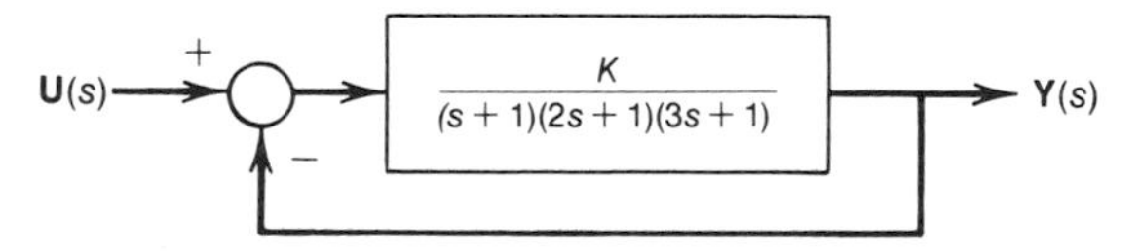

FIGURE 13.9 Block diagram of the system considered in Example 13.6.

Now, in order to satisfy the phase margin requirement, the phase angle of the open-loop transfer function for the frequency where the magnitude is one, should be

$$\angle \mathbf{T}_{OL}(j\omega_g) = -180 \text{ deg} + 45 \text{ deg} = -135 \text{ deg} \tag{13.31}$$

From the expression for the open-loop transfer function (13.26), the phase angle of $\mathbf{T}_{OL}(j\omega)$ for ω_g is

$$\angle \mathbf{T}_{OL}(j\omega_g) = \tan^{-1}(0) - \tan^{-1}\left(\frac{6\omega_g(1 - \omega_g^2)}{1 - 11\omega_g^2}\right) \tag{13.32}$$

Comparing the right-hand sides of Equations (13.31) and (13.32) yields

$$-\tan^{-1}\left(\frac{6\omega_g(1 - \omega_g^2)}{1 - 11\omega_g^2}\right) = -135 \text{ deg} \tag{13.33}$$

Taking the tangent of both sides of Equation (13.33) gives a cubic equation for ω_g

$$6\omega_g^3 + 11\omega_g^2 - 6\omega_g - 1 = 0 \tag{13.34}$$

which has the solution

$$\omega_g = 0.54776 \text{ rad/sec}$$

Now use Equation (13.24) to obtain

$$|\mathbf{T}_{OL}(j\omega_g)| = 1 \tag{13.35}$$

The magnitude of $\mathbf{T}_{OL}(j\omega)$ for this system is

$$|\mathbf{T}_{OL}(j\omega_g)| = \frac{K}{\sqrt{(1 - 11\omega_g^2)^2 + 36\omega_g^2(1 - \omega_g^2)^2}} \tag{13.36}$$

Combine Equations (13.35) and (13.36) to obtain the solution for K that satisfies the phase condition.

$$K = 3.27$$

This value is smaller than the value of K obtained from the gain condition; therefore, the open-loop gain must be equal to 3.27 (or less) to meet both, namely the gain and the phase angle conditions. ■

13.6 Root Locus Method

In Chapter 5 and also in the preceding part of this chapter, a system transient performance and ultimately the system stability are shown to be governed by the locations of the roots of the system characteristic equation in the s plane. This fact has important implications for methods used in both analysis and design of dynamic systems. In the analysis of system dynamics, it is desirable to know the locations of the characteristic roots in order to be able to predict basic specifications of the transient performance. In the design process, the system parameters are selected to obtain desired locations of the roots in the s plane. The knowledge of the locations of the characteristic roots and of the paths of their migration in the s plane due to variations of system parameters is therefore extremely important for a system engineer.

A very powerful and relatively simple technique, called root locus method, was developed by W. R. Evans[7,8] to assist in determining locations of roots of the characteristic equation of feedback systems. The graphs generated using the root locus method show the migration paths of the characteristic roots in the s plane resulting from variations of selected system parameters. The parameter that is of particular interest in the analysis of feedback systems is an open-loop gain. When the open-loop gain increases, the system response may become faster but if the increase is too great it may lead to very oscillatory or even unstable behavior.

Although the root locus method can be used to determine the migration of roots caused by variation of any of the system parameters, it is most often used to examine the effect of varying the open-loop gain.

The significance of the root locus method will be illustrated by Example 13.7.

EXAMPLE 13.7

Consider the unity feedback system shown in Figure 13.10. The system open-loop transfer function is

$$\mathbf{T}_{OL}(s) = \frac{K}{s(s+2)} \tag{13.37}$$

where $K \geq 0$ is the open-loop gain. The closed-loop transfer function is

$$\mathbf{T}_{CL}(s) = \frac{K}{(s^2 + 2s + K)} \tag{13.38}$$

Hence, the closed loop system characteristic equation is

$$s^2 + 2s + K = 0 \tag{13.39}$$

and hence the roots are

$$s_1 = -1 - \sqrt{1-K} \quad \text{and} \quad s_2 = -1 + \sqrt{1-K} \tag{13.40}$$

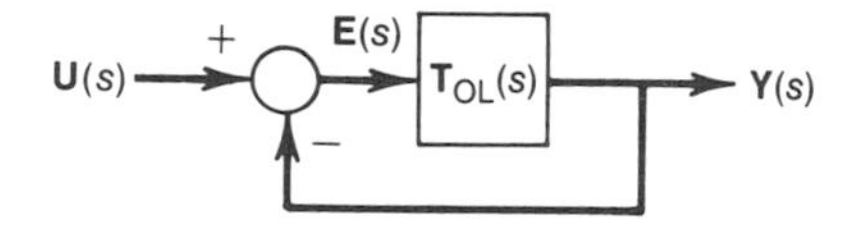

FIGURE 13.10 Block diagram of unity feedback system.

[7] W. R. Evans, "Graphical Analysis of Control Systems," *Trans. AIEE*, **67**; 547–551 (1948)

[8] W. R. Evans, *Control-System Dynamics*, McGraw-Hill Book Co., New York, 1954, pp. 96–121.

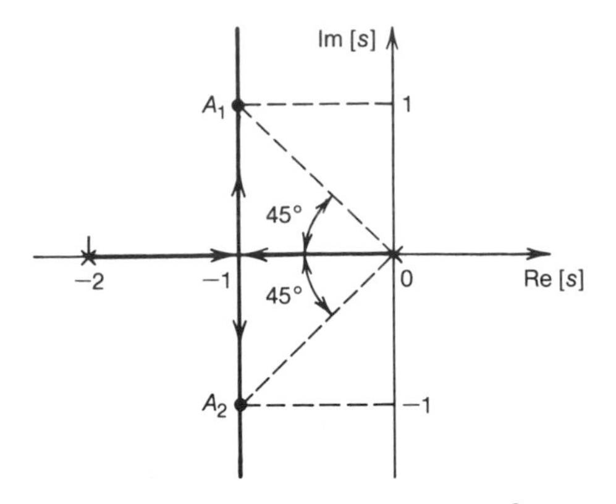

FIGURE 13.11 Root locus for $\mathbf{T}(s) = K/(s^2 + 2s + K)$.

As the open-loop gain varies from zero to infinity, the roots will change as follows:

- For $K = 0$: $s_1 = -2$ and $s_2 = 0$.
- For K increasing from 0 to 1: both roots remain real with s_1 moving from -2 to -1 and s_2 moving from 0 to -1.
- For K between 1 and infinity: the roots are complex conjugate, $s_1 = -1 - j\sqrt{K - 1}$ and $s_2 = -1 + j\sqrt{K - 1}$.

The migration of roots for $0 \leq K < \infty$ is shown in Figure 13.11. Once the root locus is plotted, it is easy to determine the locations of the characteristic roots necessary for a desired system performance. For instance, if the desired damping ratio for the system considered in this example is $\zeta = 0.7$, the roots must be located at points A_1 and A_2, which correspond to

$$s_1 = -1 + j \quad \text{and} \quad s_2 = -1 - j$$

The characteristic equation for these roots takes the form

$$(s + 1 - j)(s + 1 + j) = 0$$

or

$$s^2 + 2s + 2 = 0 \tag{13.41}$$

By comparing Equation (13.41) with the characteristic equation in terms of K (13.39), the required value of the open-loop gain is found

$$K = 2$$

■

The direct procedure for plotting loci of characteristic roots employed in Example 13.7 becomes impractical for higher-order systems. The method developed by Evans, consisting of several simple rules, greatly simplifies the process of plotting root loci. A block diagram of the system to which the method applies is shown in Figure 13.10. The characteristic equation of this system is

$$1 + \mathbf{T}_{\mathrm{OL}}(s) = 0 \tag{13.42}$$

where $\mathbf{T}_{OL}(s)$ is an open-loop transfer function that can be expressed as

$$\mathbf{T}_{OL}(s) = \frac{K\mathbf{B}(s)}{\mathbf{A}(s)} \tag{13.43}$$

where K is an open-loop gain and $\mathbf{A}(s)$ and $\mathbf{B}(s)$ are polynomials in s of nth and mth order, respectively. It should be noted that $m \leq n$. The open-loop transfer function is, in general, a complex quantity and, therefore, Equation (13.42) is equivalent to two equations representing magnitude and phase angle conditions, i.e.

$$|\mathbf{T}_{OL}(s)| = 1 \tag{13.44}$$

and for $K > 0$,

$$\underline{/\mathbf{T}_{OL}(s)} = \pm(2k + 1)\pi \quad \text{for } k = 0, 1, 2, \ldots \tag{13.45a}$$

or for $K < 0$,

$$\underline{/\mathbf{T}_{OL}(s)} = \pm 2\,k\pi \quad \text{for } k = 0, 1, 2, \ldots \tag{13.45b}$$

All roots of the characteristic equation, s_i, $i = 1, 2, \ldots, n$, must satisfy both the magnitude condition (13.44) and the phase angle condition (13.45). The following rules for plotting root loci are derived from these two conditions.

1. The root locus is symmetric about the real axis of the s plane.
2. The number of branches of root loci is equal to the number of roots of the characteristic equation or, equally, to the order of $\mathbf{A}(s)$.
3. The loci start at open-loop poles for $K = 0$ and terminate either at open-loop zeros (m branches) or at infinity along asymptotes for $K \to \infty$ ($n - m$ branches).
4. The loci exist on sections of the real axis between neighboring real poles and/or zeros, if the number of real poles and zeros to the right of this section is odd, for $K > 0$ (or if the number is even, for $K < 0$).
5. In accordance with rule 3, $n - m$ loci terminate at infinity along asymptotes. The angles between the asymptotes and the real axis are given by

$$\alpha = \pm 180 \frac{2k + 1}{n - m} \text{ deg.} \qquad k = 0, 1, 2, \ldots \tag{13.46a}$$

for $K > 0$, and for $K < 0$ the angles are given by

$$\alpha = \pm 180 \frac{2k}{n - m} \text{ deg,} \qquad k = 0, 1, 2, \ldots \tag{13.46b}$$

6. All the asymptotes intersect the real axis at the same point. The abscissa of that point is

$$\sigma_a = \frac{\sum^{n} \text{poles} - \sum^{m} \text{zeros}}{n - m} \tag{13.47}$$

where the summations are over finite poles and zeros of $\mathbf{T}_{OL}(s)$.

TABLE 13.1 Examples of root loci for common system transfer functions for $K > 0$.

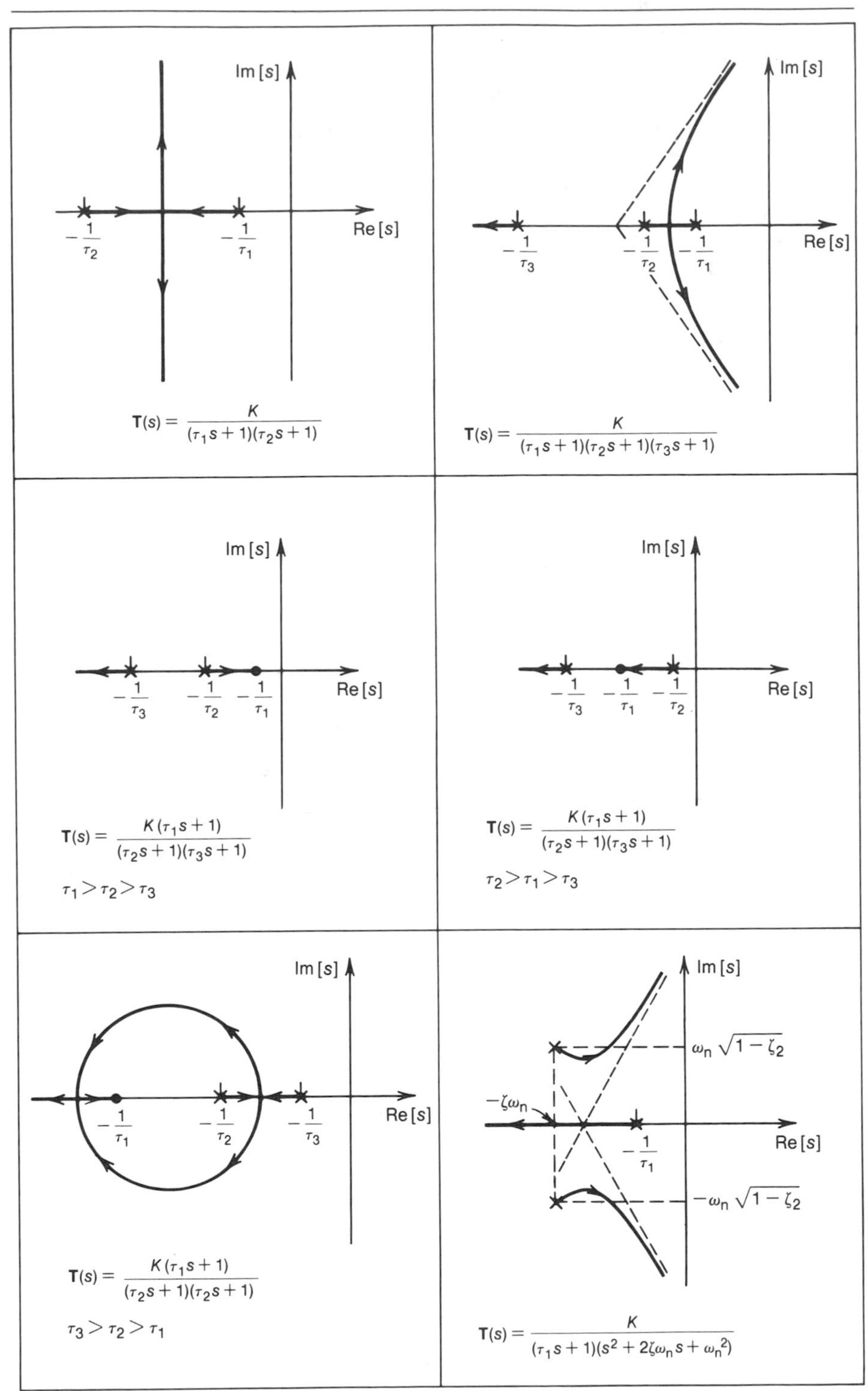

7. The loci depart the real axis at break-away points and enter the real axis at break-in points. The locations of these points can be found by solving for "min-max" values of s using[9]

$$\frac{d}{ds}\left[-\frac{\mathbf{A}(s)}{\mathbf{B}(s)}\right] = 0 \qquad (13.48)$$

8. The angles of departure from complex poles at $K = 0$ and the angles of arrival at complex zeros at $K \to \infty$ are found by applying the angle condition given by Equations (13.45) to a point infinitesimally close to the complex pole or zero in question.
9. The points where the loci cross the imaginary axis in the s plane are determined by the solution of the system characteristic equation for $s = j\omega$ or by employing Routh stability criterion.

Table 13.1 shows examples of root loci for common transfer functions for $K > 0$.

13.7 Synopsis

Along with many beneficial effects that can be attributed to the presence of feedback in engineering systems, there are also some unwanted side effects of feedback on system performance. One of such side effects is system instability which is of utmost importance for design engineers. In this chapter the conditions for stability of feedback systems have been formulated. First, the stability conditions were stated in general, descriptive terms to enhance understanding of the problem. A more rigorous definition of stability for linear systems, developed by Lyapunov, was then presented. It was shown that for stability of linear systems it is necessary and sufficient if all roots of the closed-loop system characteristic equation lie strictly in the left half of the complex plane. Two algebraic stability criteria developed by Hurwitz and Routh were introduced. The Hurwitz criterion uses determinants built on coefficients of the characteristic equation to determine if there are any roots located in the left half of the complex plane. The Routh method leads to the same result and, in addition, gives the number of unstable roots. The Nyquist stability criterion was also presented. The practical importance of the Nyquist method lies in that it determines stability of a closed-loop system based on the frequency response of the system components with the feedback loop open. Two quantitative measures of stability, gain margin and phase margin, derived from the Nyquist criterion were introduced. Finally, a root locus method for design of feedback systems, based on migration paths of the system characteristic roots in the complex plane, was described.

[9] J-C Gille, M. J. Pelegrin, P. DeCaulne, *Feedback Control Systems,* McGraw-Hill Book Co., New York, 1959, pp. 242–244.

Problems

13.1. A closed-loop transfer function of a dynamic system is

$$\mathbf{T}_{\mathrm{CL}}(s) = \frac{s + 10}{10s^4 + 10s^3 + 20s^2 + s + 1}$$

Use the Hurwitz criterion to determine stability of this system.

13.2. The transfer functions of the system represented by the block diagram shown in Figure P13.2 are

$$\mathbf{G}(s) = \frac{2s + 1}{3s^3 + 2s^2 + s + 1}$$

$$\mathbf{H}(s) = 10$$

(a) Determine stability of the open-loop system.
(b) Determine stability of the closed-loop system.

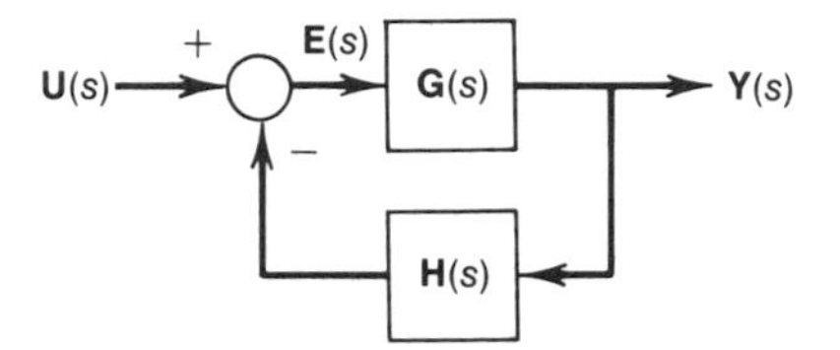

FIGURE P13.2 Block diagram of the feedback system considered in Problem 13.2.

13.3. Figure P13.3 shows a block diagram of a control system. The transfer functions of the controller $\mathbf{T}_{\mathrm{C}}(s)$ and of the controlled process $\mathbf{T}_{\mathrm{P}}(s)$ are

$$\mathbf{T}_{\mathrm{C}}(s) = K\left(1 + \frac{1}{4s}\right)$$

$$\mathbf{T}_{\mathrm{P}}(s) = \frac{5}{100s^2 + 20s + 1}$$

Using Hurwitz criterion, determine stability conditions for the open-loop and closed-loop systems in terms of the controller gain, K.

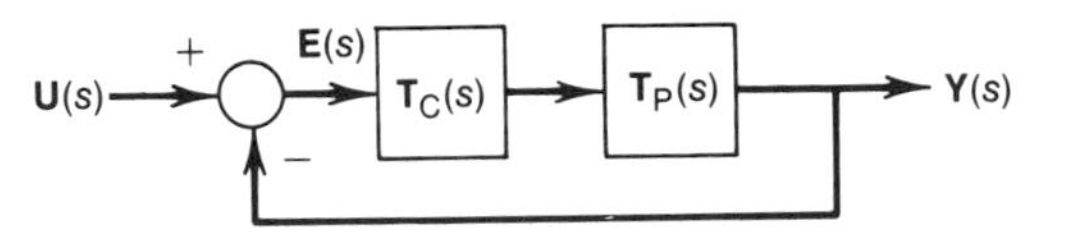

FIGURE P13.3 Block diagram of a control system.

13.4. The transfer functions of the system shown in Figure P13.3 are

$$\mathbf{T}_C(s) = K$$

$$\mathbf{T}_P(s) = \frac{2}{s(\tau s + 1)^2}$$

Determine the stability condition for the closed-loop system in terms of the controller gain K and the process time constant τ. Show the area of the system stability in the (τ, K) coordinate system.

13.5. Examine the stability of the systems whose characteristic equations are given below. Determine the number of roots of the characteristic equation having positive real parts for unstable systems.
(a) $s^3 + 12s^2 + 41s + 42 = 0$
(b) $400s^3 + 80s^2 + 44s + 10 = 0$
(c) $s^4 + s^3 - 14s^2 + 26s - 20 = 0$

13.6. The transfer functions of the system whose block diagram is shown in Figure P13.3 are

$$\mathbf{T}_C(s) = K$$

$$\mathbf{T}_P(s) = \frac{1}{s(0.2s + 1)(0.08s + 1)}$$

Sketch the polar plot of the open-loop system $\mathbf{T}_C\mathbf{T}_P$, and determine the stability condition for the closed-loop system in terms of the controller gain K using the Nyquist criterion.

13.7. The block diagram for a control system has been developed as shown in Figure P13.7. The system parameters are

$$k_C = 3.0 \text{ v/v} \qquad k_P = 4.6 \text{ m/v}$$

$$\tau_i = 3.5 \text{ sec} \qquad \tau_P = 1.4 \text{ sec}$$

$$\tau_C = 0.1 \text{ sec} \qquad k_f = 1.0 \text{ v/m}$$

(a) Determine whether the system is stable.
(b) If the system is stable, find the stability gain and phase margins.

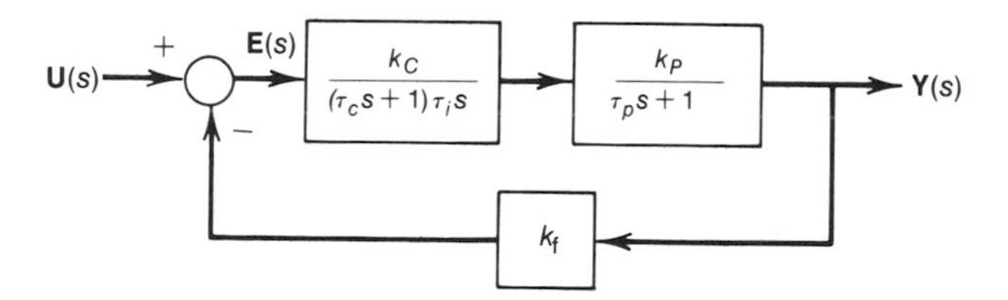

FIGURE P13.7 Block diagram of the system considered in Problem 13.7.

13.8. A simplified block diagram of a servomechanism used to control the angular position of an antenna dish, Θ, is shown in Figure P13.8. A potentiometer having gain k_p is used to produce a voltage signal proportional to the angular position of the antenna. This voltage signal is compared with voltage $u(t)$ which is proportional to the desired position of the antenna. The difference between the desired and actual position is amplified to produce a driving signal for the electrical d.c. motor.

(a) Determine the stability of the closed-loop system if the potentiometer gain is $k_p = 1.5$ v/rad

(b) Find the potentiometer gain, for which the closed-loop system will be stable with the gain margin $k_g = 1.2$.

(c) Find the potentiometer gain, for which the closed-loop system will be stable with the phase margin, $\gamma = 45$ deg.

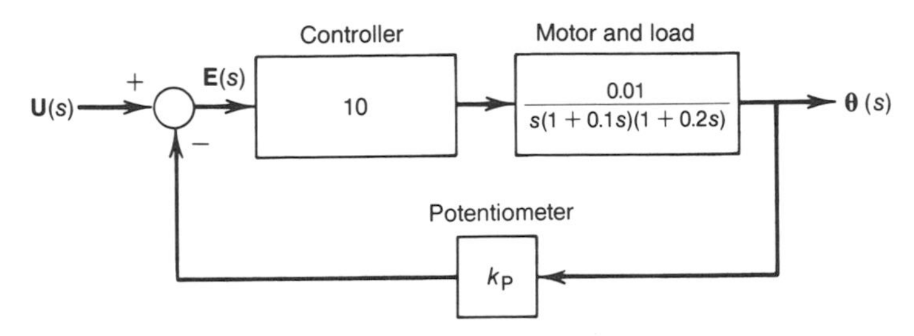

FIGURE P13.8 Simplified block diagram of a servomechanism for Problem 13.8.

13.9. Consider the feedback system represented by the block diagram shown in Figure P13.3 with the following transfer functions:

$$\mathbf{T}_C(s) = K$$

$$\mathbf{T}_P(s) = \frac{10}{(s + 5)(s + 0.2)}$$

(a) Construct the root locus for this system.

(b) Determine the locations of the roots of the system characteristic equation required for 20 percent overshoot in the system step response. Find the value of K necessary for the roots to be at the desired locations. What will be the period of damped oscillations T_d in the system step response?

13.10. The transfer functions for the system represented by the block diagram shown in Figure P13.3 are

$$\mathbf{T}_C(s) = K$$

$$\mathbf{T}_P(s) = \frac{1}{(s + 1)(s + 2)(s + 5)}$$

(a) Construct the root locus for this system.

(b) Use the constructed root locus to determine the value of K for which the closed-loop system is marginally stable.

14

Control Systems

14.1 Introduction

In Chapter 1, it was pointed out that a negative feedback is present in nearly all existing engineering systems. In control systems, which are introduced in this chapter, the negative feedback is included intentionally as a means for obtaining specified performance of the system.

Control is an action undertaken in order to obtain a desired behavior of a system, and it can be applied in an open-loop or a closed-loop configuration. In an open-loop system, shown schematically in Figure 14.1, a process is controlled in a certain prescribed manner regardless of the actual state of the process. A washing machine performing a predefined sequence of operations without any information and "with no concern" regarding the results of its operation is an example of an open-loop control system.

In a closed-loop control system, shown in Figure 14.2, the controller produces a control signal based on the difference between the desired and the actual process output. The washing machine considered earlier as an open-loop system would operate in a closed-loop mode if it was equipped with a measuring device capable of generating a signal related to the degree of cleanness of the laundry being washed.

Open-loop control systems are simpler and less expensive (at least by the cost of the measuring device necessary to produce the feedback signal); however, their performance can be satisfactory only in applications involving highly repeatable

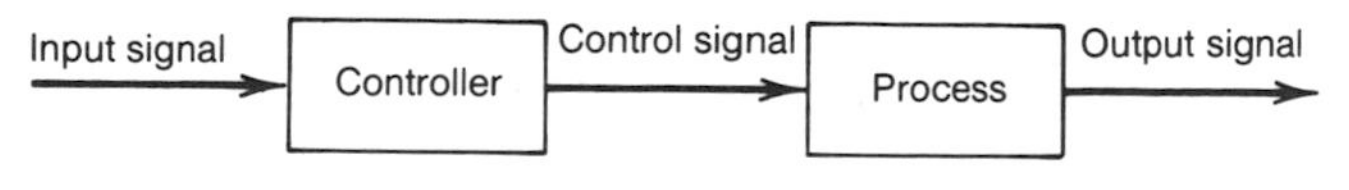

FIGURE 14.1 Open-loop control system.

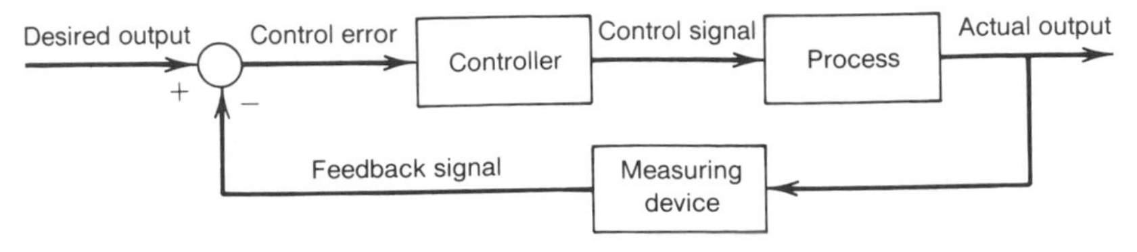

FIGURE 14.2 Closed-loop control system.

processes, having well established characteristics, and not exposed to disturbances. The methods for analysis of open-loop control systems are the same as the methods developed for analysis of dynamic systems in general, discussed in earlier chapters. In this chapter, the basic characteristics of linear closed-loop systems will be investigated.

In Chapter 13 the conditions under which a linear system is stable were derived. It was also shown how to design a linear closed-loop system for a specified gain and phase margin. Equally important to stability characteristics is the knowledge of the expected steady-state performance of the system. In Section 14.2 a control error (that is the difference between the desired and the actual system output) at steady state is evaluated for various types of systems and various types of input signals. Another aspect of steady-state performance of control systems, namely the sensitivity to disturbances, is discussed in Section 14.3. The problem of designing a control system that will meet specified steady-state and transient performance criteria is presented in Section 14.4. In Section 14.5 the most common algorithms employed in industrial controllers are described. More specialized control devices, called compensators, are the subject of Section 14.6.

14.2 Steady-State Control Error

For most closed-loop control systems the primary goal is to produce an output signal that follows an input signal as closely as possible. The steady-state performance of a control system is therefore judged by the steady-state difference between the input and output signals, that is, the steady-state error. Any physical control system inherently suffers steady-state error in response to certain types of inputs due to inadequacies in the system components, such as insufficient gain, output limiting, static friction, amplifier drift, or aging. In general, the steady-state performance depends not only on the control system itself but also on the type of input signal applied. A system may have no steady-state error to a step input, yet the same system may exhibit nonzero steady-state error in response to a ramp input. This error can usually be reduced by increasing the open-loop gain. Increasing the open-loop gain should, however, be done with care because it usually has other effects on the system performance. Such effects may include increasing the speed of response, that is reducing the time to reach steady-state, which is usually welcome, and increasing the system tendency to oscillate, eventually reducing stability margins, which is unwelcome.

The steady-state error that occurs in control systems owing to their incapability of following particular types of inputs will now be discussed.

Consider the closed-loop system shown in Figure 14.3. The open-loop transfer function of this system, $\mathbf{G}(s)\mathbf{H}(s)$ is (assuming all roots are real)

$$\mathbf{G}(s)\mathbf{H}(s) = \frac{K(\tau_1 s + 1)(\tau_2 s + 1) \cdots (\tau_m s + 1)}{s^r(\tau_{m+1} s + 1)(\tau_{m+2} s + 1) \cdots (\tau_{m+n} s + 1)} \tag{14.1}$$

where m is the number of zeros, n is the number of nonzero poles, and r is the multiplicity of poles at the origin. The "error transfer function" of the system relating the error signal $e(t)$ and the input signal $u(t)$ in the domain of complex variable s is defined as

$$\mathbf{T}_E(s) = \frac{\mathbf{E}(s)}{\mathbf{U}(s)} \tag{14.2}$$

For the system shown in Figure 14.3, $\mathbf{T}_E(s)$ takes the form

$$\mathbf{T}_E(s) = \frac{\mathbf{U}(s) - \mathbf{Y}(s)\mathbf{H}(s)}{\mathbf{U}(s)} \tag{14.3}$$

The closed-loop transfer function is

$$\frac{\mathbf{Y}(s)}{\mathbf{U}(s)} = \frac{\mathbf{G}(s)}{1 + \mathbf{G}(s)\mathbf{H}(s)} \tag{14.4}$$

Substituting Equation (14.4) into (14.3) yields

$$\mathbf{T}_E(s) = \frac{1}{1 + \mathbf{G}(s)\mathbf{H}(s)} \tag{14.5}$$

The error signal is thus given by

$$\mathbf{E}(s) = \mathbf{T}_E(s)\mathbf{U}(s) \tag{14.6}$$

or

$$\mathbf{E}(s) = \frac{\mathbf{U}(s)}{1 + \mathbf{G}(s)\mathbf{H}(s)} \tag{14.7}$$

Applying the final value theorem from Laplace transform theory (Appendix II), the steady-state error e_{ss} can be calculated as

$$e_{ss} = \lim_{t\to\infty} e(t) = \lim_{s\to 0} s\mathbf{E}(s) = \lim_{s\to 0} \frac{s\mathbf{U}(s)}{1 + \mathbf{G}(s)\mathbf{H}(s)} \tag{14.8}$$

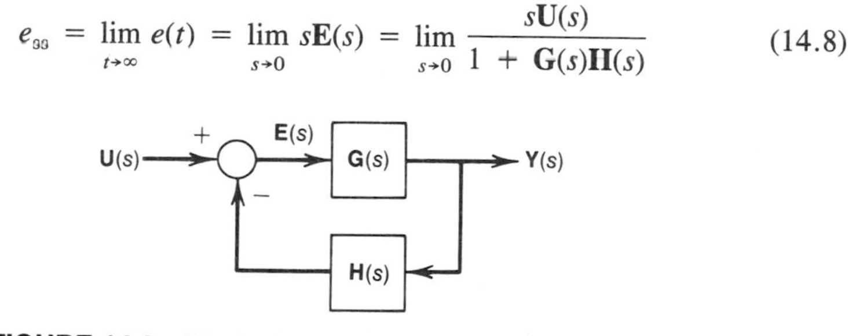

FIGURE 14.3 Block diagram of a closed-loop system.

This result will now be used in evaluating the steady-state error in response to step inputs and ramp inputs. The results obtained for those two types of input signals can be applied to more general cases involving linear systems where actual inputs may be considered combinations of such inputs.

Unit Step Input, $u(t) = U_s(t)$. The unit step function in the domain of complex variable s is represented by

$$\mathbf{U}(s) = \frac{1}{s} \tag{14.9}$$

where $\mathbf{U}(s) = \mathcal{L}[u(t)]$ with $u(t)$ being a unit step at $t = 0$. The steady-state error for a unit step is calculated using Equation (14.8).[1]

$$e_{ss} = \lim_{s \to 0} \frac{s(1/s)}{1 + \mathbf{G}(s)\mathbf{H}(s)} \tag{14.10}$$

or

$$e_{ss} = \frac{1}{1 + K_p} \tag{14.11}$$

where K_p is the static position error coefficient defined as

$$K_p = \lim_{s \to 0} \mathbf{G}(s)\mathbf{H}(s) = \mathbf{G}(0)\mathbf{H}(0) \tag{14.12}$$

or, using Equation (14.1),

$$K_p = \lim_{s \to 0} \frac{K(\tau_1 s + 1)(\tau_2 s + 1) \cdots (\tau_m s + 1)}{s^r(\tau_{m+1} s + 1)(\tau_{m+2} s + 1) \cdots (\tau_{m+n} + 1)} \tag{14.13}$$

Now assume that the system has no poles at the origin, $r = 0$. Such systems are called type 0 systems. The static position error coefficient for the type 0 system is

$$K_p = K \tag{14.14}$$

and the steady-state error is

$$e_{ss} = \frac{1}{1 + K} \tag{14.15}$$

If there is at least one pole at the origin, $r \geqq 1$, the static position error coefficient K_p is infinite and the steady-state error is zero, $e_{ss} = 0$. Systems with $r = 1, 2, \ldots$ are called type 1, 2, . . . systems, respectively. Therefore, it can be said that

$$e_{ss} = 0 \quad \text{for type 1 or higher systems} \tag{14.16}$$

[1]Note that s cancels $1/s$ in Equation 14.10. This leads to the notion of using a simplified final value expression for a unit step input, i.e.

$$e_{ss} = \lim_{s \to 0} \mathbf{T}_E(s) = \lim_{s \to 0} \frac{1}{1 + \mathbf{G}(s)\mathbf{H}(s)}$$

The use of this simplified final value expression can be substantiated by using $u(t) = \lim_{s \to 0} e^{st}$ to represent a unit step and finding the system response as $y(t) = \lim_{s \to 0} \mathbf{Y}(s)e^{st} = \lim_{s \to 0} \mathbf{T}_E(s)e^{st}$.

Unit Ramp Input, $u(t) = t$. The steady-state error for a unit ramp input is given by[2]

$$e_{ss} = \lim_{s \to 0} \frac{s \dfrac{1}{s^2}}{1 + \mathbf{G}(s)\mathbf{H}(s)} = \lim_{s \to 0} \frac{1}{s\mathbf{G}(s)\mathbf{H}(s)} \tag{14.17}$$

The static velocity error coefficient is defined as

$$K_v = \lim_{s \to 0} s\mathbf{G}(s)\mathbf{H}(s) \tag{14.18}$$

The steady-state error can thus be expressed as

$$e_{ss} = \frac{1}{K_v} \tag{14.19}$$

For a type 0 system, $r = 0$, the static velocity error coefficient is

$$K_v = \lim_{s \to 0} sK \frac{(\tau_1 s + 1)(\tau_2 s + 1) \cdots (\tau_m s + 1)}{(\tau_{m+1} s + 1)(\tau_{m+2} s + 1) \cdots (\tau_{m+n} s + 1)} = 0 \tag{14.20}$$

and the steady-state error[3]

$$e_{ss} = \frac{1}{K_v} = \infty \quad \text{for type 0 systems} \tag{14.21}$$

For a type 1 system, $r = 1$

$$K_v = \lim_{s \to 0} sK \frac{(\tau_1 s + 1)(\tau_2 s + 1) \cdots (\tau_m s + 1)}{s(\tau_{m+1} s + 1)(\tau_{m+2} s + 1) \cdots (\tau_{m+n} s + 1)} = K \tag{14.22}$$

and the steady state error

$$e_{ss} = \frac{1}{K_v} = \frac{1}{K} \quad \text{for type 1 systems} \tag{14.23}$$

For a type 2 or higher system, $r \geqq 2$

$$K_v = \lim_{s \to 0} sK \frac{(\tau_1 s + 1)(\tau_2 s + 1) \cdots (\tau_m s + 1)}{s^r(\tau_{m+1} s + 1)(\tau_{m+2} s + 1) \cdots (\tau_{m+n} s + 1)} = \infty \tag{14.24}$$

and the steady-state error

$$e_{ss} = \frac{1}{K_v} = 0 \quad \text{for type 2 or higher systems} \tag{14.25}$$

[2]Note again that a simplified final value expression like that stated in the last footnote but for the case of a unit ramp input yields an identical result as obtained from Equation 14.17, i.e.,

$$e_{ss} = \lim_{s \to 0} (1/s)\mathbf{T}_E(s) = \lim_{s \to 0} (1/s) \frac{1}{1 + \mathbf{G}(s)\mathbf{H}(s)} = \lim_{s \to 0} \frac{1}{s\mathbf{G}(s)\mathbf{H}(s)}$$

[3]Although the system does not reach a steady value for y, the time rate of change of y does reach a steady-state if the system is otherwise stable.

TABLE 14.1. Values of the steady-state error

	Type of System				
Input Signal	0	1	2	. . .	N
$U_s(t) = 1$	$1/(1+K)$	0	0		0
$U_r(t) = t$	∞	$1/K$	0		0
t^2	∞	∞	$1/K$		0
⋮	⋮	⋮	⋮		⋮
t^N	∞	∞	∞	. . .	$1/K$

Table 14.1 summarizes the steady-state errors for systems of types 0 to N when they are subjected to inputs of order 0 to N. When the order of the input signal is the same as the type of the system, the steady-state error is finite, as depicted by the values on the diagonal line of Table 14.1. When the input signal varies faster than t raised to a power equal to the type of the system, the error approaches infinity. Finally, when the input varies like t but raised to a power smaller than the type of the system, the steady-state error is zero.

14.3 Steady-State Disturbance Sensitivity

Another important aspect of a control system performance is its sensitivity to disturbances. Disturbances are all those other inputs that are not directly controlled by feedback to a separate summing point in the system. Very often disturbances are difficult to measure; so their presence can only be detected by observations of variations in the process output signal that take place while the control input signal is unchanged. Some of the most common disturbances are load force or torque in mechanical systems, variations of ambient temperature in thermal systems, and variations of load pressure or load flow-rate in fluid systems. A control system should be capable of maintaining the process output signal at the desired level in the presence of disturbances. In fact, as pointed out at the beginning of this chapter, it is the presence of disturbances in the system environment that provides the rationale for many closed-loop control systems.

When a disturbance changes suddenly, the system output signal will deviate at least temporarily from its desired value, even in the best designed system. What is expected from a well-designed control system is that, after the transients die out, the output signal will return to its previous level. A system's ability to compensate for the steady-state effects of disturbances is determined quantitatively in terms of the steady-state disturbance sensitivity S_D, defined as the ratio of the change of the output Δy versus the change of the disturbance Δv at steady-state.

$$S_D = \frac{\Delta y_{ss}}{\Delta v_{ss}} \tag{14.26}$$

In the open-loop system shown in Figure 14.4a, the disturbance sensitivity can be calculated using final value theorem, assuming that $\mathbf{U}(s) = 0$

$$S_{DO} = \frac{\lim_{s \to 0} s\mathbf{Y}(s)}{\lim_{s \to 0} s\mathbf{V}(s)} \tag{14.27}$$

Assuming that $v(t)$ is a step function, Equation (14.27) leads to the simplified final value expression

$$S_{DO} = \frac{\lim_{s \to 0} s \frac{1}{s} \mathbf{G}_v(s)}{\lim_{s \to 0} s \frac{1}{s}} = \lim_{s \to 0} \mathbf{G}_v(s) = \mathbf{G}_v(0) = K_D \tag{14.28}$$

where K_D is the static gain of $\mathbf{G}_v(s)$.

In the closed-loop system shown in Figure 14.4b, the output $\mathbf{Y}(s)$ for $\mathbf{U}(s) = 0$ is

$$\mathbf{Y}(s) = \frac{\mathbf{V}(s)\mathbf{G}_v(s)}{1 + \mathbf{G}_{\mathrm{OL}}(s)} \tag{14.29}$$

hence, the sensitivity of the closed-loop system, S_{DC}, for a step disturbance is

$$S_{DC} = \frac{\mathbf{G}_v(0)}{1 + \mathbf{G}_{\mathrm{OL}}(0)} = \frac{K_D}{1 + K} \tag{14.30}$$

where K is the steady-state gain of the open-loop transfer function $\mathbf{G}_{\mathrm{OL}}(s)$. By comparing Equations (14.28) and (14.30), it can be seen that the closed-loop system is less sensitive to disturbance, $S_{DC} < S_{DO}$, if the steady-state gain K is positive.

Example 14.1 illustrates the mathematical considerations presented above.

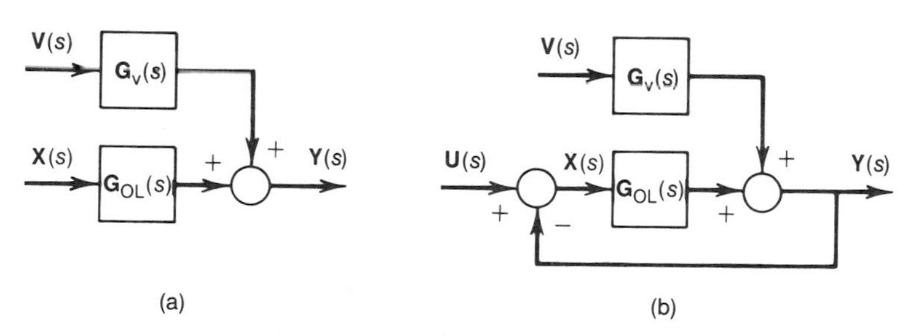

FIGURE 14.4 (a) Open-loop and (b) closed-loop systems subjected to disturbance $\mathbb{V}(s)$.

EXAMPLE 14.1

Consider a heating system for a one-room sealed up house shown schematically in Figure 14.5. The house is modeled as a lumped system having contents of mass m and average specific heat c. The rate of heat supplied by an electric heater is $Q_{in}(t)$. The spatial average temperature inside the house is T_1 and the ambient air temperature is T_2. Determine the effect of variation of T_2 on T_1 at steady state.

Solution

The heat balance equation is

$$mc\frac{dT_{1r}(t)}{dt} = Q_{in}(t) - Q_{loss}(t) \tag{14.31}$$

The rate of heat losses, Q_{loss}, is given by

$$Q_{loss}(t) = U_o(T_{1r}(t) - T_{2r}(t)) \tag{14.32}$$

where U_o is a heat loss coefficient. The heat balance equation becomes

$$mc\frac{dT_{1r}(t)}{dt} = Q_{in}(t) - U_o(T_{1r}(t) - T_{2r}(t)) \tag{14.33}$$

Transferring Equation (14.33) from the time domain into the domain of complex variable s yields

$$(mcs + U_o)\mathbf{T}_{1r}(s) = \mathbf{Q}_{in}(s) + U_o\mathbf{T}_{2r}(s)$$

or

$$\mathbf{T}_{1r}(s) = \frac{\mathbf{Q}_{in}(s) + U_o\mathbf{T}_{2r}(s)}{mcs + U_o} \tag{14.34}$$

The block diagram of the system represented Equation (14.34) is shown in Figure 14.6.

By comparing Figures 14.4a and 14.6, the disturbance and process transfer functions for the open-loop system can be identified as

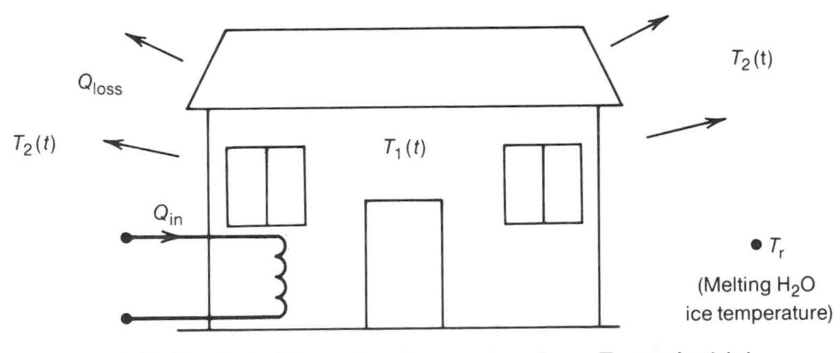

FIGURE 14.5 House heating system from Example 14.1.

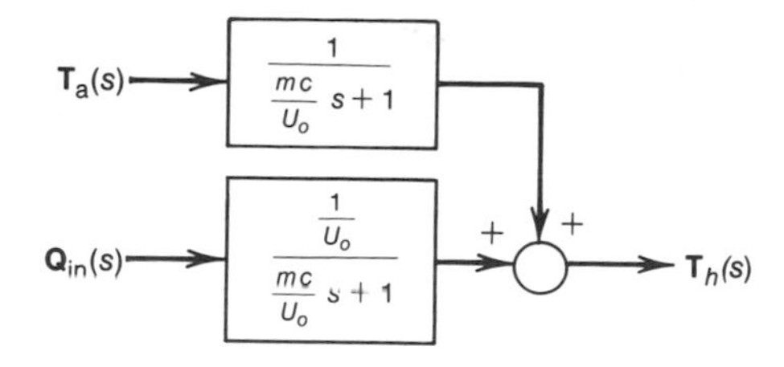

FIGURE 14.6 Block diagram of the open-loop thermal system considered in Example 14.1.

$$\mathbf{G}_v(s) = \frac{1}{\dfrac{mc}{U_o}s + 1}$$

and

$$\mathbf{G}_{\mathrm{OL}}(s) = \frac{\dfrac{1}{U_o}}{\dfrac{mc}{U_o}s + 1}$$

Using Equation (14.28) the steady-state disturbance sensitivity can be calculated as

$$S_{DO} = \lim_{s \to 0} \frac{1}{\dfrac{mc}{U_o}s + 1} = 1 \tag{14.35}$$

Equation (14.35) indicates that the change of ambient temperature by ΔT will cause the change in the house temperature by the same value ΔT.

Now consider a closed-loop system where the rate of heat supply is controlled to cause the house temperature to approach the desired level, T_d. The block diagram of the closed-loop temperature control system is shown in Figure 14.7. The rate of heat supply $Q_{\mathrm{in}}(t)$ is assumed proportional to the temperature deviation

$$Q_{\mathrm{in}}(t) = k_f(T_d(t) - T_h(t))$$

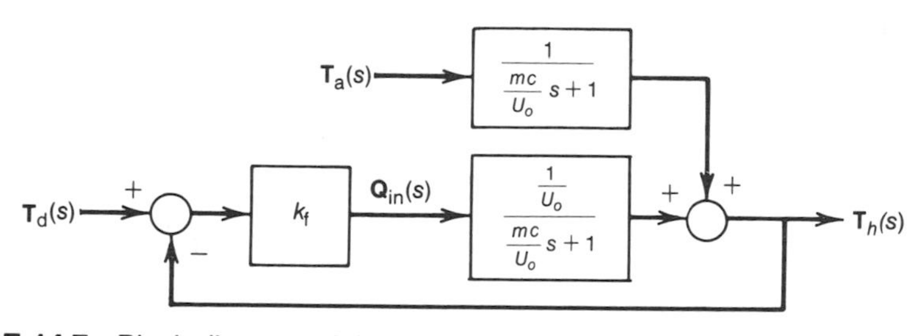

FIGURE 14.7 Block diagram of the closed-loop temperature control system considered in Example 14.1.

Using Equation (14.30), the disturbance sensitivity in this closed-loop system is found to be

$$S_{DC} = \frac{1}{1 + \dfrac{k_f}{U_o}}$$

where k_f/U_o is the open-loop gain of the system. Since both k_f and U_o are positive, the sensitivity of the closed-loop system to variations of ambient air temperature is smaller than that of the open-loop system. The greater the open-loop gain, the less sensitive is the closed-loop system to disturbances. ■

14.4 Interrelation of Steady-State and Transient Considerations

In Sections 14.2 and 14.3 it was shown that in order to improve the system steady-state performance, the open-loop gain has to be increased or an integration has to be added to the open-loop transfer function. Both remedies will, however, aggravate the stability problem. In general, the design of a system with more than two integrations in the feedforward path is very difficult. A compromise between steady-state and transient system characteristics is thus necessary. Example 14.2 illustrates this problem.

EXAMPLE 14.2

Examine the effect of the open-loop gain K on stability and steady-state performance of the system shown in Figure 14.8, which is subjected to unit ramp input signals.

Solution

The open-loop sinusoidal transfer function is

$$\mathbf{G}(j\omega)\mathbf{H}(j\omega) = \mathbf{T}_{OL}(j\omega) = \frac{K}{j\omega(j\omega + 1)(j\omega + 5)}$$

The real and imaginary parts of $\mathbf{T}_{OL}(j\omega)$ are

$$\text{Re}\,[\mathbf{T}_{OL}(j\omega)] = \frac{-6K}{(\omega^4 + 26\omega^2 + 25)}$$

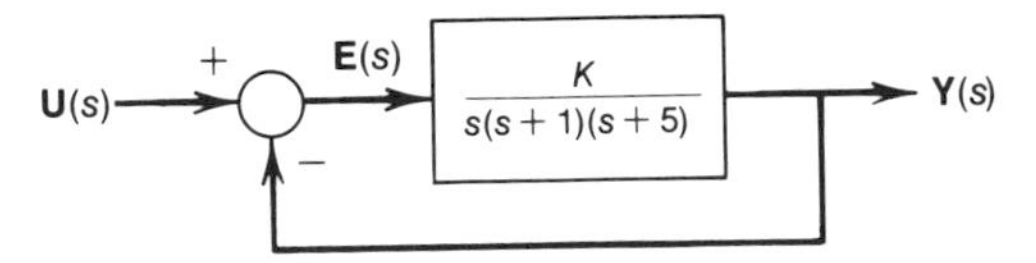

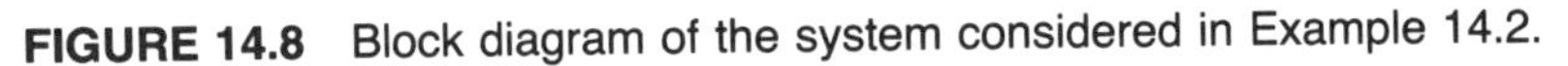

FIGURE 14.8 Block diagram of the system considered in Example 14.2.

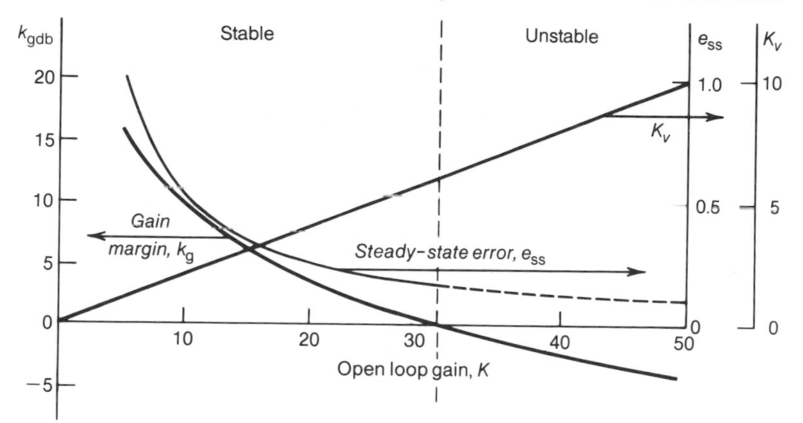

FIGURE 14.9 Effect of the open-loop gain on stability gain margin, k_g, velocity error coefficient, K_v, and steady-state error, e_{ss}.

$$\text{Im}\,[\mathbf{T}_{OL}(j\omega)] = \frac{K(\omega^2 - 5)}{(\omega^5 + 26\omega^3 + 25\omega)}$$

The stability gain margin k_g was defined in Chapter 13 as

$$k_{gdb} = 20 \log \frac{1}{|\mathbf{T}_{OL}(j\omega_p)|}$$

where ω_p is such that

$$\angle \mathbf{T}_{OL}(j\omega_p) = -180 \text{ deg}, \qquad \text{or } \mathbf{Im}[\mathbf{T}_{OL}(j\omega)] = 0$$

For this system $\omega_p = \sqrt{5}$ rad/sec, and the gain margin in decibels is

$$k_{gdb} = 20 \log \frac{30}{K}$$

Now, in order to examine the steady-state performance of the system subjected to a unit ramp input $u(t) = t$, the static velocity error coefficient must be determined. Using Equation (14.22) for a type 1 system yields

$$K_v = \lim_{s \to 0} \left(\frac{sK}{s(s + 1)(s + 5)} \right) = \frac{K}{5}$$

and hence the steady-state error is

$$e_{ss} = \frac{5}{K}$$

Figure 14.9 shows the system gain margin k_{gdb}, static velocity error coefficient K_v, and steady-state error e_{ss} as functions of the open-loop gain K. Note that the system is marginally stable for $K = 30$, for which the minimum steady-state error approaches 0.1667. Selecting values of K less than 30 will improve system stability at the cost increasing steady state error. ■

14.5 Industrial Controllers

In most process control applications standard "off-the-shelf" devices are used to obtain desired system performance. These devices, commonly called industrial controllers, compare the actual system output with the desired value and produce a signal to reduce the output signal deviation to zero or to a small value (Figure 14.2). The manner in which the controller produces the control signal in response to the control error is referred to as a control algorithm or a control law. Five of the most common control algorithms implemented by typical industrial controllers will now be described.

Two-Position or On-Off Control. The control algorithm of the two-position controller is

$$u(t) = \begin{cases} M_1 \text{ for } e(t) > 0 \\ M_2 \text{ for } e(t) < 0 \end{cases} \tag{14.36}$$

The relationship between the control signal $u(t)$ and the error signal $e(t)$ is shown in Figure 14.10. In many applications the control signal parameters M_1 and M_2 correspond to "on" and "off" positions of the actuating device. In an on-off temperature control system, a heater power is turned on or off, depending on whether the process temperature is below or above the desired level. In an on-off liquid level control system, the supply valve is either opened or closed, depending on the sign of the control error. In every two-position control system, the process output oscillates as the control signal is being switched between its two values, M_1 and M_2. Figure 14.11 shows the output and control signal of an on-off temperature control system. However this system is only piece-wise linear.

Proportional Control. The signal produced by a proportional, P, controller is, as its name implies, proportional to the control error

$$u(t) = k_p e(t) \tag{14.37}$$

The transfer function of the P controller is

$$\mathbf{T}_c(s) = \frac{\mathbf{U}(s)}{\mathbf{E}(s)} = k_p \tag{14.38}$$

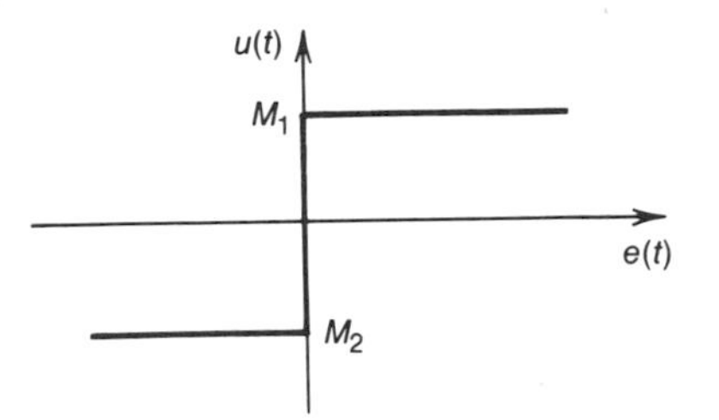

FIGURE 14.10 Control signal versus error signal in two-position control.

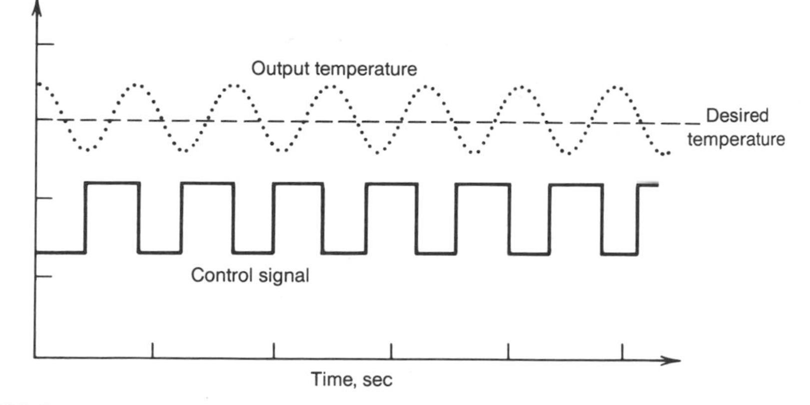

FIGURE 14.11 Output temperature and control signal in on-off control system.

The steady-state performance of the system and speed of response with P controller improve with increased gain k_p. Increasing the controller gain may, however, decrease stability margins.

Proportional-Integral Control. The steady-state performance can also be improved by adding an integral action to the control algorithm. The ideal porportional-integral, or PI, controller produces a control signal given by the equation

$$u(t) = k_p \left(e(t) + \frac{1}{T_i} \int_0^t e(\tau)\, d\tau \right) \tag{14.39}$$

The ideal controller transfer function is

$$\mathbf{T}_c(s) = k_p \left(1 + \frac{1}{T_i s} \right) \tag{14.40}$$

Adding integral control, while improving the steady-state performance, may lead to oscillatory response, (that is reduced potential stability margin) which is usually undesirable.

Proportional-Derivative Control. The stability of a system can be improved by adding a derivative action to the control algorithm. The control signal produced by an ideal proportional-derivative, or PD, controller is

$$u(t) = k_p \left(e(t) + T_d \frac{de(t)}{dt} \right) \tag{14.41}$$

The controller transfer function is[4]

$$\mathbf{T}_c(s) = k_p(1 + T_d s) \tag{14.42}$$

[4]The mathematical models given in Equations (14.41), (14.42), (14.43), (14.44) for ideal controllers are not physically realizable. All real controllers have a transfer function incorporating at least a fast first order parasitic lag term $(1 + \tau_p s)$ in the denominator. Inclusion of this lag term becomes important in order to program the controller for digital computer simulation.

The derivative action provides an anticipatory effect which results in a dampening of the system response. By increasing the stability margin in this way, it becomes possible to use greater loop gain, thus improving speed of response and reducing steady-state error.

Proportional-Integral-Derivative Control. All three control actions are incorporated in a proportional-integral-derivative control algorithm. The control signal generated by an ideal PID controller is

$$u(t)_o = k_p \left(e(t) + \frac{1}{T_i} \int_0^t e(\tau)\, d\tau + T_d \frac{de(t)}{dt} \right) \tag{14.43}$$

The transfer function of the ideal PID controller is

$$\mathbf{T}_C(s) = k_p \left(1 + \frac{1}{T_i s} + T_d s \right) \tag{14.44}$$

Determining optimal adjustments of the control parameters k_p, T_i, and T_d is one of the basic problems faced by control engineers. The tuning rules of Ziegler and Nichols[5] provide one of the simplest procedures developed for this purpose. There are two versions of this method: one is based on the process step response and the other on characteristics of sustained oscillations of the system under P control at the stability limit. The first method, based on a delay-lag model of the process, can be applied if process step response data are available in the form shown in Figure 14.12. The transfer function e^{-Ls} for the time delay L is discussed in Chapter 15.

The controller parameters are calculated using the values of slope R and delay time L of the unit step response as follows:

- For P controller

$$k_p = \frac{1}{RL}$$

- For PI controller

$$k_p = \frac{0.9}{RL} \qquad T_i = 3.3L$$

- For PID controller

$$k_p = \frac{1.2}{RL} \qquad T_i = 2L \qquad T_d = 0.5L$$

In the other method, the gain of the P controller in a closed-loop system test, shown in Figure 14.13a, is increased until a stability limit is reached with a test controller

[5] J. G. Ziegler and N. B. Nichols, ''Optimum Settings for Automatic Controllers,'' *Trans. ASME*, Vol. 64, 1942, p. 759.

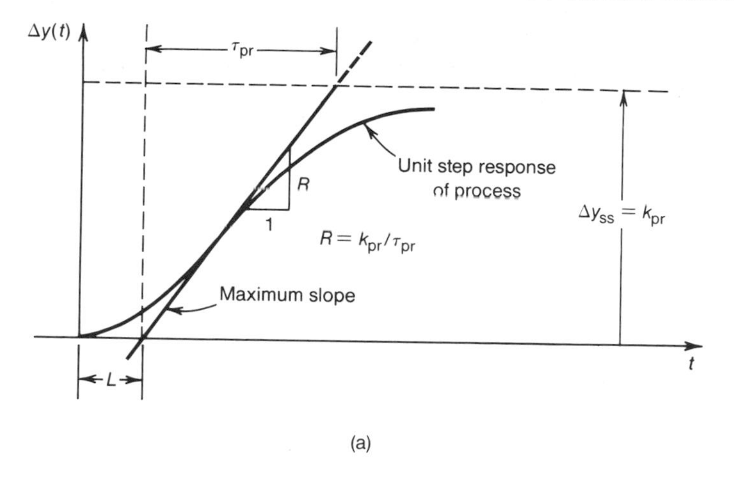

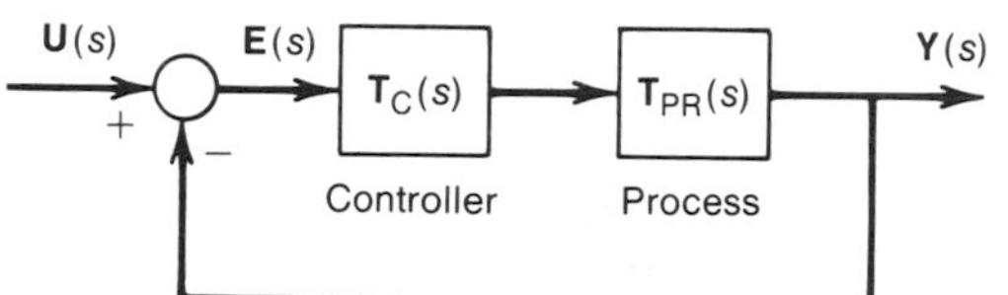

$$\mathbf{T}_C(s) = k_P(1 + \frac{1}{T_i s} + T_d s)$$

$$\mathbf{T}_{PR}(s) = k_{PR} \frac{e^{-Ls}}{\tau_{PR} s + 1}$$

$$\mathbf{T}_{PR}(o) = k_{PR}$$

(b)

FIGURE 14.12 Determining controller parameters on the basis of the process step response. (a) Process response graph, (b) Transfer function block diagram of system.

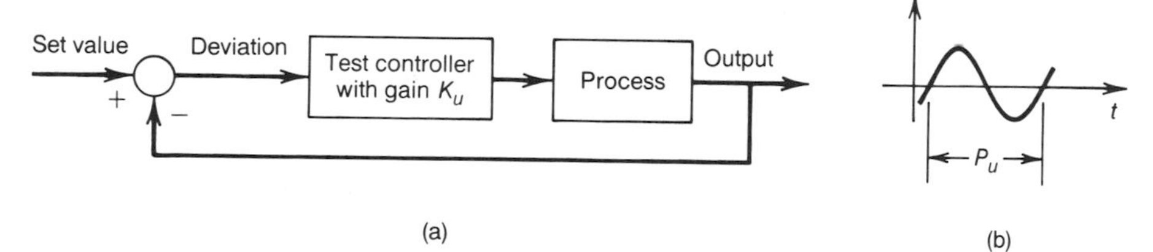

FIGURE 14.13 Determining controller parameters on the basis of stability limit oscillations.

gain K_u. The control parameters are then calculated based on this critical value of gain K_u and on the resulting period of sustained oscillations P_u using the following relations

- For P controller

$$k_p = 0.5K_u$$

- For PI controller

$$k_p = 0.45K_u \qquad T_i = 0.83P_u$$

- For PID controller

$$k_p = 0.6K_u \qquad T_i = 0.5P_u \qquad T_d = 0.125P_u$$

It has to be emphasized that the rules of Ziegler and Nichols were developed empirically, and the control parameters provided by these rules are not optimal. However, they do give a good starting point from which further tuning can be performed to obtain satisfactory system performance.

More advanced industrial controllers, available on the market today, are capable of self-tuning, that is, of automatically adjusting the values of their control settings to obtain the best performance with a given process.[6]

14.6 System Compensation

In some applications it may be very difficult or, sometimes, impossible to obtain both desired transient and steady-state system performance by adjusting parameters of PID controllers. In such cases, additional devices are inserted into the system to modify the open-loop characteristics and enhance the system performance. This technique is called system compensation and the additional devices inserted into the system are called compensators. Unlike controllers, compensators are usually designed for a specific application and their parameters are not adjustable.

There are several ways of inserting a compensator into a control system. Figure 14.14 shows block diagrams of control systems with (a) series compensation, (b) feedback compensation, and (c) feedforward disturbance compensation. The series structure is the most common and usully the simplest compensator to design. A typical example of feedback compensation is a velocity feedback in a position control system. The feedforward compensation is used to improve the system speed of response to disturbance when the disturbance is measurable.

There are two types of series compensators: lead compensators and lag compensators. The transfer function of the lead compensator is

$$\mathbf{T}_{\text{lead}}(s) = \alpha \frac{\tau s + 1}{\alpha \tau s + 1}, \quad \alpha < 1 \tag{14.45}$$

[6]K. J. Åström and T. Hagglund, "Automatic Tuning of PID Regulators", Instrument Society of America, Research Triangle Park, N.C., 1988.

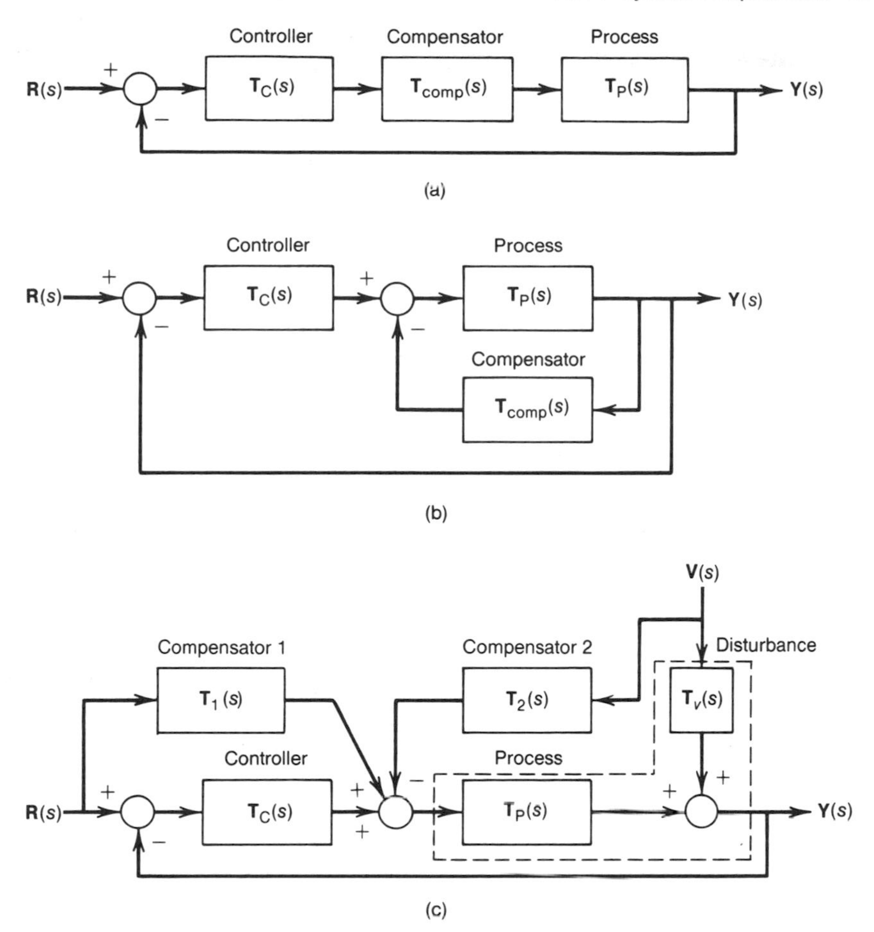

FIGURE 14.14 Block diagram of systems with (a) series, (b) feedback, and (c) feedforward compensation.

Compensators can be made of mechanical, electrical, and fluid components. An electrical lead compensator is shown in Figure 14.15. The transfer function relating the output voltage $\mathbf{E}_{2g}(s)$ to the input voltage $\mathbf{E}_{1g}(s)$, for this circuit when $i_L = 0$ is

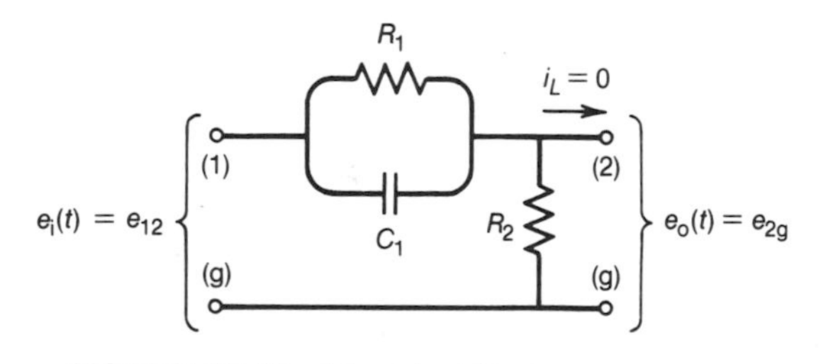

FIGURE 14.15 Electrical lead compensator.

$$\mathbf{T}(s) = \frac{\mathbf{E}_{2g}(s)}{\mathbf{E}_{1g}(s)} = \frac{R_2}{R_1 + R_2} \frac{R_1 C_1 s + 1}{\dfrac{R_1 R_2}{R_1 + R_2} C_1 s + 1} \tag{14.46}$$

By comparing Equations (14.45) and (14.46), the parameters α and τ are found.

$$\alpha = \frac{R_2}{R_1 + R_2} \tag{14.47}$$

$$\tau = R_1 C_1 \tag{14.48}$$

It can be noticed that the value of α given by Equation (14.47) is always smaller than one, as $R_2 < (R_1 + R_2)$. The lead compensator is used primarily to improve system stability. As the name indicates, this type of compensator adds a positive phase angle (phase lead) to the open-loop system frequency characteristics in a critical range of frequencies and thus increases the potential stability phase margin. Also, by increasing the potential stability margin, the lead compensation allows for further increasing the open-loop gain to achieve good dynamic and steady-state performance.[7]

The transfer function of the lag compensator when $i_L = 0$ is

$$\mathbf{T}_{lag}(s) = \frac{\tau s + 1}{\beta \tau s + 1} \quad \beta > 1 \tag{14.49}$$

Figure 14.16 shows an electrical lag compensator. The transfer function of this circuit is

$$\mathbf{T}(s) = \frac{\mathbf{E}_{2g}(s)}{\mathbf{E}_{1g}(s)} = \frac{R_2 C_2 s + 1}{(R_1 + R_2) C_2 s + 1} \tag{14.50}$$

or, equally,

$$\mathbf{T}(s) = \frac{R_2 C_2 s + 1}{s \dfrac{R_1 + R_2}{R_2} R_2 C_2 s + 1} \tag{14.51}$$

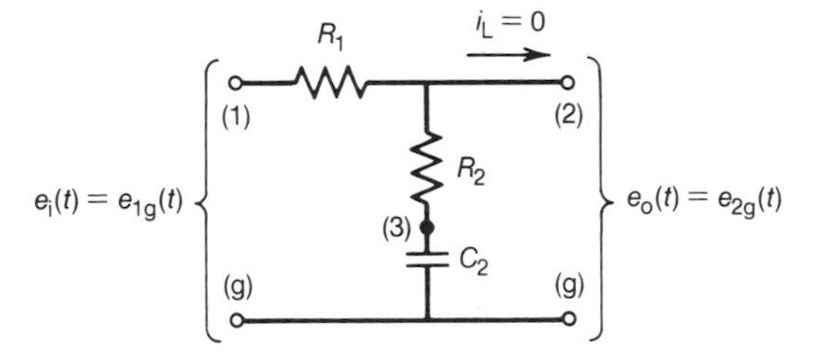

FIGURE 14.16 Electrical lag compensator.

[7]Thus using a lead compensator accomplishes somewhat the same effect as using a PD controller. Thus the use of lead compensation is sometimes referred to as using a "poor man's" PD controller.

By comparing Equations (14.49) and (14.51), the compensator parameters are found

$$\beta = \frac{R_1 + R_2}{R_2} \tag{14.52}$$

$$\tau = R_2C_2 \tag{14.53}$$

where β is always greater than one since $(R_1 + R_2) > R_2$.

The lag compensator improves steady-state performance of the system. System stability may, however, be seriously degraded by the lag compensator.[8]

The advantages of each of the two types of compensators are combined in a lag/lead compensator. An electrical lag/lead compensator is shown in Figure 14.17. The transfer function of this circuit for zero load, $i_L = 0$, is

$$\begin{aligned}\mathbf{T}_{\text{lag/lead}}(s) &= \frac{\mathbf{E}_{2g}(s)}{\mathbf{E}_{1g}(s)} \\ &= \frac{(R_1C_1s + 1)(R_2C_2s + 1)}{[(R_1C_1s + 1)(R_2C_2s + 1) + sR_1C_2]}\end{aligned} \tag{14.54}$$

or, equally,[9]

$$\mathbf{T}_{\text{lag/lead}}(s) = \left(\frac{\tau_1 s + 1}{\tau_2 s + 1}\right)\left(\frac{\gamma\tau_2 s + 1}{\gamma\tau_1 s + 1}\right) \tag{14.55}$$

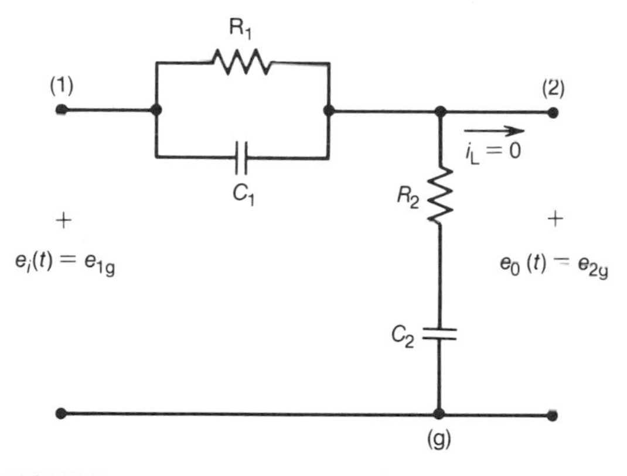

FIGURE 14.17 Electrical lag/lead compensator.

[8]Thus using a lag compensator, especially when $\beta\tau$ is very large, accomplishes somewhat the same effect as using a PI controller. Using such a properly designed lag compensator is sometimes referred to as using a "poor man's" PI controller.

[9]F. H. Raven, *Automatic Control Engineering*, McGraw-Hill Book Co., New York, 1987, pp 530–532.

where

$$\tau_1 = R_1C_1 \tag{14.56}$$

$$\tau_2 = R_2C_2 \tag{14.57}$$

$$\gamma = 1 + \frac{R_1C_2}{\tau_1 - \tau_2}, \ \tau_1 > \tau_2 \tag{14.58}$$

The lag/lead compensator is designed to improve both the transient and steady-state performance and its characteristics are similar to those of the PID controller.

14.7 Synopsis

In earlier chapters the analysis of system transient performance, including stability, was emphasized. In this chapter two important aspects of the steady-state performance have been addressed. First, the steady-state control error was considered. It was shown that the steady-state error depends on both the system transfer function and the type of input signal. A system is said to be of type r if there are r poles of the system open-loop transfer function located at the origin of the complex plane, $s = 0$. A type r system will produce zero steady-state error if the input signal is a function of time of order less than r, $u(t) = t^p$, where $p < r$. If $r = p$, the steady-state error has a finite value that depends on the gain of the open-loop transfer function; the greater the gain, thc smaller the steady-state error. If the input signal varies like time raised to a power higher than the type of the system, $p > r$, the steady-state error is infinity. The second aspect of steady-state performance addressed in this chapter was steady-state sensitivity to disturbances. It was shown that feedback reduces the effect of external disturbances on the system output at steady state.

The steady-state performance and the speed of response of feedback systems both usually improve when the system open-loop gain increases. On the other hand, the system stability margins usually decrease and the system may eventually become unstable when the open-loop gain is increased. A compromise is therefore necessary in designing feedback systems to ensure satisfactory steady-state and transient performance at the same time.

The most common algorithms employed with industrial controllers: on-off, proportional, proportional-integral, proportional-derivative, and proportional-integral-derivative, and their basic characteristics were presented. In general, the integral action improves steady-state performance, whereas the derivative action improves transient performance of the control system. Compensators, custom-designed control devices that complement the typical controllers to enhance the control system performance, were also introduced.

Problems

14.1. The open-loop transfer function of a system was found to be

$$\mathbf{T}_{\mathrm{OL}}(s) = \frac{K}{(s + 5)(s + 2)^2}$$

Determine the range of K for which the closed-loop system meets the following performance requirements: (1) the steady-state error for a unit step input is less than 10 percent of the input signal, and (2) the system is stable.

14.2. A simplified block diagram of the engine-speed control system, known as a flyball governor, invented by James Watt in the eighteenth century is shown in Figure P14.2.

(a) Develop the closed-loop transfer function $\mathbf{T}_{\mathrm{CL}}(s) = \mathbf{\Omega}_{\mathrm{o}}(s)/\mathbf{\Omega}_{\mathrm{d}}(s)$ and write the differential equation relating the actual speed of the engine $\Omega_{\mathrm{o}}(t)$ to the desired speed $\Omega_d(t)$ in the time domain.

(b) Find the gain of the hydraulic servo necessary for the steady-state value of the error in the system, $e(t) = \Omega_{\mathrm{o}}(t) - \Omega_d(t)$, to be less than 1% of the magnitude of the step input.

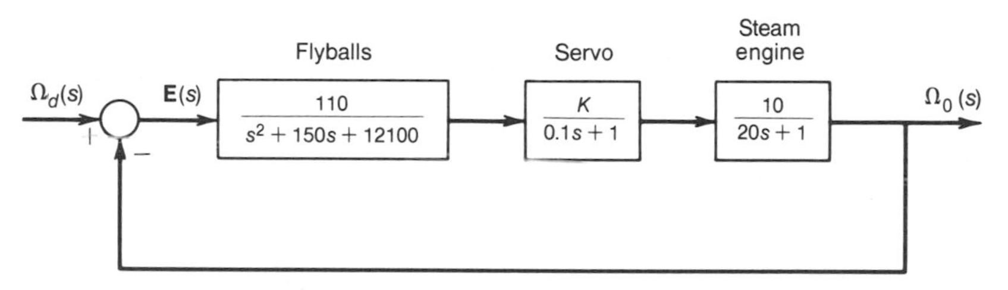

FIGURE P14.2 Block diagram of the steam engine speed control system.

14.3. The block diagram of the system designed to control angular velocity of the motor shaft Ω is shown in Figure P14.3. The system parameters are

$L = 100$ mH $\qquad R = 12$ ohms $\qquad \alpha = 68$ v-sec/rad (or N-m/amp)

$J = 4$ N-m-sec^2 $\qquad B = 15$ N-m-sec

where L and R are the series inductance and resistance of the armature of a DC motor, J is the combined motor and load inertia, B is the combined motor and load friction coefficient, and α is the electromechanical coupling coefficient.

(a) Determine the value of the tachometer gain k_T for which the damping ratio of the closed-loop system is greater than 0.5 and the system sensitivity to the load torque T_L is less than 2.0×10^{-3} rad/sec-N-m.

(b) Find the steady-state control error for a unit step change in the input voltage, $e_i(t) = U_s(t)$, using the value of k_T calculated in part (a).

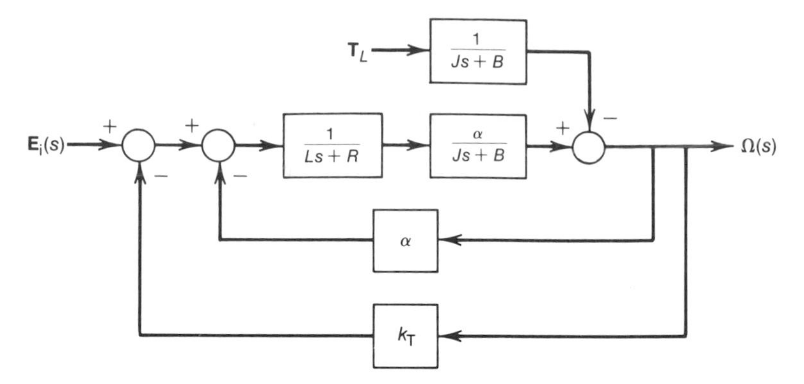

FIGURE P14.3 Block diagram of angular velocity control system.

14.4. The block diagram of the control system developed for a thermal process is shown in Figure P14.4.

(a) Determine the gain of the proportional controller k_p necessary for the stability gain margin $k_g = 1.2$.

(b) Find the steady-state control error in the system when the input temperature changes suddenly by 10°C, $T_i(t) = 10U_s(t)$, using the value of the proportional gain obtained from the stability requirement in part (a).

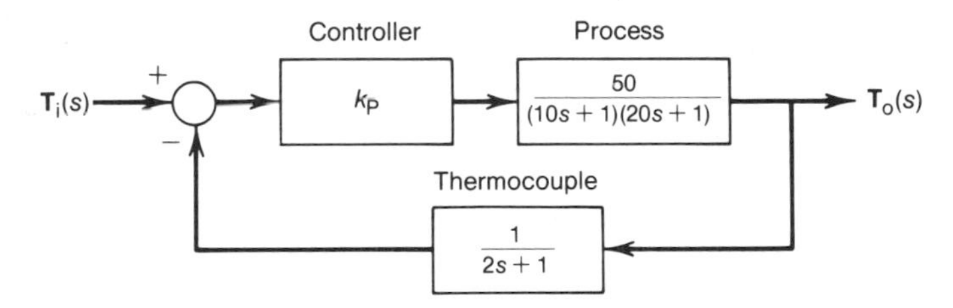

FIGURE P14.4 Block diagram of temperature control system.

14.5. A system open-loop transfer function is

$$\mathbf{T}_{OL}(s) = \frac{k}{s^2(\tau s + 1)}$$

Find the steady-state control error in the closed-loop system subjected to input $u(t) = t^2$. Express the steady-state error in terms of the static acceleration error coefficient K_a, defined as

$$K_a = \lim_{s \to 0} s^2 \mathbf{T}_{OL}(s)$$

14.6. The block diagram of a control system is shown in Figure P14.6. The process transfer function, $\mathbf{T}_P(s)$, is

$$\mathbf{T}_P(s) = \frac{1}{10s + 1}$$

Compare the performance of the control system with P and PI controllers. The controller transfer functions are

$$\mathbf{T}_C(s) = 9$$

for the P controller, and

$$\mathbf{T}_C(s) = 9\left(1 + \frac{1}{1.8s}\right)$$

for the PI controller. In particular, compare the percent overshoot of the step responses and the steady-state errors for a unit step input obtained with the two controllers.

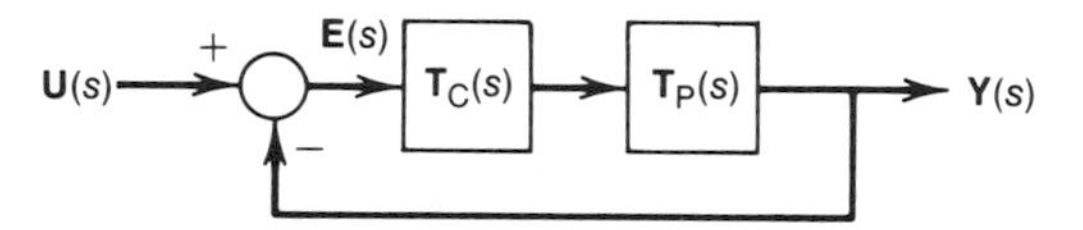

FIGURE P14.6 Block diagram of control system.

14.7. The process transfer function in the control system shown in Figure P14.6 has been found to be

$$\mathbf{T}_P(s) = \frac{k}{(\tau_1 s + 1)(\tau_2 s + 1)}$$

Compare the damping ratios and the steady-state errors for a step input obtained in this system with P and PD controllers. The controllers transfer functions are

$$\mathbf{T}_C(s) = k_p$$

for the P controller, and

$$\mathbf{T}_C(s) = k_p(1 + T_d s)$$

for the ideal PD controller.

14.8. A position control system, shown in Figure P14.8a, is being considered for a large turn-table. The turn-table is to be driven by a "torque motor" that provides an output torque that is proportional to its input signal, using power supplied from a dc source to achieve its inherent power amplification. One special requirement for the control system is to minimize the effects of an external load torque that may act from time to time on the turn-table. The system has been modeled as shown in Figure P14.8b to investigate the use of a P controller for this task. The values of the system parameters are:

$$k_m = 1.0 \text{ N-m/volt} \qquad k_a = 2.0 \text{ volt/rad}$$

$$J = 3.75 \text{ N-m-sec}^2 \qquad B = 1.25 \text{ N-m sec}$$

The general requirements for the steady-state performance of this system are:

(i) The steady-state error after a step input must be zero.
(ii) The steady-state load sensitivity resulting from a step change in the load torque T_L is to be less than 0.1 rad/N-m.

Find the value of the controller gain k_p required to achieve the desired steady-state performance. Calculate the damping ratio and the natural frequency of the system with this value of k_p.

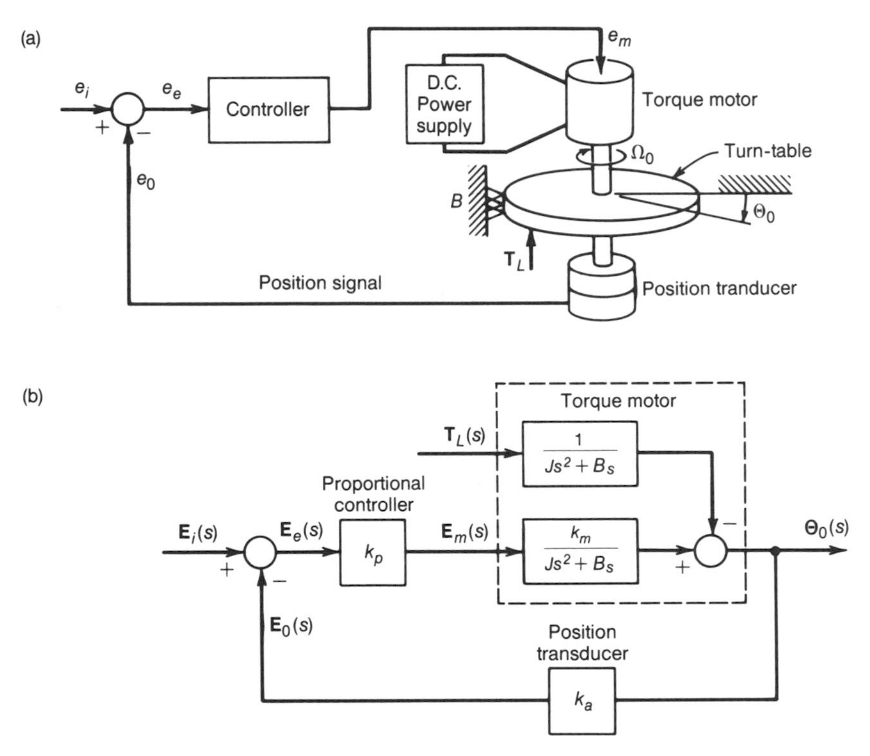

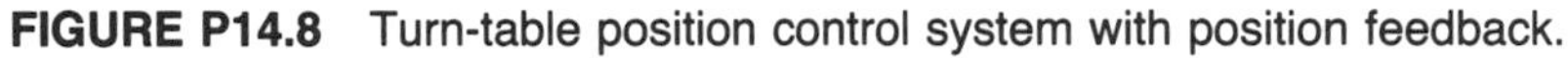
FIGURE P14.8 Turn-table position control system with position feedback.

14.9. This problem is a continuation of Problem 14.8 which must be solved first. Consider again the turn-table control system shown in Figure P14.8a. In order to improve the degree of damping of this system with the parameter values found in Problem 14.8, it is now proposed to employ the velocity transducer of gain k_v. Figure P14.9 shows a block diagram of the system developed to investigate the performance attainable with a P controller augmented with an inner loop being closed by velocity feedback to the torque motor. In addition to the steady-state performance requirements stated in Problem 14.8, the system must be stable, having a damping ratio of at least 0.5. Using the value of the controller gain, k_p, obtained in Problem 14.8, find the value of k_v needed to achieve the required damping ratio of 0.5. Check if all the steady-state and transient performance requirements specified in Problems 14.8 and 14.9 are satisfied in the redesigned system. Also, compare the speeds of response obtained with the two systems designed in this and in the preceeding problem.

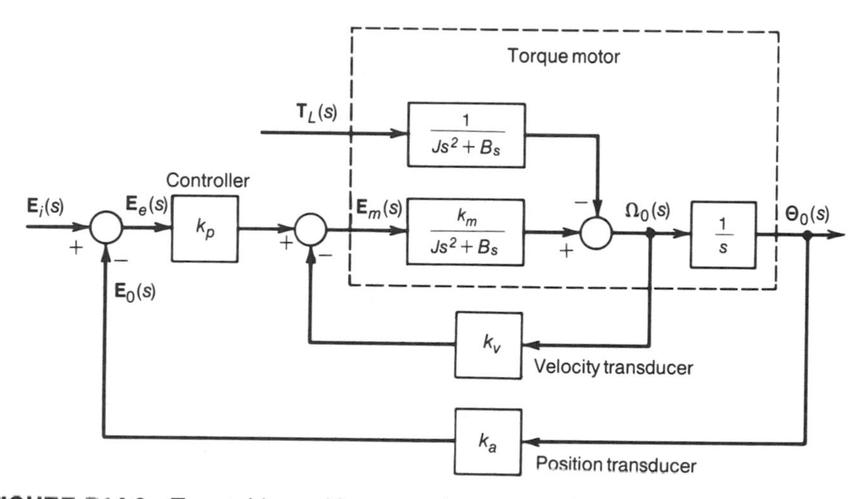

FIGURE P14.9 Turn-table position control system with position and velocity feedback.

15

Analysis of Discrete-Time Systems

15.1 Introduction

In almost all existing engineering systems, the system variables (input, output, state) are continuous functions of time. The first fourteen chapters of this book deal with this category of systems, classified in Chapter 1 as continuous dynamic systems. The last two chapters are devoted to discrete-time systems in which, according to the definition given in Chapter 1, the system variables are defined only at distinct instants of time. It may seem that there are not many such systems and, indeed, very few examples of intrinsically discrete engineering systems come to mind. There are, however, many systems involving continuous subsystems that are classified as discrete because of the discrete-time elements used to observe and control the continuous processes. Any system in which a continuous process is measured and/or controlled by a digital computer is considered discrete. Although some variables in such systems are continuous functions of time, they are known only at distinct instants of time determined by the computer sampling frequency, and therefore are treated as discrete-time variables.

The number of digital computer applications in data acquisition and control of continuous processes has been growing rapidly in recent years; thus the knowledge of basic methods available for analysis and design of discrete-time systems is an increasingly important element of engineering education. The next two chapters provide introductory material on analysis and control of linear discrete-time systems.

In Section 15.2, a problem of mathematical modeling of discrete-time systems is presented. Both input-output and state forms of system models are introduced. The process of discretization of continuous systems as a result of sampling at discrete time intervals is described in Section 15.3. Theoretical and practical criteria for selecting the sampling frequency to assure that no information relevant to the dynamics of the continuous process is lost due to sampling are discussed. In Section 15.4, the z transform is introduced. The concept of transfer function of discrete-time systems defined in the domain of complex variable z is presented in Section 15.5. A procedure for calculation of a response of a discrete-time system to an arbitrary input is also outlined.

15.2 Mathematical Modeling

A discretized model of a continuous system employs a sequence of values of each continuous variable taken only at carefully chosen[1] distinct increments of time (usually equal increments). A continuous variable $x(t)$, for instance, is represented in a discretized model by a sequence $\{x(k)\}$, $k = 0, 1, 2, \ldots$, consisting of the values $x(0)$, $x(T)$, $x(2T)$, . . . or, simply, $x(0)$, $x(1)$, $x(2)$, ... Furthermore, the amplitude of a signal in a discrete-time system may be quantized; that is, it may only take a finite number of values, and in such case, the signal is called a digital signal. If the signal amplitude is not quantized, such a signal is referred to as a sampled-data signal. In this introductory treatment of discrete-time systems, the effect of quantization will be neglected; thus, no distinction will be made between digital and sampled-data signals.

Mathematical discrete-time models can be derived in either the form of an input-output equation or in a state form. Just as in the case of continuous systems, both forms of mathematical models of discrete-time systems are equivalent in terms of information incorporated in them. The decision on which form should be used in modeling discrete-time systems depends on a particular application. The input-output form is usually preferred in modeling low-order linear systems. The state variable form is used primarily in modeling higher-order systems and in solving optimal control problems.

15.2.1 Input-Output Model

As mentioned in the previous section, discrete-time models are often derived by descretizing continuous models. It may, therefore, be expected that a certain correspondence exists between the two types of models. The connection between the corresponding continuous and discrete-time models can best be illustrated by considering simple first- and second-order models.

A linear first-order continuous input-output model equation is

$$a_1 \frac{dy}{dt} + a_0 y = b_1 \frac{du}{dt} + b_0 u \tag{15.1}$$

A frequently employed discrete-time model results from an approximate discretization based on finite difference approximation of the state-variable time deriva-

[1]The choice of T will be discussed later in Chapters 15 and 16. See also discussion of choice of Δt in Chapter 6.

tives. Equation (15.1) can be discretized by replacing the continuous derivatives of input and output variables by appropriate approximating quotients at the selected time increments. A backward difference approximation scheme for a derivative of a continuous variable $x(t)$ is

$$\left.\frac{dx}{dt}\right|_{t=kT} \approx \frac{x(kT) - x((k-1)T)}{T} \tag{15.2}$$

A simplified notation, i.e., $x(k)$ instead of $x(kT)$, will be used from now on for all discrete-time variables. Using this notation, Equation (15.2) can be rewritten as

$$\left.\frac{dx}{dt}\right|_{t=kT} \approx \frac{x(k) - x(k-1)}{T} \tag{15.3}$$

By applying the backward difference approximation defined by Equation (15.3) to Equation (15.1), the following first-order difference equation is obtained, which will be used here as an approximate discretized model.

$$y(k) = \frac{a_1}{a_1 + a_0 T} y(k-1) + \frac{b_1 + b_0 T}{a_1 + a_0 T} u(k) - \frac{b_1}{a_1 + a_0 T} u(k-1) \tag{15.4}$$

or, simply,

$$y(k) + g_1 y(k-1) = h_0 u(k) + h_1 u(k-1) \tag{15.5}$$

where g_1, h_0, and h_1 are the parameters of this discrete-time model. By comparison of Equations (15.4) and (15.5), these parameters can be expressed in terms of the parameters of the continuous model

$$\begin{aligned} g_1 &= \frac{-a_1}{a_1 + a_0 T} \\ h_0 &= \frac{b_1 + b_0 T}{a_1 + a_0 T} \\ h_1 &= \frac{-b_1}{a_1 + a_0 T} \end{aligned} \tag{15.6}$$

A similar procedure can be applied to a linear second-order model of a continuous system described by the following differential equation

$$a_2 \frac{d^2 y}{dt^2} + a_1 \frac{dy}{dt} + a_0 y = b_2 \frac{d^2 u}{dt^2} + b_1 \frac{du}{dt} + b_0 u \tag{15.7}$$

A discrete approximation of a second derivative of a continuous variable x is

$$\left.\frac{d^2 x}{dt^2}\right|_{t=kT} \approx \frac{x(k) - 2x(k-1) + x(k-2)}{T^2} \tag{15.8}$$

which leads to the following discrete-time model.

$$y(k) + g_1 y(k-1) + g_2 y(k-2) = h_0 u(k) + h_1 u(k-1) + h_2 u(k-2) \tag{15.9}$$

where

$$g_1 = \frac{-(2a_2 + a_1T)}{a_2 + a_1T + a_0T^2}$$

$$g_2 = \frac{a_2}{a_2 + a_1T + a_0T^2}$$

$$h_0 = \frac{b_2 + b_1T + b_0T^2}{a_2 + a_1T + a_0T^2} \tag{15.10}$$

$$h_1 = \frac{-(2b_2 + b_1T)}{a_2 + a_1T + a_0T^2}$$

$$h_2 = \frac{b_2}{a_2 + a_1T + a_0T_2}$$

From Equations (15.5) and (15.9), it can be deduced that a nth-order approximate discrete-time model can be presented in the following form

$$y(k) + g_1y(k-1) + \cdots + g_ny(k-n) = h_0u(k) + h_1u(k-1) + \cdots + h_nu(k-n) \tag{15.11}$$

where some of the coefficients g_i $(i = 1, 2, \ldots, n)$ and h_j $(j = 0, 1, \ldots, n)$ may be equal to zero.

It should be realized that the connection between the corresponding continuous and discrete-time models is rather symbolic. Although the parameters of the discrete-time model can be expressed in terms of the parameters of the corresponding continuous model for a given discretization method, that relationship is not unique. By choosing different values of the sampling time T, different sets of discrete-time model parameters are obtained and each of these approximate discrete-time models can be considered to be "corresponding" to the continuous model. The selection of time T is not uniquely determined either and is usually based on a rule of thumb. One such rule states that T should be smaller than one fourth of the smallest time constant of the continuous model. Another rule of thumb, developed for models producing oscillatory step response, recommends the value of T smaller than about one twenty fourth of the period of the highest frequency oscillation of interest. Obviously, none of these rules are very precise in determining the value of the sampling time.

Thus the order of the model is the same, regardless of whether the modeling is performed in continuous or in discrete-time domain. And an output variable of a discrete-time system at time $t = kT$, $y(kT)$, can be expressed as a function of n previous values of the output and $m + 1$ present and past values of the input variable, u, which can be written mathematically as

$$y(k) = f[y(k-1), y(k-2), \ldots, y(k-n), \ldots, u(k), u(k-1), \ldots, u(k-m)] \tag{15.12}$$

For linear stationary systems, the input-output model takes the form of a linear difference Equation (15.11). A recursive solution of Equation (15.11) can be obtained for given initial conditions and a specified input sequence, $u(k)$, $k = 0, 1, \ldots,$ as illustrated in Example 15.1.

EXAMPLE 15.1

Find the solution of the following input-output difference equation for $k = 0, 1, \ldots, 10$.

$$y(k) + 0.6y(k-1) + 0.05y(k-2) = 0.25u(k-1) + 0.2u(k-2)$$

The input signal, $u(k)$, is a unit step sequence given by

$$u(k) = \begin{cases} 0 \text{ for } k < 0 \\ 1 \text{ for } k = 0, 1, 2, \ldots \end{cases}$$

The output sequence $y(k)$ is initially zero

$$y(k) = 0 \quad \text{for } k < 0$$

Solution

The recursive solution of the given difference equation can be obtained in a step by step manner, starting from $k = 0$ and progressing towards the final value of $k = 10$ as follows

$$\text{For } k = 0, \quad y(0) = 0.6y(-1) - 0.05y(-2) + 0.25u(-1) + 0.2u(-2) = 0$$

$$\text{For } k = 1, \quad y(1) = 0.6y(0) - 0.05y(-1) + 0.25u(0) + 0.2u(-1) = 0.25$$

In a similar manner, the following values of $y(k)$ are calculated from the equation

$$y(k) = 0.6y(k-1) - 0.05y(k-2) + 0.25u(k-1) + 0.2u(k-2) \quad \text{for } k = 1, 2, \ldots, 10$$

A listing of a very simple BASIC computer program for solving the difference equation considered in this example is given below.

```
10  Y0=0
20  Y1=.25
30  U=1
40  Y2=.6*Y1-.05*Y0+.25*U+.2*U
50  YY=Y0
60  Y0=Y1
70  Y1=Y2
80  PRINT I,YY
90  I=I+1
100 IF I<11 GOTO 40
110 END
```

The plot of $u(k)$ and $y(k)$ are shown in Figure 15.1. ■

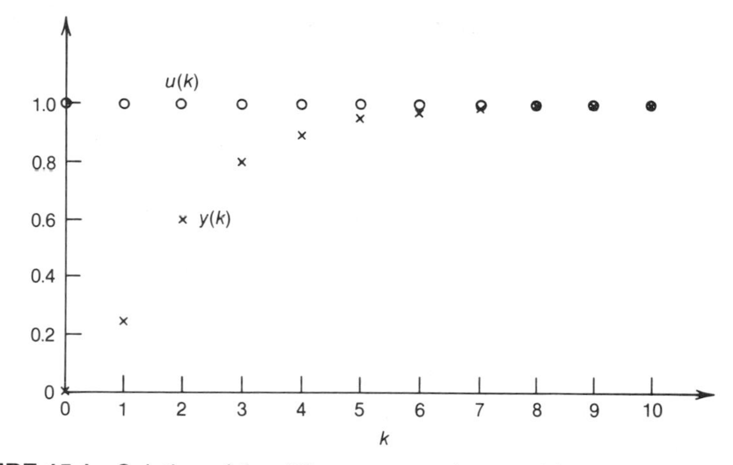

FIGURE 15.1 Solution of the difference equation considered in Example 15.1.

15.2.2. State Model

The basic concept and definitions associated with state models of discrete-time systems are the same as those used in modeling continuous systems.

In order to derive a state model for a linear system described by an nth-order input-output Equation (15.11), first an auxiliary discrete-time variable x is introduced; this satisfies a simplified input-output equation in which all coefficients on the right-hand side except h_0 are assumed to be zero, that is

$$x(k + n) + g_1x(k + n - 1) + \cdots + g_nx(k) = u(k) \tag{15.13}$$

The following set of n discrete-time state variables is then selected

$$\begin{aligned} q_1(k) &= x(k) \\ q_2(k) &= x(k + 1) = q_1(k + 1) \\ q_3(k) &= x(k + 2) = q_2(k + 1) \\ &\vdots \\ q_n(k) &= x(k + n - 1) = q_{n-1}(k + 1) \end{aligned} \tag{15.14}$$

Equations (15.14) yield $n - 1$ state equations

$$\begin{aligned} q_1(k + 1) &= q_2(k) \\ q_2(k + 1) &= q_3(k) \\ &\vdots \\ q_{n-1}(k + 1) &= q_n(k) \end{aligned} \tag{15.15}$$

To obtain the nth state equation, first notice that the last equation in (15.14) for $k = k + 1$ becomes

$$q_n(k + 1) = x(k + n) \tag{15.16}$$

Substituting $x(k + n)$ from Equation (15.13) gives the nth state equation

$$q_n(k + 1) = -g_1q_n(k) - g_2q_{n-1}(k) - \cdots - g_nq_1(k) + u(k) \tag{15.17}$$

Hence, a complete state model for a discrete system is

$$\begin{bmatrix} q_1(k+1) \\ q_2(k+1) \\ \vdots \\ q_n(k+1) \end{bmatrix} = \begin{bmatrix} 0 & 1 & 0 & \cdots & 0 \\ 0 & 0 & 1 & \cdots & 0 \\ & & & & \\ -g_n & -g_{n-1} & & \cdots & -g_1 \end{bmatrix} \begin{bmatrix} q_1(k) \\ q_2(k) \\ \vdots \\ q_n(k) \end{bmatrix} + \begin{bmatrix} 0 \\ 0 \\ \vdots \\ 1 \end{bmatrix} u(k) \tag{15.18}$$

or in a more compact form

$$\mathbf{q}(k + 1) = \mathbb{G}\mathbf{q}(k) + \mathbf{h}u(k) \tag{15.19}$$

where $\mathbf{q}$ is the state vector, $\mathbb{G}$ is the system matrix, $\mathbf{h}$ is the input vector, and $u(k)$ is an input signal.

The preceding state model was derived neglecting all but one of the terms on the right-hand side of Equation (15.11). In order to incorporate these terms in the system representation, an output model is derived relating the output variable $y(k)$ to the discrete-time state vector $\mathbf{q}(k)$, and the input variable $u(k)$. The procedure followed in the derivation is very similar to that used in Chapter 4 to derive the output model for continuous systems described by input-output equations that involve derivatives of input variables. The resulting output equation for a discrete-time model is

$$y(k) = [(h_n - h_0g_n) \quad (h_{n-1} - h_0g_{n-1}) \cdots (h_1 - h_0g_1)] \begin{bmatrix} q_1(k) \\ q_2(k) \\ \vdots \\ q_n(k) \end{bmatrix} + h_0u(k) \tag{15.20}$$

A complete state model of a single-input, single-output discrete system, described by Equation (15.11), can now be presented in the following form

$$\begin{aligned} \mathbf{q}(k + 1) &= \mathbb{G}\mathbf{q}(k) + \mathbf{h}u(k) \\ y(k) &= \mathbf{p}^T\mathbf{q}(k) + ru(k) \end{aligned} \tag{15.21}$$

where $r = h_0$ and column vector $\mathbf{p}$ is

$$\mathbf{p} = \begin{bmatrix} (h_n - h_0 g_n) \\ (h_{n-1} - h_0 g_{n-1}) \\ \vdots \\ (h_1 - h_0 g_1) \end{bmatrix} \tag{15.22}$$

It is interesting to note that the set of discrete-time state variables is *not* the same as the set of discretized continuous system state variables. The discrete-time state variables are successively advanced versions of the output $y(k)$, whereas the continuous system state variables are successively differentiated versions of the output $y(t)$. Corresponding solution techniques thus involve successive time delays between the discrete-time variables, and successive integrations between the continuous system variables chosen in this way. The procedure presented above will be illustrated by Example 15.2.

EXAMPLE 15.2

Derive a state model for the discrete system considered in Example 15.1.

Solution

The input-output equation of the system can be written in the following form

$$y(k+2) - 0.6y(k+1) + 0.05y(k) = 0.25u(k+1) + 0.2u(k)$$

Define state variables as

$$q_1(k) = y(k)$$
$$q_2(k) = y(k+1)$$

An auxiliary variable $x(k)$ is introduced such that

$$x(k+2) - 0.6x(k+1) + 0.05x(k) = u(k)$$

The state equations take the form

$$\begin{bmatrix} q_1(k+1) \\ q_2(k+1) \end{bmatrix} = \begin{bmatrix} 0 & 1.0 \\ -0.05 & 0.6 \end{bmatrix} \begin{bmatrix} q_1(k) \\ q_2(k) \end{bmatrix} + \begin{bmatrix} 0 \\ 1 \end{bmatrix} u(k)$$

The output equation is

$$y(k) = [0.2 \quad 0.25] \begin{bmatrix} q_1(k) \\ q_2(k) \end{bmatrix}$$

■

15.3 Sampling and Holding Devices

Most systems classified as discrete-time systems involve discrete as well as continuous components. An example of such a system is a digital control system, shown in block diagram form in Figure 15.2. In this system a digital device is used to control a continuous process. Figure 15.3 shows the form of signals appearing in the system using the digital PID algorithm to control a third-order process.

The digital controller generates a discrete-time control signal and accepts only discrete input signals. The continuous process produces a continuous output signal and can be effectively manipulated by a continuous input signal. Interface devices capable of transforming continuous signals into discrete signals and vice versa are therefore necessary to create compatibility between a digital controller and a continuous process. The two devices, a sampling device that converts a continuous signal into a discrete-time signal, and a holding device, which performs the opposite signal conversion, will now be described.

The sampling device allows the continuous input signal to pass through at distinct instants of time. In an actual sampler, the path for the input signal remains open for a finite period of time, Δ, as illustrated in Figure 15.4a. It is usually assumed that Δ is much smaller than the sampling period T; as a result, the output signal from a sampling device is presented by the strengths of a sequence of impulses shown in Figure 15.4b. The strength of an impulse is defined here as the area of the impulse.

Think of obtaining the sampled signal $x^*(t)$ from a continuous signal $x(t)$ as being the result of modulation of $x(t)$ with a gating function $G_s(t)$. Then

$$x^*(t) = G_s(t)x(t)$$

when the function $G_s(t)$ is equal to one at $t = 0, \mathrm{T}, 2\mathrm{T}, \ldots$, and zero elsewhere.

Some authors[2] employ Dirac's delta function to describe $x^*(t)$ mathematically. The mathematical formula representing an idealized sampled signal shown in Figure 15.4b is

$$x^*(t) = \sum_{k=0}^{\infty} x(kT)U_i(t - kT) \tag{15.24}$$

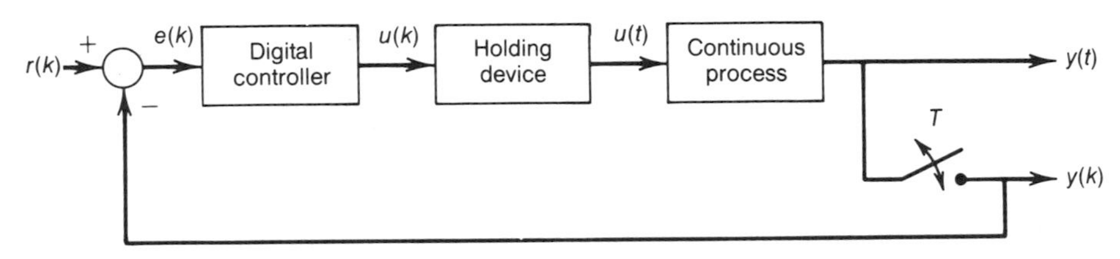

FIGURE 15.2 Block diagram of a digital control system with a continuous process.

[2]G. F. Franklin and J. D. Powell, *Digital Control of Dynamic Systems*, Addison-Wesley Publishing Co., Reading, Mass., 1980.

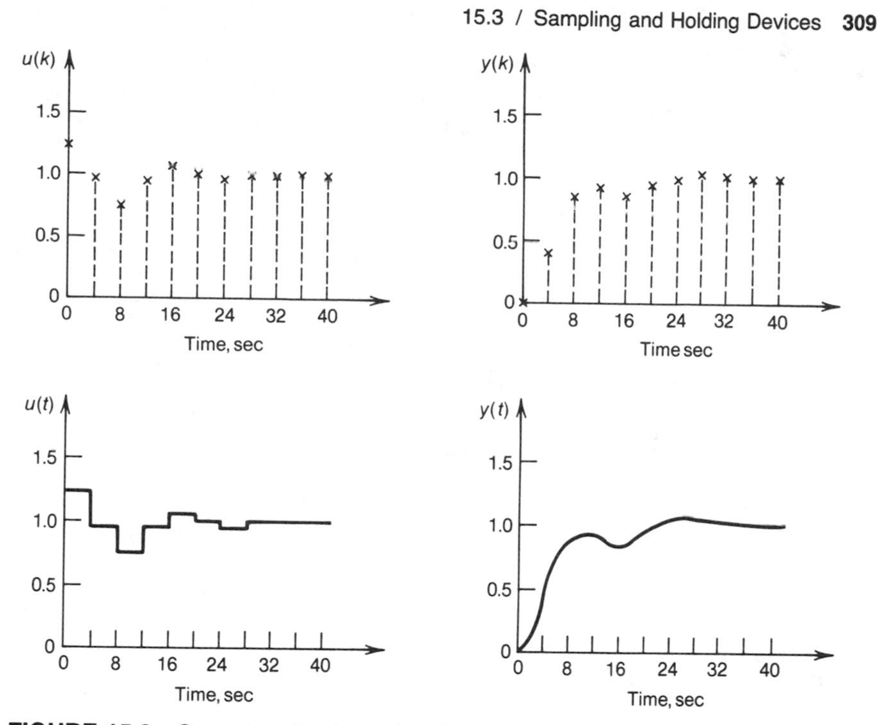

FIGURE 15.3 Control and output signals in the system shown in Figure 15.2.

where Dirac's delta function, introduced earlier in Section 5.4, is defined as a unit impulse having area of unity

$$U_i(t - kT) = \begin{cases} 0 \text{ for } t \neq kT \\ \infty \text{ for } t = kT \end{cases} \tag{15.25a}$$

and

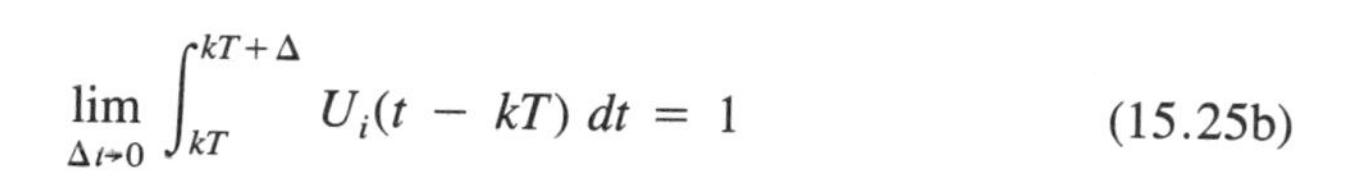

$$\lim_{\Delta t \to 0} \int_{kT}^{kT+\Delta} U_i(t - kT)\, dt = 1 \tag{15.25b}$$

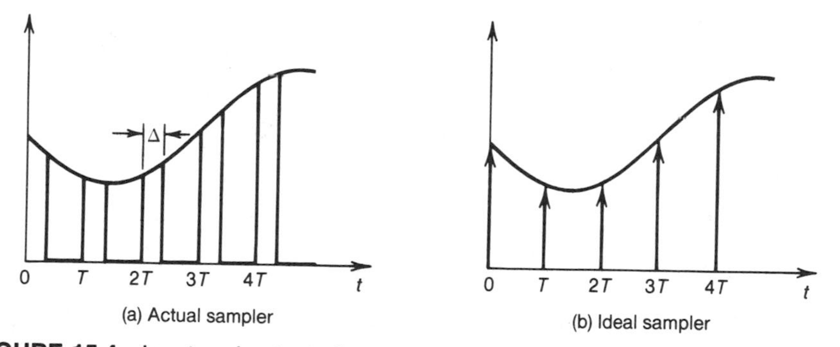

FIGURE 15.4 Input and output signals in an (a) actual and (b) ideal sampling device.

Equation (15.24) describes a sampled signal $x^*(t)$ in terms of a series of impulses such that the strength of each impulse is equal to the magnitude of the continuous signal at the corresponding instant of time, $t = kT$, $k = 0, 1, 2, \ldots$, that is

$$\int_{-\infty}^{+\infty} x(t)U_i(t - kT)\,dt = x(kT) \tag{15.26}$$

This is usually referred to as the "sifting integral."[3] Remember that, as pointed out in Chapter 5, although Dirac's delta function represents a useful mathematical idealization, it cannot be generated physically.

Intuitively, it seems that a certain loss of information must occur when a continuous signal is replaced by a discrete signal. However, according to Shannon's theorem[4], a continuous, band-limited signal of maximum frequency ω_{max} can be recovered from a sample signal if the sampling frequency ω_s, is greater than twice the maximum signal band frequency; that is, if

$$\omega_s > 2\omega_{max} \tag{15.27}$$

or, in terms of the sampling period T, if

$$T < \pi/\omega_{max} \tag{15.28}$$

In digital control practice the value for T should be less than $1/2\omega_{max}$. Half of the frequency that satisfies inequality (15.27), $\omega_s/2$, is often referred to as the Nyquist frequency[5]. Selecting sampling frequency on the basis of condition (15.27) is difficult in practice because real signals in engineering systems have unlimited frequency spectra; hence, determining the value of ω_{max} involves some uncertainty. Note too that the sampling frequency used in digital control systems can be higher than the frequency determined from Shannon's theorem because it is selected based on different criteria. The problem of selecting sampling frequency in digital control systems will be discussed in more detail in Chapter 16.

A serious problem, called aliasing, occurs as a result of sampling if the sampling frequency is too small. Aliasing is manifested by the presence of harmonic components in the sampled signal that are not present in the original continuous signal. The frequency of the alias harmonics is

$$\omega_a = \omega_s - \omega_c \tag{15.29}$$

where ω_c is a frequency of the continuous signal. Aliasing is illustrated in Figure 15.5 where a sinusoidal signal of frequency ω_c (and period T_c) is sampled with frequency $\omega_s = \frac{4}{3}\omega_c$ which corresponds to period $T_s = \frac{3}{4}T_c$ and is less than the sampling frequency required by the Shannon theorem. As a result, an alias sinusoidal signal is generated with a frequency of $\omega_a = \omega_s - \omega_c = \frac{1}{3}\omega_c$ and a period $T_a = 3T_c$

In order to prevent aliasing from affecting the sampled signal, the sampling

[3]Franklin and Powell.

[4]C. E. Shannon and W. Weaver, *The Mathematical Theory of Communication,* University of Illinois Press, Urbana, Il., 1949.

[5]Franklin and Powell.

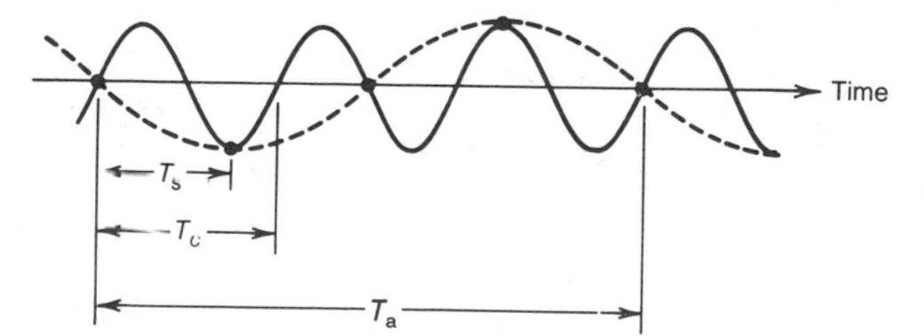

FIGURE 15.5 Aliasing due to sampling frequency $\omega_s < 2\omega_c$.

frequency should be high enough, which is easier to accomplish with a continuous signal if it is processed by a low-pass filter before sampling. The bandwidth of such an antialiasing low-pass filter, often called a guard filter, should be higher than the bandwidth of the sampled signal.

A holding device is used in order to convert sampled values (such as a control sequence generated by a digital controller) into signals that can be applied to continuous systems. Operation of a holding device can be thought of as an extrapolation of past values of the discrete-time signal over the next sampling period. Mathematically, for nth-order extrapolation, the output signal from a holding device, $x_h(t)$ for time t such that $kT < t < (k + 1)T$, can be expressed as

$$x_h(t) = x(kT) + a_1\tau + a_2\tau^2 + \cdots + a_n\tau^n \tag{15.30}$$

where $0 < \tau < T$. The coefficients $a_1, a_2, \ldots, a_n$ have to be estimated, using n past values of $x(k)$. In digital control practice the simplest holding device, the one that maintains the last last value over the next sampling period, is used. This device, a zero-order hold (ZOH), is described by the equation

$$x_h(kT + \tau) = x(kT) \quad \text{for } 0 \leqq \tau < T \tag{15.31}$$

In general, a staircase signal $x_h(t)$ that has the same form as $u(t)$ shown in Figure 15.3, can be expressed mathematically as the summation

$$x_h(t) = \sum_{k=0}^{\infty} x(kT)[U_s(t - kT) - U_s(t - (k + 1)T)] \tag{15.32}$$

where U_s is a unit step function. The transfer function of the zero-order hold can be found from Equation (15.32).

$$\mathbf{T}_h(s) = \frac{\mathbf{X}_h(s)}{\mathbf{X}^*(s)} = \frac{1 - e^{-sT}}{s} \tag{15.33}$$

where $\mathbf{X}_h(s)$ and $\mathbf{X}^*(s)$ are Laplace transforms of output and input signals of the zero-order hold, respectively. It can be seen from Equation (15.33) that the zero-order hold involves nontrivial dynamics that may have a significant effect on performance of the discrete system incorporating a continous process. Very often the dynamics of the holding device is combined with the dynamics of the process to obtain a single-block "equivalent" process.

15.4 The z Transform

As shown in Section 15.2, difference equations describing discrete-time systems are usually solved recursively, that is in a step-by-step manner, starting from the initial conditions and progressing towards the final time. This method, although quite effective when applied with a computer, does not produce a closed-form solution, which is often needed in analysis of system dynamics. The z transform provides a useful tool that allows for transforming difference equations derived in the time domain into equivalent algebraic equations in the domain of a complex variable z. The algebraic equations in the z domain are usually much easier to solve than the original difference equations. By applying the inverse z transform, closed-form solutions of the model difference equations can be obtained. Moreover, many powerful methods exist for analysis of discrete-time systems in the z domain. Most of these methods, by the way, are analogous to corresponding methods for analysis of continuous systems in the s domain.

15.4.1. Definition and Basic z Transforms

The z transform of sequence $x(kT)$ such that $x(kT) = 0$ for $k < 0$ is defined as follows.

$$\mathbf{X}(z) = \mathcal{Z}\{x(kT)\} = \sum_{k=0}^{\infty} x(kT)z^{-k} \tag{15.34}$$

Because of the assumption that $x(kT)$ is equal to zero for negative k, the z transform defined in Equation (15.34) is referred to as the one-sided z transform. It can be noted that $\mathbf{X}(z)$ is a power series of z^{-1}.

$$\mathbf{X}(z) = x(0) + x(T)z^{-1} + x(2T)z^{-2} + \cdots \tag{15.35}$$

Example 15.3 will demonstrate how Equation (15.34) can be used to find z transforms of functions representing typical input and output signals in linear discrete-time systems.

EXAMPLE 15.3

Find z transforms of the following functions.

(a) Unit step function defined as

$$x(k) = U_s(kT) = \begin{cases} 0 \text{ for } k < 0 \\ 1 \text{ for } k > 0 \end{cases}$$

Using the definition of the z transform given by Equation (15.34)

$$\mathbf{X}(z) = \sum_{k=0}^{\infty} 1 \cdot z^{-k} = \frac{1}{1 - z^{-1}} = \frac{z}{z - 1}$$

Note that $\mathbf{X}(z)$ converges if $|z| > 1$. In general, in calculating z transforms it is not necessary to determine the region of convergence of $\mathbf{X}(z)$; it is sufficient to know that such a region does exist.

(b) Unit ramp function defined as

$$x(k) = \begin{cases} 0 \text{ for } k < 0 \\ kT \text{ for } k \geqq 0 \end{cases}$$

Using Equation (15.34)

$$\mathbf{X}(z) = \sum_{k=0}^{\infty} kT\, z^{-k} = T(z^{-1} + 2z^{-2} + 3z^{-3} + \cdots)$$

$$= Tz^{-1}(1 + 2z^{-1} + 3z^{-2} + \cdots)$$

$$= \frac{Tz^{-1}}{(1 - z^{-1})^2} = \frac{Tz}{(z - 1)^2}$$

(c) Kronecker delta function defined as

$$x(k) = \begin{cases} 1 \text{ for } k = 0 \\ 0 \text{ for } k \neq 0 \end{cases}$$

Using Equation (15.34)

$$\mathbf{X}(z) = (1 \cdot z^{-0}) + (0 \cdot z^{-1}) + (0 \cdot z^{-2}) + \cdots = 1$$

(d) Power function defined as

$$x(k) = \begin{cases} 0 \text{ for } k < 0 \\ a^{kT} \text{ for } k \geqq 0 \end{cases}$$

Again, using Equation (15.34), the z transform is found

$$\mathbf{X}(z) = \sum_{k=0}^{\infty} a^{kT} z^{-k} = \sum_{k=0}^{\infty} (a^{-T} z)^{-k} = \frac{1}{1 - a^{T} z^{-1}} = \frac{z}{z - a^{T}}$$

A special case of the power function is the exponential function defined as

$$x(k) = \begin{cases} 0 \text{ for } k < 0 \\ e^{-bkT} \text{ for } k \geqq 0 \end{cases}$$

for which the z transform is

$$\mathbf{X}(z) = \frac{1}{1 - e^{-bT} z^{-1}} = \frac{z}{z - e^{-bT}}$$

■

More z transforms of selected discrete functions can be found in Appendix 2.

15.4.2. z Transform Theorems

Several basic theorems of the z transform, those that are most useful in analysis of discrete-time systems, are introduced in this section. For proofs of these theorems, the reader is referred to the textbook by Ogata[6].

(a) Linearity

$$\mathfrak{Z}\{a_1 f_1(kT) + a_2 f_2(kT)\} = a_1 \mathfrak{Z}\{f_1(kT)\} + a_2 \mathfrak{Z}\{f_2(kT)\} \quad (15.36)$$

(b) Delay of argument

$$\mathfrak{Z}\{f(k - n)T\} = z^{-n}\mathfrak{Z}\{f(kT)\} \quad \text{for } n > 0 \quad (15.37)$$

(c) Advance of argument

$$\mathfrak{Z}\{f(k + n)T\} = z^n\left[\mathfrak{Z}\{f(kT)\} - \sum_{k=0}^{n-1} f(kT)z^{-k}\right] \quad (15.38)$$

(d) Initial value theorem

$$f(0^+) = \lim_{z\to\infty} \mathbf{F}(z) \quad (15.39)$$

provided $\lim_{z\to\infty} \mathbf{F}(z)$ exists.

(e) Final value theorem

$$\lim_{k\to\infty} f(kT) = \lim_{z\to 1} (z - 1)\mathbf{F}(z) \quad (15.40)$$

provided $f(kT)$ remains finite for $k = 0, 1, 2, \ldots$.

15.4.3. Inverse z Transform

It was mentioned earlier that one of the main applications of the z transform is in solving difference equations. Once the closed-form solution in the z domain is found, it must be transformed back into the discrete-time domain using the inverse z transform. The following notation is used for the inverse z transform

$$f(k) = f(kT) = \mathfrak{Z}^{-1}\{\mathbf{F}(z)\} \quad (15.41)$$

In analysis of discrete-time systems, the function to be inverted is usually in the form of the ratio of polynomials in z^{-1}

$$\mathbf{F}(z) = \frac{b_0 + b_1 z^{-1} + \cdots + b_m z^{-m}}{1 + a_1 z^{-1} + \cdots + a_n z^{-n}} \quad (15.42)$$

By direct division of the numerator and denominator polynomials, a series is obtained

$$\mathbf{C}(z) = c_0 + c_1 z^{-1} + c_2 z^{-2} + \cdots \quad (15.43)$$

[6]K. Ogata, *Discrete-Time Control Systems*, Prentice-Hall Press, New York, 1987.

By comparing this form with the definition of the z transform, Equation (15.34), the values of the sequence $f(k)$ can be found

$$f(0) = c_0, \quad f(1) = c_1, \quad f(2) = c_2, \quad \ldots \tag{15.44}$$

In general, the direct method does not give a closed-form solution and it is practical only if no more than the first several terms of the sequence $f(k)$ are to be found.

The most powerful method for calculation of inverse z transforms is the partial fraction expansion method. In this method, which is parallel to the method used in the inverse Laplace transformation, function $\mathbf{F}(z)$ is expanded into a sum of simple terms, which are usually included in tables of common z transforms. Because of linearity of the z transform, the corresponding function $f(k)$ is obtained as a sum of the inverse z transforms of the simple terms resulting from the partial fraction expansion. The use of the partial fraction expansion method will be demonstrated in Examples 15.4 and 15.5.

EXAMPLE 15.4

Find the inverse z transform of the function

$$\mathbf{F}(z) = \frac{z(z+1)}{(z^2 - 1.4z + 0.48)(z-1)} \tag{15.45}$$

First, the denominator of $\mathbf{F}(z)$ must be factored. The roots of the quadratic term in the denominator are 0.6 and 0.8; hence the factored form is

$$\mathbf{F}(z) = \frac{z(z+1)}{(z-0.6)(z-0.8)(z-1)} \tag{15.46}$$

When $\mathbf{F}(z)$ has a zero at the origin, $z = 0$, it is convenient to find a partial fraction expansion for $\mathbf{F}(z)/z$. In this case

$$\frac{\mathbf{F}(z)}{z} = \frac{z+1}{(z-0.6)(z-0.8)(z-1)} \tag{15.47}$$

and the expanded form is

$$\frac{\mathbf{F}(z)}{z} = \frac{c_1}{z-0.6} + \frac{c_2}{z-0.8} + \frac{c_3}{z-1} \tag{15.48}$$

If all poles of $\mathbf{F}(z)$, equation (15.46), are of multiplicity one, the constants c_i are calculated as

$$c_i = \left. \frac{(z-z_i)\mathbf{F}(z)}{z} \right|_{z=z_i} \tag{15.49}$$

where z_i are the poles of $\mathbf{F}(z)$. Using Equation (15.49), the expanded form of $\mathbf{F}(z)/z$ is found

$$\frac{\mathbf{F}(z)}{z} = \frac{20}{z-0.6} - \frac{45}{z-0.8} + \frac{25}{z-1} \tag{15.50}$$

Multiplying both sides of Equation (15.50) by z gives

$$\mathbf{F}(z) = \frac{20z}{z - 0.6} - \frac{45z}{z - 0.8} + \frac{25z}{z - 1} \tag{15.51}$$

The inverse z transforms of each of the three terms on the right-hand side of Equation (15.51) can be found easily to obtain Equation (15.52), the solution in the discrete-time domain.

$$f(k) = (20 \cdot 0.6^k) - (45 \cdot 0.8^k) + 25 \tag{15.52}$$

The final value theorem can be used to verify the solution for k approaching infinity. When k approaches infinity, the first two terms on the right-hand side of Equation (15.52) are approaching zero, and thus the final value of $f(k)$ is 25. Applying the final value theorem to Equation (15.45) gives

$$\lim_{k \to \infty} f(k) = \lim_{z \to 1} (z - 1)\mathbf{F}(z) = \lim_{z \to 1} \frac{z(z + 1)}{z^2 - 1.4z + 0.48} = 25$$

which verifies the final value of the solution. ■

EXAMPLE 15.5

Find the inverse z transform of the following function

$$\mathbf{F}(z) = \frac{z(z + 2)}{(z - 1)^2}$$

Since $\mathbf{F}(z)$ has a zero at $z = 0$, it is convenient to expand $\mathbf{F}(z)/z$ rather than $\mathbf{F}(z)$. The expanded form is

$$\frac{\mathbf{F}(z)}{z} = \frac{c_1}{z - 1} + \frac{c_2}{(z - 1)^2} \tag{15.54}$$

where the constants c_1 and c_2 are

$$c_1 = \left(\frac{d}{dz} \frac{(z - 1)^2\mathbf{F}(z)}{z}\right)_{z=1} = 1$$

$$c_2 = \left(\frac{(z - 1)^2\mathbf{F}(z)}{z}\right)_{z=1} = 3$$

Thus the expanded form of $\mathbf{F}(z)$ is

$$\mathbf{F}(z) = \frac{z}{(z - 1)} + \frac{3z}{(z - 1)^2}$$

or

$$\mathbf{F}(z) = \frac{1}{1 - z^{-1}} + \frac{3z^{-1}}{(1 - z^{-1})^2} \tag{15.55}$$

The inverse transforms of the terms on the right-hand side of Equation (15.55) are found from the table of z transforms, given in Appendix II, to give the solution for $f(k)$

$$f(k) = 1(k) + 3k \tag{15.56}$$

The final value theorem cannot be applied in this case to verify the solution because $\mathbf{F}(z)$ in Equation (15.53) has a double pole at $z = 1$ and thus $f(k)$ does not remain finite for $k = 0, 1, 2, \ldots$. As will be shown in Chapter 16, for $f(k)$ to remain finite for $k = 0, 1, 2, \ldots$, it is necessary that all poles of $\mathbf{F}(z)$ lie inside the unit circle in the z plane, with the possible exception of a single pole at $z = 1$. ■

15.5 Pulse Transfer Function

Another form of mathematical model of a linear discrete-time system, in addition to input-output and state models introduced in Section 15.2, is a pulse transfer function. For a system with input $u(k)$ and output $y(k)$, the pulse transfer function is defined as the ratio of z transforms of $y(k)$ and $u(k)$ for zero initial conditions

$$\mathbf{T}(z) = \frac{\mathbf{Y}(z)}{\mathbf{U}(z)} \tag{15.57}$$

where $\mathbf{U}(z)$ and $\mathbf{Y}(z)$ are z transforms of $u(k)$ and $y(k)$, respectively. The pulse transfer function for a system described by Equation (15.11) is

$$\mathbf{T}(z) = \frac{h_0 + h_1 z^{-1} + \cdots + h_m z^{-m}}{1 + g_1 z^{-1} + \cdots + g_n z^{-n}} \tag{15.58}$$

In general, the pulse transfer function of an engineering discrete-time system takes the form of the ratio of polynomials in z^{-1}. Equation (15.58) can be rewritten as

$$\mathbf{T}(z) = \frac{\mathbf{H}(z^{-1})}{\mathbf{G}(z^{-1})} \tag{15.59}$$

where

$$\mathbf{G}(z^{-1}) = 1 + g_1 z^{-1} + \cdots + g_n z^{-n} \tag{15.60}$$

$$\mathbf{H}(z^{-1}) = h_0 + h_1 z^{-1} + \cdots + h_m z^{-m} \tag{15.61}$$

The pulse transfer function is obtained by applying the z transform to the system input-output equation, as depicted in Figure 15.6. Note that the form of the pulse

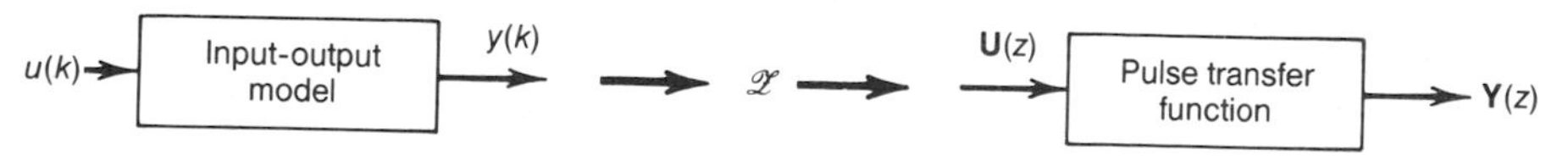

FIGURE 15.6 Obtaining pulse transfer function from input-output model.

transfer function is not affected by shifting the argument of the system input-output equation, provided the system is initially at rest. In particular, the transfer function obtained from Equation (15.11) is the same as the transfer function derived from Equation (15.62)

$$\begin{aligned} y(k+n) + g_1 y(k+n-1) + \cdots + g_n y(k) \\ = h_0 u(k+n) + h_1 u(k+n-1) + \cdots + h_m u(k+n-m) \end{aligned} \tag{15.62}$$

provided both $u(k)$ and $y(k)$ are zero for $k < 0$. This property of the pulse transfer function will be illustrated by the following simple example.

EXAMPLE 15.6

Consider the system from Example 15.1. Find the pulse transfer function $\mathbf{T}(z)$ for the system having the following input-output equation

Solution

$$y(k) - 0.6y(k-1) + 0.05y(k-2) = 0.25u(k-1) + 0.2u(k-2)$$

Taking the z transform of both sides of this equation gives

$$\mathbf{Y}(z) - 0.6z^{-1}\mathbf{Y}(z) + 0.05z^{-2}\mathbf{Y}(z) = 0.25z^{-1}\mathbf{U}(z) + 0.2z^{-2}\mathbf{U}(z) \tag{15.63}$$

Hence, the system pulse transfer function is

$$\mathbf{T}(z) = \frac{\mathbf{Y}(z)}{\mathbf{U}(z)} = \frac{0.25z^{-1} + 0.2z^{-2}}{1 - 0.6z^{-1} + 0.05z^{-2}} \tag{15.64}$$

Now, shift the argument of the original input-output equation by 2 steps to yield

$$y(k+2) - 0.6y(k+1) + 0.05y(k) = 0.25u(k+1) + 0.2u(k) \tag{15.65}$$

Transforming Equation (15.65) into the z domain gives

$$\begin{aligned} z^2\mathbf{Y}(z) - z^2y(0) - y(1)z - 0.6z\mathbf{Y}(z) + 0.6zy(0) + 0.05\mathbf{Y}(z) \\ = 0.25z\mathbf{U}(z) - 0.25zu(0) + 0.2\mathbf{U}(z) \end{aligned} \tag{15.66}$$

To determine $y(0)$ and $y(1)$, substitute first $k = -2$ and then $k = -1$ into the Equation (15.65). For $k = -2$

$$y(0) - 0.6y(-1) + 0.05y(-2) = 0.25u(-1) + 0.2u(-2)$$

It was assumed that both $u(k)$ and $y(k)$ are zero for $k < 0$, which gives

$$y(0) = 0$$

For $k = -1$, the Equation (15.65) is

$$y(1) - 0.6y(0) + 0.05y(-1) = 0.25u(0) + 0.2u(-1)$$

and hence

$$y(1) = 0.25u(0) = 0.25$$

Now substitute $y(0) = 0$ and $y(1) = 0.25$ into Equation (15.66) to obtain

$$z^2\mathbf{Y}(z) - (z^2 \cdot 0) - 0.25z - 0.6z\mathbf{Y}(z) + 0.6 \cdot z \cdot 0 + 0.05\mathbf{Y}(z)$$
$$= 0.25z\mathbf{U}(z) - (0.25 \cdot z \cdot 1) + 0.2\mathbf{U}(z)$$

The resulting transfer function is

$$\mathbf{T}(z) = \frac{\mathbf{Y}(z)}{\mathbf{U}(z)} = \frac{0.25z + 0.2}{z^2 - 0.6z + 0.05)} \tag{15.67}$$

Multiplying numerator and denominator of Equation (15.67) by z^{-2} gives

$$\mathbf{T}(z) = \frac{0.25z^{-1} + 0.2z^{-2}}{1 - 0.6z^{-1} + 0.05z^{-2}}$$

which is the same as Equation (15.64), obtained from the original input-output equation. ■

A response of a linear discrete-time system to a discrete impulse function defined in Section 15.4.1 is called a weighting sequence. The z transform of the discrete impulse function is equal to one and thus the z transform of the weighting sequence is

$$\mathbf{Y}(z) = \mathbf{T}(z) \cdot 1 = \mathbf{T}(z) \tag{15.68}$$

The weighting sequence is thus given by an inverse z transform of the system pulse transfer function $\mathbf{T}(z)$

$$\mathcal{Z}^{-1}\{\mathbf{T}(z)\} = w(k) \tag{15.69}$$

The weighting sequence of a discrete-time system, as most of other terms introduced in this chapter, has its analogous term in the area of continuous systems—the impulse response. Although this analogy between various aspects of continuous and discrete-time systems is, in most cases, clearly drawn and very useful, it should be taken with caution. One such example is an analogy between the relationships involving a continuous function of time $f(t)$ and its Laplace transform $\mathbf{F}(s)$ versus a discrete function $f(k)$ and its z transform $\mathbf{F}(z)$. By applying an inverse Laplace transformation to $\mathbf{F}(s)$, the same continuous function $f(t)$ is obtained. A discrete function $f(k)$ obtained by sampling a continuous function $f(t)$ having sampling period T is transformed into $\mathbf{F}(z)$ in the domain of complex variable z. Application of the inverse z transform to $\mathbf{F}(z)$ will result in the same discrete function $f(k)$; however, there is no basis for considering $f(k)$ as a sampled version of any specific continuous function of time. In other words, the function $f(k)$ obtained from the inverse z transform is defined only at distinct instants of time $0, T, 2T, \ldots$, and it would be entirely meaningless to deduce what values it might take between the sampling instants of time. After a continuous function of time is sampled, there is no unique transformation that would allow for return from discrete-time domain to the original function.

15.6 Synopsis

In this chapter, basic methods for modeling and analysis of discrete-time systems were introduced. Although most engineering systems are continuous, more and more of those systems are observed and/or controlled by digital computers which result in overall systems that are considered discrete. In such situations the continuous system variables are known only at distinct, usually equal, increments of time. As in the modeling of systems that include continuous elements only, input-output and state models are employed in modeling discrete-time systems. Difference equations describing discrete-time models can be solved using simple computer programs based on recursive algorithms or using the z transform which leads to closed form solutions. General forms of the discrete-time models were presented and compared with the corresponding continuous system models. It was shown that the correspondence between continuous and discrete-time models is rather elusive. A continuous system can be approximated by many different discrete-time models resulting from different discretization methods or from different time intervals selected for the discrete-time approximation. In general, there are many similarities between continuous and discrete-time systems that are very helpful for those who have considerable experience in the area of continuous systems in their introductory studies of discrete-time systems. However, as your knowledge of discrete-time systems progresses, you will notice many distinct characteristics of these systems that open new and attractive opportunities for analysis and, more importantly, for applications in process control, robotics, and so forth.

Problems

15.1. A linear discrete-time model is described by the following input-output equation

$$y(k) - 1.2y(k-1) + 0.6y(k-2) = u(k-1) + u(k-2)$$

Find the output sequence, $y(k)$, $k = 0, 1, \ldots, 25$, assuming that $y(k) = 0$ for $k < 0$, for the following input signals:

(a) $u(k) = \begin{cases} 0 \text{ for } k < 0 \\ 1 \text{ for } k \geq 0 \end{cases}$

(b) $u(k) = \begin{cases} 1 \text{ for } k = 0 \\ 0 \text{ for } k \neq 0 \end{cases}$

15.2. Select sampling frequency for a digital data acquisition system measuring velocity v of mass m in the mechanical system considered in Example 5.1. The mechanical system parameters are $m = 5$ kg and $b = 2$ N-sec/m.

15.3. Determine sampling frequency for digital measurement of position of mass m in the mechanical system shown in Figure 5.15a and described by the input-output equation

$$9\ddot{x} + 4\dot{x} + 4x = F(t)$$

where $F(t)$ varies in a stepwise manner.

15.4. A sinusoidal signal $y(t) = \sin 200t$ is to be recorded using a computer data acquisition system. It is expected that the measuring signal may be contaminated by an electric noise of frequency 60 Hz. Select the sampling frequency and determine the value of the time constant of the guard filter to prevent aliasing. The transfer function of the filter is $\mathbf{T}_f(s) = 1/(\tau_f s + 1)$.

15.5. Find z transforms of the following discrete functions of time defined for $k = 0, 1, 2, \ldots$:

(a) $a(1 - e^{-bkT})$

(b) $(1 - akT)e^{-bkT}$

(c) $e^{-bkT} \sin \omega kT$

(d) $\cos \omega(k - 2)T$

15.6. Obtain z transforms of the sequences $x(k)$ described below. Express the solutions as ratios of polynomials in z.

(a) $x(k) = \begin{cases} 0 \text{ for } k \leqslant 1 \\ 0.5 \text{ for } k = 2 \\ 1 \text{ for } k \geq 3 \end{cases}$

(b) $x(k) = \begin{cases} 0 \text{ for } k < 0 \\ e^{0.5k} + U_s(k - 2) \text{ for } k \geq 0 \end{cases}$

15.7. Find a closed-form solution for a unit step response of the system having pulse transfer function, $\mathbf{T}(z) = (z + 1)/(z^2 - 1.1z + 0.28)$. Verify the steady-state solution using final value theorem.

15.8. A weighting sequence of a linear discrete-time system was measured at equally spaced instants of time, and the results are tabulated below. After time $10T$ the measured output signal was zero.

Time	0	T	$2T$	$3T$	$4T$	$5T$	$6T$	$7T$	$8T$	$9T$	$10T$
$w(kT)$	1.0	0.5	0.25	0.125	0.063	0.031	0.016	0.008	0.004	0.002	0.001

Find the response of this system to a unit step input $U_s(k)$.

15.9. Obtain the weighting sequence for the pulse transfer function, $\mathbf{T}(z) = (z^2 + z)/(z^2 - 0.988z + 0.49)(z - 0.6)$.

15.10. Obtain the weighting sequence for the system represented by the input-output equation

$$y(k + 2) + 0.7y(k + 1) + 0.1y(k) = u(k + 1)$$

15.11. Find the output sequences for the discrete-time system of the transfer function $\mathbf{T}(z) = z/(z^2 - 1.125z + 0.125)$ for the following input signals:

(a) Kronecker delta defined as

$$u(k) = \begin{cases} 1 \text{ for } k = 0 \\ 0 \text{ for } k \neq 0 \end{cases}$$

(b) Sequence $u(k)$, shown in Figure P15.11.

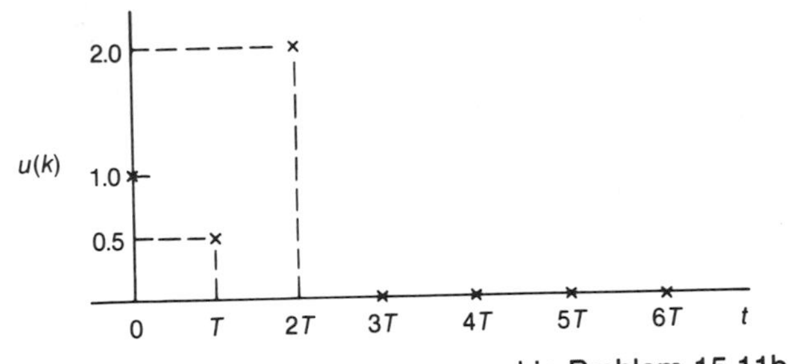

FIGURE P15.11 Input sequence used in Problem 15.11b.

16

Digital Control Systems

16.1 Introduction

Unprecedented advances in electronics have revolutionized control technology in recent years. Digital controllers, built around microcomputer chips as stand-alone units or implemented using ubiquitous personal computers, have dominated modern industrial process control applications. The computational power and operational speed of digital controllers allow for performing much more sophisticated algorithms than were possible with analog controllers. Even in relatively simple control tasks digital controllers are superior to analog controllers due to their improved flexibility, reliability, and, more and more often, lower cost.

The main objective of this chapter is to introduce the very basic concepts of analysis of digital control systems. Only linear, stationary models will be considered. In Section 16.2 a pulse transfer function block diagram of a single-loop digital control system is presented. Section 16.3 deals with transient characteristics determined by the location of roots of the system characteristic equation; methods for determining stability are also briefly discussed. In Section 16.4 steady-state performance characteristics of digital control systems are reviewed. Section 16.5 provides introductory material on digital control algorithms. A digital version of the PID controller is given special attention because of its popularity in industrial process control applications.

16.2 Single-Loop Control System

At the time of the first digital process control applications, which took place in the late 1950s, the cost of computers used to perform control functions was relatively high. In order to make these systems economic and obtain reasonable pay-back time, at least 100 or more individual control loops had to be included in a single installation. Hundreds of measuring signals were transmitted from the process to the computer, often over very long distances. The control signals were sent back from the central computer to the process over the same long distances. As a result, an excessive network of wire and tubing was necessary to transmit electrical and pneumatic signals back and forth between the process and the computer. In addition to the obviously high cost of such installations, they were also very vulnerable to damage and interference from all kinds of industrial disturbances. Moreover, every failure of the central computer affected the entire process being controlled, which caused serious reliability problems.

In the 1970s when microprocessors became available, distributed digital control systems were introduced. In these systems controllers, built around microprocessors, are responsible for only local portions of the process. Each digital controller is thus handling only a few (usually just one) control loops. Thus the controllers may be located in close proximity to the process, reducing cable and tubing cost as compared with the centralized systems. The local controllers can then be connected with a supervisory controller through a data highway bus, as illustrated in Figure 16.1. During start-up and shut-down of complex multiloop process control systems it is often advantageous to maintain autonomy of local control loops under supervision by experienced personnel in order to achieve the transition between dead-start and normal operation. A failure of any of the local controllers or even of the supervisory controller has a limited impact on the performance of the rest of the system, which results in much greater reliability than was possible in centralized systems. Also, it is usually easier and more economical to provide redundant digital controllers than to provide redundant continuous controllers when the need for reliability is very great. Moreover, the modular structure of distributed systems allows for gradual (piece by piece, controller by controller) implementation of new systems and easier expansion of existing systems. In summary, distributed control systems have proven to be the most efficient and reliable structure for industrial process control today. The material presented in this chapter is limited

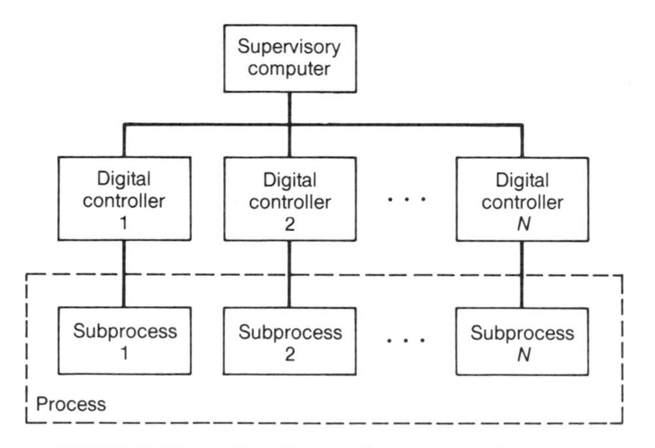

FIGURE 16.1 Distributed digital control system.

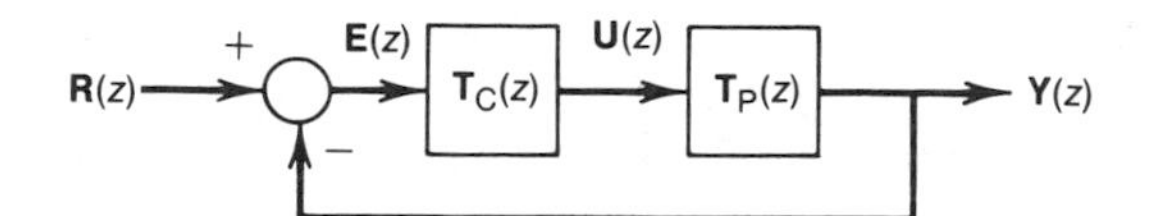

FIGURE 16.2 Block diagram of a single-loop digital control system.

to single-loop systems such as those implemented at the lowest level of distributed control systems.

A block diagram of a single-loop digital control system is shown in Figure 16.2. The controller pulse transfer function is $\mathbf{T}_C(z)$. The other block, $\mathbf{T}_P(z)$, represents a controlled process together with a preceding zero-order hold and can be expressed as

$$\mathbf{T}_P(z) = \mathcal{Z}\{\mathbf{T}_h(s)\mathbf{T}_P(s)\} \tag{16.1}$$

where $\mathbf{T}_h(s)$ is the transfer function of the zero order hold and $\mathbf{T}_P(s)$ represents a continuous model of the process. Using Equation (15.33), $\mathbf{T}_P(z)$ can be expressed as

$$\mathbf{T}_P(z) = \mathcal{Z}\left\{\frac{1 - e^{-sT}}{s}\mathbf{T}_P(s)\right\} \tag{16.2}$$

and hence

$$\mathbf{T}_P(z) = (1 - z^{-1})\mathcal{Z}\left\{\frac{\mathbf{T}_P(s)}{s}\right\} \tag{16.3}$$

A closed-loop pulse transfer function of the digital control system shown in Figure 16.2 is

$$\mathbf{T}_{CL}(z) = \frac{\mathbf{T}_C(z)\mathbf{T}_P(z)}{[1 + \mathbf{T}_C(z)\mathbf{T}_P(z)]} \tag{16.4}$$

In the next two sections basic transient and steady-state characteristics of linear single-loop digital control systems will be examined.

16.3 Transient Performance

Just as in the case of continuous systems, transient performance of discrete-time systems is determined by location of the poles of the system transfer function. The poles of the transfer function are the roots of the system characteristic equation, which in the case of the system shown in Figure 16.2 takes the form

$$1 + \mathbf{T}_C(z)\mathbf{T}_P(z) = 0 \tag{16.5}$$

For linear systems represented by the input-output Equation (15.11), the transfer function is of the form

$$\mathbf{T}(z) = \frac{h_0 + h_1 z^{-1} + \cdots + h_n z^{-n}}{1 + g_1 z^{-1} + \cdots + g_n z^{-n}} \tag{16.6}$$

or, equally,

$$\mathbf{T}(z) = \frac{h_0 z^n + h_1 z^{n-1} + \cdots + h_n}{z^n + g_1 z^{n-1} + \cdots + g_n} \tag{16.7}$$

The characteristic equation is thus an nth-order algebraic equation in z

$$z^n + g_1 z^{n-1} + \cdots + g_n = 0 \tag{16.8}$$

This equation can be written in a factored form as

$$(z - p_1)(z - p_2) \cdots (z - p_n) = 0 \tag{16.9}$$

where $p_1, p_2, \ldots, p_n$ are the poles of the transfer function. In order to investigate the effect of the locations of the poles in the domain of complex variable z on the system transient performance, a first-order system will be considered. The system pulse transfer function is

$$\mathbf{T}(z) = \frac{1}{1 + g_1 z^{-1}} \tag{16.10}$$

or, equally,

$$\mathbf{T}(z) = \frac{1}{1 - p_1 z^{-1}} \tag{16.11}$$

where p_1 is a single real pole. The corresponding input-output equation is

$$y(k) - p_1 y(k - 1) = u(k) \tag{16.12}$$

The homogenous equation is

$$y(k) - p_1 y(k - 1) = 0 \tag{16.13}$$

For a nonzero initial condition, $y(0) \neq 0$, the output sequence is

$$y(k) = y(0) p_1^k \tag{16.14}$$

Plots of the sequence $y(k)$, for $k = 0, 1, 2, \ldots$, given by Equation (16.14), for different values of p_1 are shown in Table 16.1. Note that the free response of the system, represented by the solution of the homogenous input-output equation, converges to zero only if the absolute value of p_1 is less than unity, $|p_1| < 1$.

Discrete-time systems of higher than first order may have real as well as complex poles, which occur in pairs of complex conjugate numbers just as they do in continuous systems. Moreover, the relation between the s-plane locations of continuous system poles and the z-plane locations of poles of a corresponding discrete-time system with sampling interval T is given by

$$z = e^{sT} \tag{16.15}$$

or, equally,

$$s = \frac{1}{\mathrm{T}} \ln z \tag{16.16}$$

TABLE 16.1. Locations of the Pole in The z Plane and Corresponding Free Response Sequences for a First-Order System.

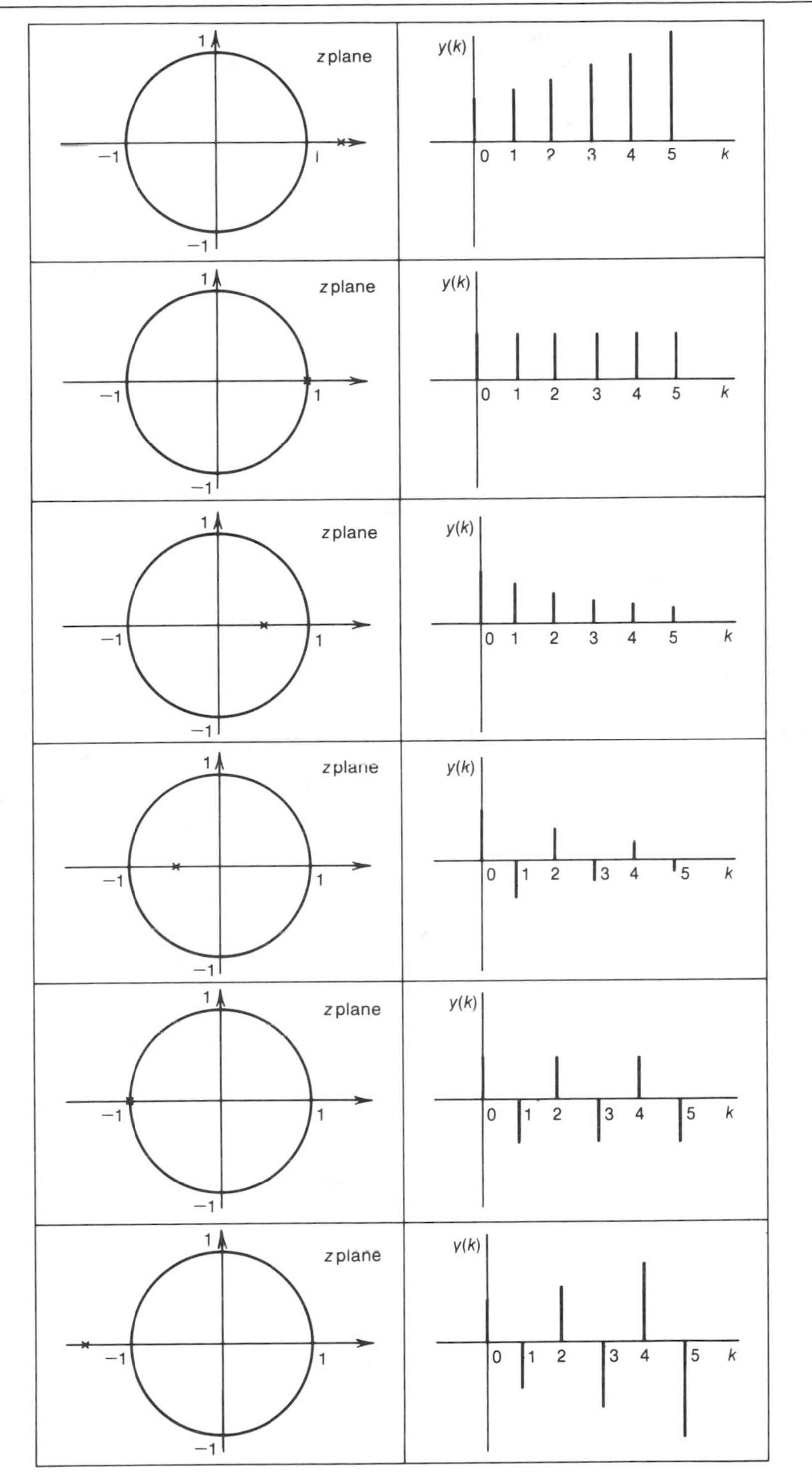

Equations (16.15) and (16.16) represent a mapping between the s plane and the z plane that applies to all poles of system transfer function, not just the complex ones.[1] The term "corresponding" used here means the relation between a continuous system and a discrete-time system involving the original continuous system together with a zero order hold and an ideal sampler, as shown in Figure 16.3.

In Chapter 5 the transient performance specifications of continuous systems were discussed in details. The mapping given by Equation (16.15) can be used to transform those specifications from continuous time to the discrete-time domain. A complex pole in the s plane can be expressed, from Equation (5.86), as

$$s = -\zeta\omega_n + j\omega_d \tag{16.17}$$

Using mapping (16.15), the corresponding pole in the z plane is

$$z = e^{(-\zeta\omega_n + j\omega_d)T} = e^{-\zeta\omega_n T} e^{j\omega_d T} \tag{16.18}$$

The real decreasing exponential term on the right-hand side of Equation (16.18) represents the distance d between the pole and the origin of the z plane coordinate system

$$d = |z| = e^{-\zeta\omega_n T} \tag{16.19}$$

whereas the complex exponential factor represents the phase angle φ associated with the pole

$$\varphi = \angle z = \omega_d T \tag{16.20}$$

Replacing the sampling period T by the sampling frequency ω_s yields

$$\varphi = \frac{2\pi\omega_d}{\omega_s} \tag{16.21}$$

For a specified sampling frequency ω_s and a constant damped frequency ω_d, a constant value of φ is obtained. The loci of the constant frequency ratio ω_d/ω_s are therefore straight lines crossing the origin of the coordinate system at the angle given by the right-hand side of Equation (16.21) with respect to the positive real axis in the z plane.

Loci of another important parameter associated with complex poles, the damping ratio ζ, can be found by combining Equations (16.19) and (16.20), which gives

$$d = e^{-(\varphi\zeta/\sqrt{1 - \zeta^2})} \tag{16.22}$$

Equation (16.22) describes spiral curves in the z plane with constant value of the damping ratio along each curve. Figure 16.4 shows loci of constant frequency ratio

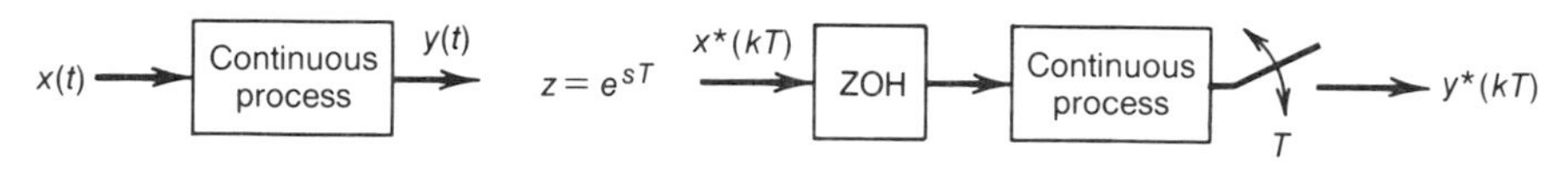

FIGURE 16.3 Illustration of mapping between the s and z planes.

[1] G. F. Franklin and J. D. Powell, *Digital Control of Dynamic Systems*, Addison-Wesley Publishing Co., Reading, Mass., 1980.

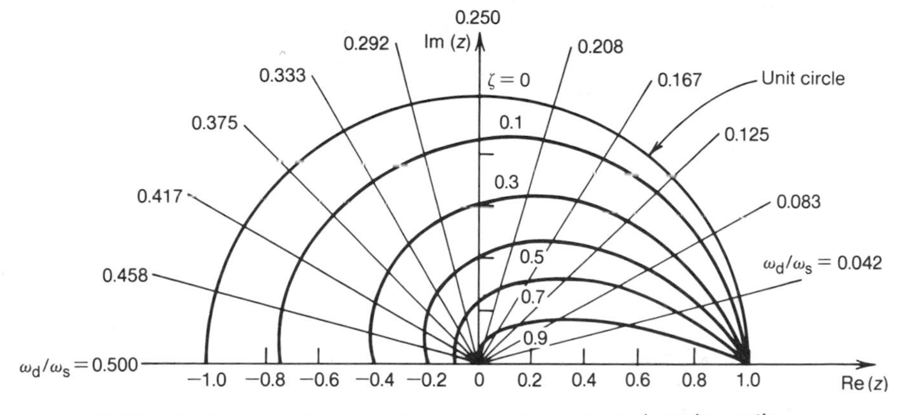

FIGURE 16.4 Loci of constant ω_d/ω_s and constant damping ratio.

given by Equation (16.21) and constant damping ratio, Equation (16.22). The loci are symmetrical with respect to the real axis; however, only the upper half of the z plane shown in Figure 16.4 is of practical significance because it represents the area where the sampling frequency satisfies the Shannon theorem condition, Equation (15.27).

The principal requirement regarding system transient performance is its stability. It was observed earlier that the free response of a system with real poles converges to zero (which indicates asymptotic stability according to the definition stated in Chapter 13) only if the absolute value of each real pole is less than unity. To determine the stability condition for discrete-time systems having both real and complex poles, the mapping defined by Equation (16.15) can be employed. Using this mapping, the left-hand side of the s plane is transformed into an inside of a unit circle in the z plane. It can thus be concluded that a linear discrete-time system is asymptotically stable if all poles of the system transfer function lie inside the unit circle in the plane of complex variable z, that is if

$$|p_i| < 1 \quad \text{for } i = 1, 2, \ldots, n$$

The area inside the unit circle in the plane of complex variable z plays, therefore, the same role in analysis of stability of discrete-time systems as the left-hand side of the s plane does in analysis of stability of continuous systems.

A direct method for analysis of stability of a discrete-time system involves calculation of all roots of the system characteristic equation to determine their location with respect to the unit circle in the z plane. This method, which is efficient only in analysis of low-order systems, will be illustrated in Example 16.1.

EXAMPLE 16.1

Determine the stability condition for a digital control system, with a proportional controller shown in Figure 16.5. The s-domain transfer function of the continuous process is

$$\mathbf{T}_P(s) = \frac{k_P}{\tau s + 1}$$

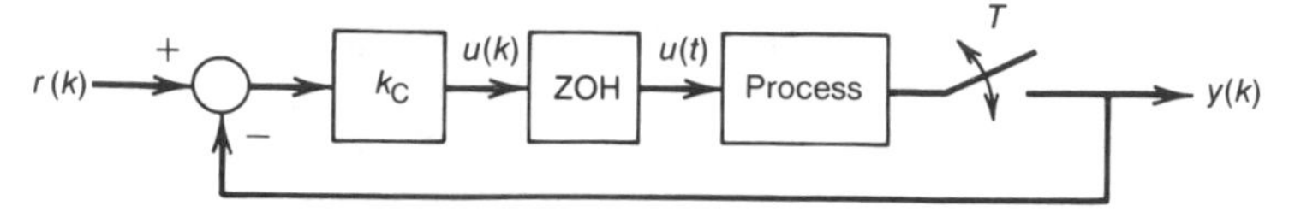

FIGURE 16.5 Digital control system from Example 16.1.

Solution

First, the equivalent process transfer function representing the dynamics of the continuous process together with the zero-order hold can be found using Equation (16.3), together with the partial fraction expansion for $\mathbf{T}_P(s)/s$, $k_P\left(\frac{1}{s} - \frac{1}{s + 1/\tau}\right)$, and Table A2.1 in Appendix 2.

$$\mathbf{T}_P(z) = (1 - z^{-1})\mathcal{Z}\left\{\frac{\mathbf{T}_P(s)}{s}\right\} = \frac{k_P z^{-1}(1 - e^{-T/\tau})}{1 - e^{-T/\tau}z^{-1}}$$

where T is the sampling time.

The transfer function of the proportional controller is

$$\mathbf{T}_C(z) = k_C$$

The closed-loop system transfer function is

$$\mathbf{T}_{CL}(z) = \frac{k_C k_P z^{-1}(1 - e^{-T/\tau})}{1 - e^{-T/\tau}z^{-1} + k_C k_P z^{-1}(1 - e^{-T/\tau})}$$

and the system characteristic equation is

$$1 - e^{-T/\tau}z^{-1} + k_C k_P z^{-1}(1 - e^{-T/\tau}) = 0$$

The single real pole of the closed-loop system is

$$p = e^{-T/\tau} - k_C k_P(1 - e^{-T/\tau})$$

To assure stability, the pole must lie between -1 and $+1$ on the real axis in the z plane. The stability condition can thus be written as

$$-1 < [e^{-T/\tau} - k_C k_P(1 - e^{-T/\tau})] < 1$$

which yields the admissible range of values for positive controller gain

$$0 < k_C < \frac{1 + e^{-T/\tau}}{k_P(1 - e^{-T/\tau})}$$

■

Several observations can be made from Example 16.1. First of all, it can be seen that a simple discrete-time system, including a first-order process and a proportional controller, can be unstable if the open-loop gain is too high, whereas a first-order continuous control system is always stable. The difference is caused by the presence of the zero-order hold in the discrete-time system. Second, from the stability condition found for the system, it can be seen that the maximum admissible value of k_C increases when the sampling time T decreases. In fact, as T approaches zero, the system becomes unconditionally stable, that is, stable for all values of gain k_C from zero to infinity. Third, note that the relation between the continuous process pole $s = -1/\tau$ and the pole of the corresponding discrete-time pole $z = e^{T/\tau}$ is indeed as given by Equation (16.15).

Of course, the direct method for analysis of stability is not the most efficient. One of the more efficient methods is based on a bilinear transformation defined by Equation (16.23).

$$w = \frac{z + 1}{z - 1} \tag{16.23}$$

or, equally,

$$z = \frac{w + 1}{w - 1} \tag{16.24}$$

The mapping defined by these equations transforms the unit circle in the z plane into the imaginary axis of the w plane, and the inside of the unit circle in the z plane into the left-hand side of the w plane. Substituting the right-hand side of Equation (16.24) for z in the system characteristic Equation (16.8) yields a nth order equation in variable w. If all roots of this equation lie in the left hand side of the w plane, the discrete-time system is stable. To determine if the roots of the transformed equation satisfy this condition, the same methods can be used as those developed for analysis of stability of linear continuous systems, such as Hurwitz or Routh criteria presented in Chapter 13.

16.4 Steady-State Performance

In Section 14.2 the steady-state performance of continuous control systems was evaluated. The two primary criteria used in this evaluation, the steady-state control error and the steady-state sensitivity to disturbances, can also be applied to discrete-time systems.

The steady-state value of the error signal $e(k)$ in a single-loop control system, shown in Figure 16.2, is defined as

$$e_{ss} = \lim_{k \to \infty} e(k) \tag{16.25}$$

Applying the final value theorem, e_{ss} can also be expressed as

$$e_{ss} = \lim_{z \to 1} (1 - z^{-1})\mathbf{E}(z) \tag{16.26}$$

where $\mathbf{E}(z)$ is a z transform of $e(k)$. An algebraic equation for the summing point in Figure 16.2 is

$$\mathbf{E}(z) = \mathbf{R}(z) - \mathbf{Y}(z) \tag{16.27}$$

The output $\mathbf{Y}(z)$ is

$$\mathbf{Y}(z) = \mathbf{E}(z)\mathbf{T}_C(z)\mathbf{T}_P(z) \tag{16.28}$$

Substituting (16.28) into (16.27) gives

$$\mathbf{E}(z) = \mathbf{R}(z) - \mathbf{E}(z)\mathbf{T}_C(z)\mathbf{T}_P(z) \tag{16.29}$$

Hence, the error pulse transfer function $\mathbf{T}_E(z)$ is defined as

$$\mathbf{T}_E(z) = \frac{\mathbf{E}(z)}{\mathbf{R}(z)} = \frac{1}{1 + \mathbf{T}_C(z)\mathbf{T}_P(z)} \tag{16.30}$$

The steady-state error can now be expressed in terms of the error transfer function $\mathbf{T}_E(z)$ and the z transform of the input signal $\mathbf{R}(z)$

$$e_{ss} = \lim_{z \to 1} (1 - z^{-1})\mathbf{R}(z)\mathbf{T}_E(z) \tag{16.31}$$

As indicated by Equation (16.31), the steady-state performance of discrete-time control systems depends not only on the system characteristics, represented by the error transfer function $\mathbf{T}_E(z)$, but also on the type of input signal, $\mathbf{R}(z)$. This, again, is true for continuous control systems also. A steady-state response to unit step and unit ramp inputs will now be derived.

Unit step input. The z transform of the unit step function $U_s(kT)$ is

$$\mathbf{R}(z) = \mathscr{Z}\{U_s(kT)\} = \frac{1}{1 - z^{-1}} \tag{16.32}$$

Substituting Equation (16.32) into Equation (16.31) gives the steady-state error for a unit step input

$$e_{ss} = \lim_{z \to 1} \mathbf{T}_E(z) = \lim_{z \to 1} \frac{1}{1 + \mathbf{T}_C(z)\mathbf{T}_P(z)} \tag{16.33}$$

A static position error coefficient is defined as a limit of the open-loop pulse transfer function for z approaching unity

$$K_P = \lim_{z \to 1} \mathbf{T}_C(z)\mathbf{T}_P(z) \tag{16.34}$$

The steady-state error can now be expressed as

$$e_{ss} = \frac{1}{1 + K_P} \tag{16.35}$$

This result shows that the steady-state error in response to a unit step input will be zero only if the static position error coefficient is infinity. From Equation (16.34) it can be seen that K_P will approach infinity if the open-loop transfer function, $\mathbf{T}_C(z)\mathbf{T}_P(z)$, has at least one pole at $z = 1$. In general, $\mathbf{T}_C(z)\mathbf{T}_P(z)$ can be presented as

$$\mathbf{T}_C(z)\mathbf{T}_P(z) = \frac{k(z - z_1)(z - z_2) \cdots (z - z_m)}{(z - 1)^r(z - p_1)(z - p_2) \cdots (z - p_n)} \tag{16.36}$$

where $z_1, z_2, \ldots, z_m$ are the open-loop zeros other than unity and $p_1, p_2, \ldots, p_n$ are the open-loop poles, none of which is equal to one. There are also r open-loop poles at $z = 1$. The multiplicity of the open-loop poles at $z = 1$ determines the type of the system. As shown earlier, the steady-state error in response to a unit step input is zero for systems of type 1 or higher.

Unit Ramp Input. The z transform of the unit ramp signal was found in Example 15.3 to be

$$\mathbf{R}(z) = \mathcal{Z}\{x(kT)\} = \frac{Tz^{-1}}{(1 - z^{-1})^2} \tag{16.37}$$

where $x(kT)$ is

$$x(k) = \begin{cases} 0 \text{ for } k < 0 \\ kT \text{ for } k \geq 0 \end{cases}$$

The steady-state error is, from Equation (16.31),

$$\begin{aligned} e_{ss} &= \lim_{z \to 1} (1 - z^{-1}) \frac{\dfrac{Tz^{-1}}{(1 - z^{-1})^2}}{1 + \mathbf{T}_C(z)\mathbf{T}_P(z)} \\ &= \lim_{z \to 1} \frac{T}{(1 - z^{-1})\mathbf{T}_C(z)\mathbf{T}_P(z)} \end{aligned} \tag{16.38}$$

A static velocity error coefficient K_v is defined in Equation (16.39).

$$K_v = \lim_{z \to 1} (1 - z^{-1})\mathbf{T}_C(z)\mathbf{T}_P(z)/T \tag{16.39}$$

Comparing Equations (16.38) and (16.39) gives

$$e_{ss} = \frac{1}{K_v} \tag{16.40}$$

From Equation (16.40) the steady-state error in response to a ramp signal can be seen to equal zero only if the system static velocity error coefficient is infinity, which occurs when the open-loop transfer function has at least a double pole at $z = 1$. In other words, using the terminology introduced earlier in this section, the steady-state error in response to a ramp input is zero if the system is of type 2 or higher.

The results derived here for unit step and ramp inputs can be extended for parabolic and higher-order input signals.

Another important aspect of performance of control systems is their ability to eliminate steady-state effects of disturbances on the output variable. The parameter introduced in Chapter 14 for evaluation of continuous systems, a steady-state disturbance sensitivity, can also be applied to discrete-time systems. According to the definition formulated in Chapter 14, the steady-state disturbance sensitivity, S_D, is

$$S_D = \frac{\Delta y_{ss}}{\Delta v_{ss}} \tag{16.41}$$

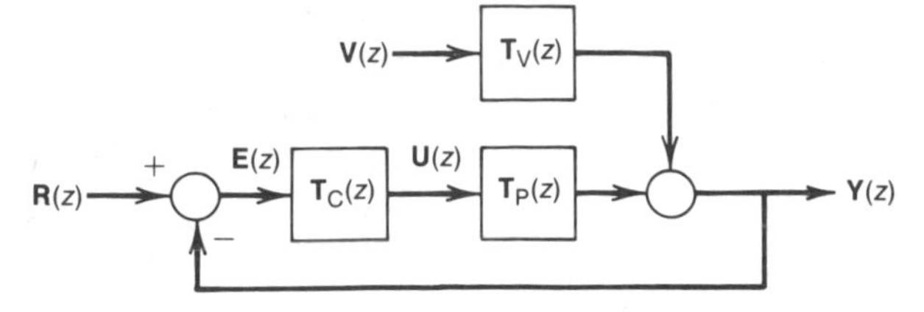

FIGURE 16.6 Block diagram of a digital control system subjected to disturbance V(z).

Figure 16.6 shows a digital control system subjected to a disturbance signal having z-transform $\mathbf{V}(z)$. The output $\mathbf{Y}(z)$ in this system consists of two components, one caused by the input signal $\mathbf{R}(z)$, and the other produced by the disturbance $\mathbf{V}(z)$; that is

$$\mathbf{Y}(z) = \mathbf{T}_{CL}(z)\mathbf{R}(z) + \mathbf{T}_D(z)\mathbf{V}(z) \tag{16.42}$$

where $\mathbf{T}_{CL}(z)$ is the system closed-loop pulse transfer function and $\mathbf{T}_D(z)$ is the disturbance pulse transfer function defined as

$$\mathbf{T}_D(z) = \left.\frac{\mathbf{Y}(z)}{\mathbf{V}(z)}\right|_{\mathbf{R}(z)=0} \tag{16.43}$$

Using block diagram algebra, the disturbance transfer function is found to be

$$\mathbf{T}_D(z) = \frac{\mathbf{T}_V(z)}{1 + \mathbf{T}_P(z)\mathbf{T}_C(z)} \tag{16.44}$$

Assuming that Δv_{ss} is unity, which does not restrict the generality of these considerations, the disturbance sensitivity can be expressed as

$$S_D = \Delta y_{ss} = \lim_{z \to 1} (1 - z^{-1})\mathbf{V}(z)\mathbf{T}_D(z) = \lim_{z \to 1} \mathbf{T}_D(z) \tag{16.45}$$

In general, in order to reduce the system disturbance sensitivity the open-loop gain has to be high, however, the extent to which the open-loop gain can be increased is usually limited by the system transient performance requirements.

16.5 Digital Controllers

One of the greatest advantages of digital controllers is their flexibility. A control algorithm performed by a digital controller is introduced in the form of a computer code, in some cases just a few lines of BASIC or FORTRAN program. All it takes to change the control algorithm is to rewrite those few lines of computer programming. In addition, there are very few limitations, especially when compared with analog controllers, on what kind of control actions can be encoded. Yet, in spite of the great flexibility and ease of implementing various control algorithms, over 90 percent of industrial digital controllers perform a classical proportional-integral-derivative (PID) algorithm. The main reason for the popularity of the PID controllers is an extensive theoretical and practical knowledge of many aspects of their performance carried over from the years of analog controllers. The presentation of digital controllers in this book will be limited to PID control algorithm.

The continuous PID control law was given by Equation (14.43)

$$u(t) = k_p \left(e(t) + \frac{1}{T_i} \int_0^t e(t_D)\, dt_D + T_d \frac{de(t)}{dt} \right) \tag{14.43}$$

In a discrete-time system the integral and derivative terms are approximated by using discretized models. Various discretization methods can be used resulting in different versions of the digital PID algorithm. The simplest method for approximating an integral is a rectangular (staircase) backward difference approximation described by Equation (16.46).

$$\int_0^t e(t_D)\, dt_D \approx \sum_{i=0}^{k-1} e(i)T \tag{16.46}$$

where T is the sampling time. The derivative of the control error is approximated by a backward difference quotient

$$\frac{de(t)}{dt} \approx \frac{e(k) - e(k-1)}{T} \tag{16.47}$$

Substituting the approximating expressions (16.46) and (16.47) into Equation (14.43) and replacing $u(t)$ and $e(t)$ with $u(k)$ and $e(k)$, respectively, yields

$$u(k) = K_p e(k) + K_i \sum_{i=0}^{k-1} e(i) + K_d[e(k) - e(k-1)] \tag{16.48}$$

where K_p is the digital proportional gain, and K_i and K_d are the digital integral and derivative gains, respectively, given by

$$K_p = k_p \tag{16.49}$$

$$K_i = k_p T / T_i \tag{16.50}$$

$$K_d = k_p T_d / T \tag{16.51}$$

Equations (16.49) through (16.51) are provided only to show the relationship of a discrete model to a continuous model of a PID controller. Thus they do not imply the existence of more than nominal correspondence between discrete model coefficients and the continuous model coefficients derived earlier by Ziegler and Nichols.

Equation (16.48) is referred to as a position PID algorithm. Another, very widely used form of digital PID, is a velocity algorithm. To derive the velocity algorithm, first consider the control signal defined by Equation (16.48) at $(k - 1)T$ instant of time

$$u(k-1) = K_p e(k-1) + K_i \sum_{i=0}^{k-2} e(i) + K_d[e(k-1) - e(k-2)] \tag{16.52}$$

Subtracting (16.52) from (16.48) yields

$$\Delta u(k) = u(k) - u(k-1) = K_p[e(k) - e(k-1)] + K_i e(k-1) + K_d[e(k) - 2e(k-1) + e(k-2)] \quad (16.53)$$

or, in more compact form,

$$\Delta u(k) = K_0 e(k) + K_1 e(k-1) + K_2 e(k-2) \quad (16.53)$$

where

$$\begin{aligned} K_0 &= K_p + K_d \\ K_1 &= K_i - 2K_d - K_p \\ K_2 &= K_d \end{aligned} \quad (16.54)$$

The velocity algorithm is usually preferred to the position algorithm because it is computationally simpler, safer (in case of controller failure the control signal remains unchanged, $\Delta u(k) = 0$), and it better handles "wind-up." A control error wind-up occurs within the controller after a control actuator, for instance for a control valve, hits a stop. When this happens the error signal to the controller persists and the integrator output continues to increase producing a "wind-up" phenomenon.

The controller gains, K_p, K_i, K_d, are selected to meet specified process performance requirements and must be adjusted according to the process transient and steady-state characteristics. A set of tuning rules, derived from the Ziegler-Nichols rules for analog controllers presented in Chapter 14, can be used to adjust the controller gains on the basis of the process step response. The Ziegler-Nichols rules for a digital PID controller are

$$\begin{aligned} K_i &= \frac{0.6T}{R(L + 0.5T)^2} \\ K_p &= \frac{1.2}{R(L + T)} \\ K_d &= \frac{0.5}{RT} \end{aligned} \quad (16.55)$$

The values of R and L are determined from the step response curve as shown in Figure 14.12. It should be stressed that, just as in the case of analog controllers, the Ziegler-Nichols rules do not guarantee optimal settings for the digital controller gains. In most cases, however, the values obtained using Equations 16.55 provide a good starting point for further fine tuning of the controller gains based on the system on-line performance.

As shown by Equation (16.55), the values of the control settings of a digital controller are also dependent on the sampling time T. Selecting a proper sampling time for a digital controller is a complex matter involving many factors. Some of the more important system characteristics affected by the sampling time are stability, bandwidth, sensitivity to disturbances, and sensitivity to parameter variations. All of these characteristics improve when the sampling interval is decreased.

However, the cost of the digital system increases when a small sampling time or, equally, a high sampling frequency is required simply because faster computers and faster process interface devices are more expensive than the slower ones. Moreover, an initial magnitude of a control signal produced by a digital PID controller in response to a step change of an input signal is greater if the sampling time is smaller for most control algorithms. Thus selecting the sampling time for a digital control system is a compromise between performance and cost (including the control effort), a familiar dilemma for a system designer.[2]

16.6 Synopsis

Digital control is a broad and still rapidly growing area. In this chapter an attempt has been made to present in simple terms some of the basic problems associated with analysis and design of digital control systems. The discussion was limited to single-loop systems, the most common structure in industrial digital control implemented either as a simple stand-alone system or as a part of a more complex distributed system. In spite of availability of many new improved control algorithms developed specifically for discrete-time systems, industrial digital controllers are usually programmed to perform conventional PID algorithms in discrete-time form. Two versions of the digital PID algorithms were introduced. Tuning rules derived from Ziegler-Nichols rules for analog PID controllers were given.

Design of digital control systems is an art of compromise between transient and steady-state performance, just as the design of continuous control systems. A method of bilinear transformation for determining stability of discrete-time systems was described. The relation between location of poles of the closed-loop transfer function and the system transient performance was examined. Two basic criteria of the steady-state performance, steady-state control error and sensitivity to disturbances, were shown to depend on the system open-loop gain and the type of input signal in a way similar to continuous feedback systems.

Although the use of digital control offers many advantages both to the designer and user of closed-loop control systems, it should be noted that the best performance achievable with a digital PID controller is dependent upon choice of a sufficiently small value of the sampling time interval T. Unless this value can become infinitesimal, the closed-loop performance of a given system with digital PID control will not be as good as the closed-loop performance of the same system attainable with continuous (analog) PID control, due to the pure delay time T inherent in the sample-and-hold process of the digital control system.

Furthermore, the presence of the derivative term in the discrete model PID controller generates a derivative component of the controller output signal which varies inversely with T, often resulting in controller output signals that are too large for the process to use (signal saturation at the process input), with the result that the beneficial effects desired from the derivative action are lost. Hence this

[2]For more information on selection of sampling time for digital control systems see Franklin and Powell, pp. 275–289.

tends to limit the smallness of T. The same saturation effect also occurs with continuous PID control, unless a time lag is also present in the controller, but the continous control system does not suffer from the presence of the finite delay time T.

For the engineer who has accumulated a body of experience working with continous control systems the design of a conventional digital controller is readily accomplished by employing characteristics (PID, compensation, and so forth) needed for continuous control in the s-domain and then finding the corresponding characteristics of a digital control system. This can be accomplished through the use of one of a number of possible transform methods, as discussed in some detail by Ogata[3] for instance.

Problems

16.1. The shaded area in Figure P16.1 represents a region of desirable locations of poles for a continuous dynamic system. The damping ratio associated with the complex conjugate poles located in this region is greater than 0.7 and the damped natural frequency is smaller than 3.93 rad/sec. Find a corresponding region in the z plane assuming that the sampling period is 0.2 sec.

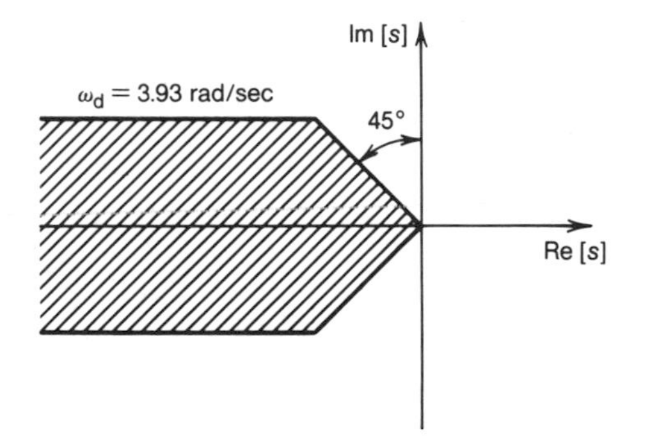

FIGURE P16.1 Region of desirable locations of poles for a continuous system.

16.2. In certain applications of continuous control systems it is desirable that the closed-loop poles of the system transfer function are located far enough to the left from the imaginary axis in the s plane. Mathematically this requirement can be written as

$$\text{Re}\,[p_i] \leq a_{\min} \qquad i = 1, 2, \ldots, n$$

where $a_{\min}$ is a negative number and p_i are closed-loop poles. This inequality is satisfied in the shaded area shown in Figure P16.2. Find a corresponding region in the z plane.

[3]K. Ogata, *Discrete-time Control Systems*, Prentice-Hall, Inc., New York, 1987, pp 306–478.

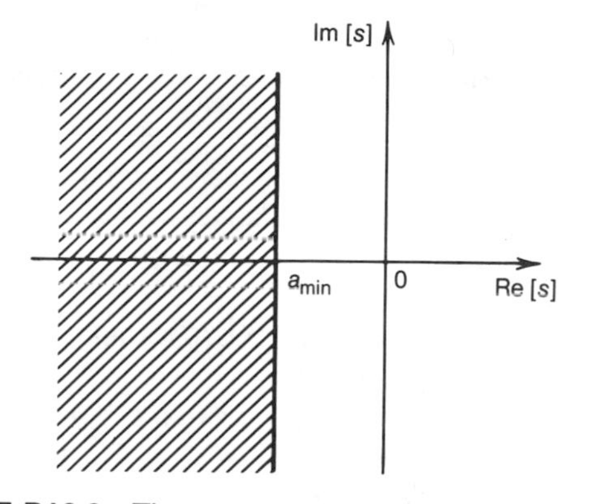

FIGURE P16.2 The region Re $[p_i] \le a_{min}$ in the s plane.

16.3. Obtain the z transform of a digital PID control algorithm given by Equation (16.53). Find a pulse transfer function of the PID controller and arrange it into a ratio of polynomials in z.

16.4. A control algorithm employed in a digital control system, shown in Figure P16.4, is given by the following difference equation

$$u(k) = u(k-1) + K_p e(k) - 0.9512 K_p e(k-1)$$

The sampling time is 0.1 sec and the continuous process transfer function is

$$\mathbf{T}_P(s) = \frac{0.25}{s+1}$$

(a) Determine the maximum value of the controller gain K_p for which the closed-loop system, consisting of the controller, zero-order hold, and the continuous process $\mathbf{T}_P(s)$, remains stable.

(b) Derive an expression for the steady-state error of the system in response to a unit ramp input kT, $k = 0, 1, \ldots$. Find the value of the steady-state error for the value of the controller gain found in part (a).

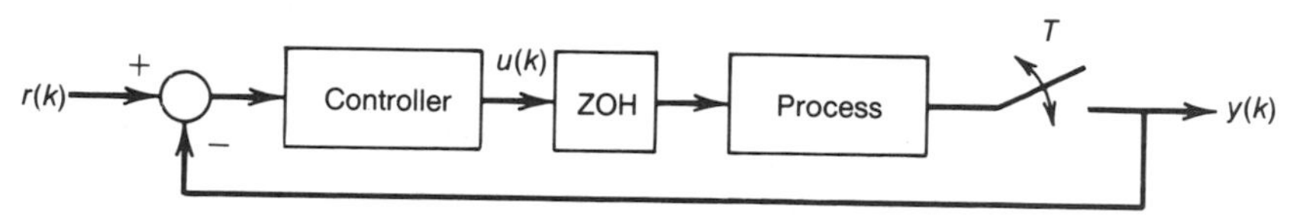

FIGURE P16.4 Digital control system considered in Problem 16.4.

16.5. A pulse transfer function of a continuous process with zero-order hold was found to be

$$\mathbf{T}_P(z) = (z-1)\mathcal{Z}\left\{\frac{\mathbf{T}_P(s)}{s}\right\} = \frac{0.025z^2 + 0.06z + 0.008}{z^3 - 1.6z^2 + 0.73z - 0.1}$$

Determine stability condition for a proportional controller of gain K_p in a digital control system with the process and zero-order hold represented by $\mathbf{T}_P(z)$.

16.6. A digital PD controller is to be designed for control of a continuous process having transfer function $\mathbf{T}_P(s) = 1/s(s + 2)$. The design specifications for the closed-loop system include a damping ratio of 0.7 and a period of step response oscillation equal to 5 sec.

(a) Find the control parameters k_p and T_d for a continuous PD controller to meet the design specifications.

(b) Find the proportional and derivative gains of a corresponding digital PD controller, assuming sampling time $T = 0.2$ sec and write the discrete control algorithm in velocity form.

(c) Obtain the pulse transfer function of the closed-loop system, including the digital PD controller and the continuous process preceded by a zero-order hold.

(d) Find a difference equation relating output and input variables of the closed-loop system.

(e) Calculate the first 15 values of the closed-loop system unit step response.

APPENDIX 1

Fourier Series and the Fourier Transform

When the input to a dynamic system is periodic, i.e., a continuously repeating function of time, having period T such as the function shown in Figure A1.1, it is often useful to describe this function in terms of an infinite series of pure sinusoids known as a Fourier series.

One form of such an infinite series is that given in Equation (A1.1).

$$x(t) = \frac{a_0}{2} + \sum_{k=1}^{\infty} a_k \cos k\omega_1 t + \sum_{k=1}^{\infty} b_k \sin k\omega_1 t \tag{A1.1}$$

where $a_0/2 = (1/T) \int_{t_0}^{t_0+T} x(t)\, dt$ is the average, or constant, value of the function, $\omega_1 = 2\pi/T$ is the radian frequency of the lowest frequency component, and the amplitudes of the series of component sinusoids at succeeding frequencies $k\omega_1$ are given by the following:

$$a_k = \frac{2}{T} \int_{t_0}^{t_0+T} x(t) \cos k\omega_1 t\, dt \tag{A1.2}$$

$$b_k = \frac{2}{T} \int_{t_0}^{t_0+T} x(t) \sin k\omega_1 t\, dt \tag{A1.3}$$

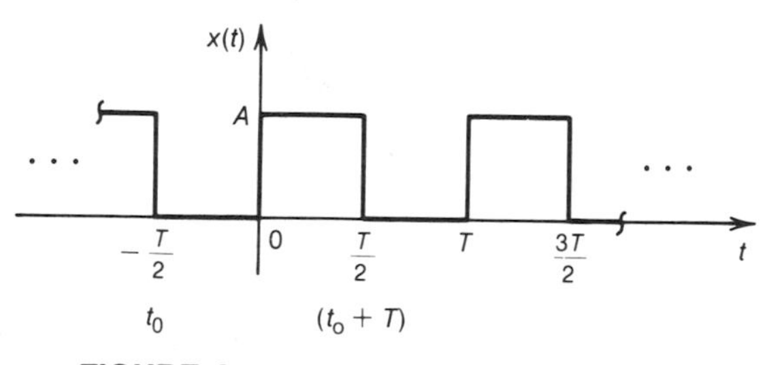

FIGURE A1.1 A typical periodic function.

Alternatively this function may be expressed as a series of only sine waves or only cosine waves, using Equation (A1.4) or Equation (A1.5).

$$x(t) = X_0 + \sum_{k=1}^{\infty} X_k \sin(k\omega_1 t + \phi_{ks}) \tag{A1.4}$$

$$x(t) = X_0 + \sum_{k=1}^{\infty} X_k \cos(k\omega_1 t + \phi_{kc}) \tag{A1.5}$$

where

$$X_0 = \frac{a_0}{2} = \frac{1}{T}\int_{t_0}^{t_0+T} x(t)\,dt$$

$$X_k = \sqrt{a_k^2 + b_k^2}$$

$$\phi_{ks} = \tan^{-1}\frac{b_k}{a_k}$$

$$\phi_{kc} = -\tan^{-1}\frac{a_k}{b_k}$$

The steady response of a dynamic system (i.e., the response remaining after all transients have decayed to zero) to an infinite series of sine waves may also be expressed as an infinite series.

$$y(t) = Y_0 + \sum_{k=1}^{\infty} Y_k \sin(k\omega_1 t + \phi_{ko}) \tag{A1.6}$$

where Y_0 is the constant component of the output and the coefficients Y_k are the amplitudes of the successive sine waves having frequencies $k\omega_1$ and phase angles ϕ_{ko}. The values of Y_0, Y_k, and ϕ_{ko} at each frequency are obtained by the methods developed in Chapter 12.

Alternatively the periodic function $x(t)$ may be expressed in terms of an infinite series of exponentials having the form $e^{jk\omega_0 t}$ through the use of Euler's equations (A1.7) and (A1.8).

$$\sin \omega t = \frac{1}{2j}(e^{j\omega t} - e^{-j\omega t}) \tag{A1.7}$$

$$\cos \omega t = \frac{1}{2}(e^{j\omega t} + e^{-j\omega t}) \tag{A1.8}$$

Substituting these expressions in Equation (A1.1) yields

$$x(t) = \frac{a_0}{2} + \sum_{k=1}^{\infty} \frac{1}{2}[(a_k - jb_k)e^{jk\omega_1 t} + (a_k + jb_k)e^{-jk\omega_1 t}]$$

or

$$x(t) = X_0 + \sum_{k=1}^{\infty} (\mathbf{X}_k e^{jk\omega_1 t} + \overline{\mathbf{X}}_k e^{-jk\omega_1 t}) \tag{A1.9}$$

where

$$X_0 = \frac{a_0}{2} = \frac{1}{T}\int_{t_0}^{t_0+T} x(t)\,dt \tag{A1.10}$$

$$\mathbf{X}_k = \frac{(a_k - jb_k)}{2} \tag{A1.11}$$

and

$$\overline{\mathbf{X}}_k = \frac{a_k + jb_k}{2} \tag{A1.12}$$

Noting that

$$\sum_{k=1}^{\infty} \overline{\mathbf{X}}_k e^{-jk\omega_1 t} = \sum_{k=-1}^{-\infty} \mathbf{X}_k e^{jk\omega_1 t}$$

the expression for $x(t)$ may be simplified and then combined with the expression for X_0 to form a single summation given in Equation (A1.13).

$$x(t) = \sum_{n=-\infty}^{+\infty} \mathbf{X}_n e^{jn\omega_1 t} \tag{A1.13}$$

where

$$\mathbf{X}_n = \frac{a_n - jb_n n}{|n|} = \frac{1}{T}\int_{t_0}^{t_0+T} x(t)e^{-jn\omega_1 t}\,dt \tag{A1.14}$$

and n represents the complete set of integers from minus infinity to plus infinity, *including zero*.

The complex coefficients $\mathbf{X}_n$ (i.e., the conjugate pairs $\mathbf{X}_k$ and $\overline{\mathbf{X}}_k$) have magnitudes and phase angles given by

$$|\mathbf{X}_n| = \frac{1}{2}\sqrt{a_n^2 + b_n^2} \tag{A1.15}$$

and

$$\angle \mathbf{X}_n = -\tan^{-1}\frac{n}{|n|}\frac{b_n}{a_n} \tag{A1.16}$$

Because the exponential components of the series in Equation (A1.13) act in conjugate pairs for each value of $|n|$, the magnitude of $\mathbf{X}_n$ is one half the amplitude of the corresponding wave in the sine wave series in Equation (A1.4), and the phase angle of $\mathbf{X}_n$ is the negative of the phase angle ϕ_{ks} of the corresponding wave in the sine wave series.

As a prelude to the discussion of Laplace transforms in Appendix 2, it can be seen that if the period T is allowed to increase and approach infinity, then the frequency, ω_0, approaches zero and the frequency interval between successive values of $n\omega_0$ becomes infinitesimal. In the limit, Equation (A1.14) becomes the expression for the Fourier transform

$$\mathbf{X}(j\omega) = \mathcal{F}\{x(t)\} = \int_{-\infty}^{\infty} x(t)e^{-j\omega t}\, dt \qquad \text{(A1.17)}$$

The Fourier transform is closely related[1] to the Laplace transform, which employs the exponential e^{-st} instead of the exponential $e^{-j\omega t}$ in the transformation integral.

EXAMPLE A1.1

Find the expression for the complex coefficients $\mathbf{X}_n$ of the infinite series of Fourier exponentials to represent the periodic function shown in Figure A1.1 and develop the terms of the exponential series into a series of sines and/or cosines. Sketch roughly to scale each of the first two sinusoidal components (i.e., for $n = 1, 3$) on a graph, together with the square-wave periodic function.

Solution

From Equation (A1.14) $\mathbf{X}_n$ is found

$$\mathbf{X}_n = \frac{1}{T}\int_{-T/2}^{+T/2} x(t)e - {}^{jn\omega_1 t}\, dt \qquad \text{(A1.18)}$$

Integrating from $-T/2$ to 0 and from 0 to $T/2$ and substituting the corresponding values of $x(t)$ yields

$$\begin{aligned}\mathbf{X}_n &= \frac{1}{T}\left[\int_{-T/2}^{0} 0 \cdot e^{-jn\omega_1 t}\, dt + \int_{0}^{+T/2} Ae^{-jn\omega_1 t}\, dt\right] \\ &= \frac{1}{T}\left[0 - \frac{A}{jn\omega_0} e^{-jn\omega_1 t}\Big|_0^{+T/2}\right] = \frac{jA}{n\omega_0 T}(e^{-jn\omega_1 T/2} - 1)\end{aligned} \qquad \text{(A1.19)}$$

Substituting the trigonometric form for the complex exponential in Equation (A1.19) and using $\omega_1 = 2\pi/T$ gives

$$\mathbf{X}_n = \frac{jA}{2n\pi}(\cos n\pi - j\sin n\pi - 1) = \frac{jA}{2n\pi}(\cos n\pi - 1) \qquad \text{(A1.20)}$$

Thus for even values of n, $\mathbf{X}_{n_{\text{even}}} = 0$, and for odd values of n

$$\mathbf{X}_{n_{\text{odd}}} = \frac{-jA}{n\pi} \qquad \text{(A1.21)}$$

For $n = 0$

$$X_0 = \frac{A}{2} \qquad \text{(A1.22)}$$

[1]See *Dynamics Of Physical Systems*, by R. H. Cannon, Jr., McGraw-Hill Book Co., New York, 1967, pp 533, 545–548.

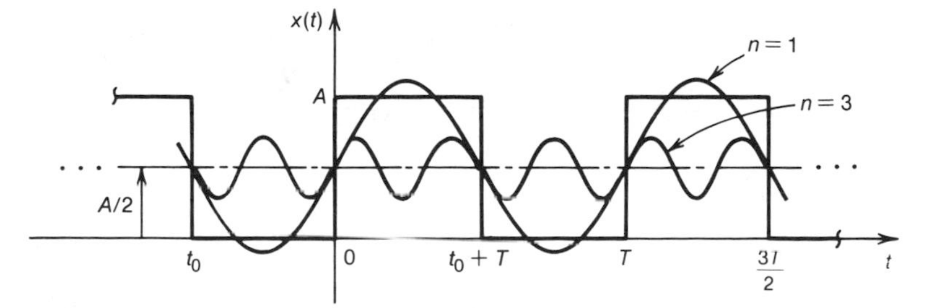

FIGURE A1.2 Sketch showing sine waves for $n = 1, 3$ superposed on square-wave function.

Substituting Equations (A1.21) and (A1.22) into Equation (A1.13) yields

$$\begin{aligned} x(t) &= \frac{A}{2} + \frac{2A}{\pi}\frac{1}{2j}\left(e^{j\omega_1 t} - e^{-j\omega_1 t}\right) \\ &+ \frac{2A}{3\pi}\frac{1}{2j}\left(e^{j3\omega_1 t} - e^{-j3\omega_1 t}\right) + \cdots \\ &+ \frac{2A}{n\pi}\frac{1}{2j}\left(e^{jn\omega_1 t} - e^{-jn\omega_1 t}\right) \end{aligned} \tag{A1.23}$$

Then using Euler's identity for sin ωt

$$\begin{aligned} x(t) &= \frac{A}{2} + \frac{2A}{\pi}\sin(\omega_1 t) + \frac{2A}{3\pi}\sin(3\omega_1 t) \\ &+ \cdots \frac{2A}{n\pi}\sin(n\omega_1 t) \end{aligned} \tag{A1.24}$$

The sine waves for $n = 1$ and $n = 3$ are shown superposed on the square-wave function in Figure A1.2. ■

APPENDIX 2

Laplace Transform

Definition of Laplace Transform. Laplace transform is a mapping between the time domain and the domain of complex variable s defined by

$$\mathbf{F}(s) = \mathcal{L}\{f(t)\} = \int_0^\infty f(t)e^{-st}\,dt \tag{A2.1}$$

where s is a complex variable, $s = \sigma + j\omega$, and $f(t)$ is a sectionally continuous function of time. Function $f(t)$ is also assumed to be equal to zero for $t < 0$. With this assumption the transform defined by Equation (A2.1) is called a one-sided Laplace transform.

The condition for the existance of a Laplace transform of $f(t)$ is that the integral in Equation (A2.1) exists, which in turn requires that there exist real numbers, A and b, such that $|f(t)| < Ae^{bt}$. Most functions of time encountered in engineering systems are Laplace transformable.

Laplace transform is commonly used in solving linear differential equations. By applying the Laplace transform, the differential equations involving variables of time t are transformed into algebraic equations in the domain of complex variable s. The solutions of the algebraic equations, which are usually much easier to obtain than the solutions of the original differential equations, are then transformed back to the time domain using the inverse Laplace transform.

Inverse Laplace Transform. The inverse Laplace transform is defined by the Riemann integral

$$f(t) = \mathcal{L}^{-1}\{\mathbf{F}(s)\} = \frac{1}{2\pi j}\int_{c-j\infty}^{c+j\infty} \mathbf{F}(s)e^{st}\,ds \tag{A2.2}$$

TABLE A2.1. Laplace and z transforms of most common functions of time.

Continuous	Discrete	Laplace Transform	z Transform
$U_i(t)$	$U_i(kT)$	1	1
$U_s(t)$	$U_s(kT)$	$1/s$	$z/(z - 1)$
t	kT	$1/s^2$	$Tz/(z - 1)^2$
t^2	$(kT)^2$	$2/s^3$	$T^2z(z + 1)/(z - 1)^3$
e^{-at}	e^{-akt}	$1/(s + a)$	$z/(z - e^{-aT})$
te^{-at}	kTe^{-akT}	$1/(s + a)^2$	$Te^{-aT}z/(z - e^{-aT})^2$
$\sin \omega t$	$\sin \omega kT$	$\omega/(s^2 + \omega^2)$	$z \sin \omega T/(z^2 - 2z \cos \omega T + 1)$
$\cos \omega t$	$\cos \omega kT$	$s/(s^2 + \omega^2)$	$z(z - \cos \omega T)/(z^2 - 2z \cos \omega T + 1)$

The Riemann integral is rarely used in practice. The most common practical method employed in inverse Laplace transform is the method of partial fraction expansion, which will be described later.

A short list of Laplace transforms and z transforms, defined in Chapter 15, of the most common functions of time is given in Table A2.1.

Basic Properties of the Laplace Transform. Following are useful basic properties of the Laplace transform.

1. Linearity

$$\mathcal{L}\{a_1 f_1(t) + a_2 f_2(t)\} = a_1\mathbf{F}_1(s) + a_2\mathbf{F}_2(s) \tag{A2.3}$$

2. Integration

$$\mathcal{L}\left\{\int_0^t f(\tau)\, d\tau\right\} = \frac{\mathbf{F}(s)}{s} \tag{A2.4}$$

3. Differentiation

$$\mathcal{L}\left\{\frac{d^n f(t)}{dt^n}\right\} = s^n\mathbf{F}(s) - \sum_{k=0}^{n-1} s^{n-k-1}\left[\frac{d^k f(t)}{dt^k}\right]\Bigg|_{t=0^-} \tag{A2.5}$$

4. Shifting argument in the time domain (multiplication by an exponential in the s domain)

$$\mathcal{L}\{f(t - a)\} = e^{-as}\,\mathbf{F}(s) \tag{A2.6}$$

5. Shifting argument in the s domain (multiplication by an exponential in the time domain)

$$\mathcal{L}\{f(t)e^{-at}\} = \mathbf{F}(s + a) \tag{A2.7}$$

In addition to these properties, two theorems are very useful in evaluating initial and steady-state values of function $f(t)$. The initial value theorem is

$$f(0^+) = \lim_{s\to\infty} s\mathbf{F}(s) \tag{A2.8}$$

The steady-state value of function $f(t)$ is given by the final value theorem

$$\lim_{t \to \infty} f(t) = \lim_{s \to 0} s\mathbf{F}(s) \tag{A2.9}$$

Partial Fraction Expansion Method. The method of partial fraction expansion is most commonly used in obtaining inverse Laplace transforms of functions that have a form of ratios of polynomials in s, such as

$$\mathbf{F}(s) = \frac{\mathbf{B}(s)}{\mathbf{A}(s)} = \frac{b_0 + b_1 s + \cdots + b_m s^m}{a_0 + a_1 s + \cdots + a_n s^n} \tag{A2.10}$$

where $m \leq n$. Transfer functions of linear systems often take such form. If the roots of the characteristic equation, $\mathbf{A}(s) = 0$, are $s_1, s_2, \ldots, s_q$, the expansion of $\mathbf{F}(s)$ takes the form

$$\mathbf{F}(s) = \frac{C_{11}}{s - s_1} + \frac{C_{12}}{(s - s_1)^2} + \cdots + \frac{C_{1p_1}}{(s - s_1)^{p_1}} + \cdots + \frac{C_{qp_q}}{(s - s_q)^{p_q}} \tag{A2.11}$$

where q is the number of roots, some of which may be multiple roots, and p_i is a multiplicity of ith root. Note that if all roots are distinct, then $q = n$. Equation (A2.11) can be rewritten in a more compact form

$$\mathbf{F}(s) = \sum_{i=1}^{n} \sum_{j=1}^{p_i} \frac{C_{ij}}{(s - s_i)^j} \tag{A2.12}$$

The constants C_{ij} are given by

$$\begin{aligned} C_{i1} &= \lim_{s \to s_i} [\mathbf{F}(s)(s - s_i)^{p_i}] \\ C_{i2} &= \lim_{s \to s_i} \left\{ \frac{d[\mathbf{F}(s)(s - s_i)^{p_i}]}{ds} \right\} \\ &\vdots \\ C_{ij} &= \frac{1}{(p_i - j)!} \lim_{s \to s_i} \frac{d^{p_i - j}[\mathbf{F}(s)(s - s_i)^{p_i - j}]}{ds^{p_i - j}} \end{aligned} \tag{A2.13}$$

In Examples A2.1 and A2.2 the partial fraction expansion method is used to find inverse Laplace transforms. In Example A2.3 the Laplace transform is employed to obtain a transfer function and a step response for a mass-spring-dashpot system.

EXAMPLE A2.1

Find an inverse Laplace transform of function $\mathbf{F}(s)$

$$\mathbf{F}(s) = \frac{2s + 4}{s^3 + 7s^2 + 15s + 9}$$

Solution

The first step is to find poles of $\mathbf{F}(s)$. There is a single pole at $s = -1$ and a pole of multiplicity 2 at $s = -3$. $\mathbf{F}(s)$ can thus be rewritten as

$$\mathbf{F}(s) = \frac{2s + 4}{(s + 1)(s + 3)^2}$$

Hence, the expanded form of $\mathbf{F}(s)$ is

$$\mathbf{F}(s) = \frac{C_{11}}{s + 1} + \frac{C_{21}}{s + 3} + \frac{C_{22}}{(s + 3)^2}$$

Using Equations (A2.13), the constants C_{11}, C_{21}, and C_{22} are found

$$C_{11} = 0.5 \qquad C_{21} = -0.5 \qquad C_{22} = 1$$

The partial fraction expansion of $\mathbf{F}(s)$ becomes

$$\mathbf{F}(s) = \frac{0.5}{s + 1} - \frac{0.5}{s + 3} + \frac{1}{(s + 3)^2}$$

The inverse Laplace transforms of the three simple terms on the right-hand side of the above equation can be found in Table A2.1 to give the solution

$$f(t) = 0.5e^{-t} - 0.5e^{-3t} + te^{-3t}$$

■

EXAMPLE A2.2

Find the inverse Laplace transform of

$$\mathbf{F}(s) = \frac{4}{s(s^2 + 2.4s + 4)}$$

Solution

In addition to the pole $s = 0$, $\mathbf{F}(s)$ has two complex conjugate poles, $s_2 = -1.2 - j1.6$ and $s_3 = -1.2 + j1.6$. The factored form of $\mathbf{F}(s)$ is

$$\mathbf{F}(s) = \frac{4}{s(s + 1.2 + j1.6)(s + 1.2 - j1.6)}$$

The partial fraction expansion in this case is

$$\mathbf{F}(s) = \frac{C_{11}}{s} + \frac{C_{21}}{(s + 1.2 + j1.6)} + \frac{C_{31}}{(s + 1.2 - j1.6)}$$

The constants C_{11}, C_{21}, C_{31} are obtained using Equations (A2.13)

$$C_{11} = 1 \qquad C_{21} = 0.625e^{-j0.6435} \qquad C_{31} = 0.625e^{j0.6435}$$

Hence

$$\mathbf{F}(s) = \frac{1}{s} + 0.625\left(\frac{e^{j0.6435}}{s + 1.2 + j1.6} + \frac{e^{-j0.6435}}{s + 1.2 - j1.6}\right)$$

Using Table A2.1 to find the inverse transforms of the terms on the right-hand side of the above equation, the function of time is obtained

$$f(t) = 1 + 0.625[e^{j0.6435}e^{(-1.2-j1.6)t} + e^{-j0.6435}e^{(-1.2+j1.6)t}]$$

In order to simplify the form of $f(t)$, first group the exponents of the exponential terms

$$f(t) = 1 + 0.625(e^{j0.6435-1.2t-j1.6t} + e^{-j0.6435-1.2t+1.6t})$$

Moving the common factor in front of the parenthesis gives

$$f(t) = 1 + 0.625e^{-1.2t}[e^{-j(-0.6435+1.6t)} + e^{-j(0.6435-1.6t)}]$$

Now, substitute equivalent trigonometric expressions for the two exponential terms to get

$$f(t) = 1 + 0.625e^{-1.2t}[\cos(-0.6435 + 1.6t) - j\sin(-0.6435 + 1.6t) + \cos(0.6435 - 1.6t) - j\sin(0.6435 - 1.6t)]$$

Using basic properties of sine and cosine functions, the final form of the solution is obtained.

$$f(t) = 1 + 1.25e^{-1.2t}\sin(0.9273 + 1.6t)$$ ■

EXAMPLE A2.3

Consider a mass, spring, dashpot system, shown in Figure 11.7. Use the Laplace transformation to find the system transfer function and obtain the step response $x(t)$ for a step change of force $F(t)$.

Solution

The system differential equations of motion are

$$m\frac{dv}{dt} = F(t) - bv - kx$$

$$\frac{dx}{dt} = v$$

Laplace transform of the two equations gives

$$m[s\mathbf{V}(s) - v(0^-)] = \mathbf{F}(s) - b\mathbf{V}(s) - k\mathbf{X}(s)$$

$$s\mathbf{X}(s) - x(0^-) = \mathbf{V}(s)$$

where $x(0^-)$ and $v(0^-)$ represent the initial conditions for displacement and velocity of mass m just before the input force was applied. Now, combining the Laplace-transformed equations yields

$$\mathbf{X}(s)(ms^2 + bs + k) = \mathbf{F}(s) + x(0^-)(ms + b) + mv(0^-)$$

Assuming zero initial conditions, $x(0^-) = 0$ and $v(0^-) = 0$, the system transfer function is found

$$\mathbf{T}(s) = \frac{\mathbf{X}(s)}{\mathbf{F}(s)} = \frac{1}{ms^2 + bs + k}$$

Note that the transfer function obtained here is the same as the transfer function derived in Section 11.4.3 using an exponential input approach.

Next, the system step response is to be found for the input given by

$$F(t) = \Delta F \cdot U_s(t)$$

The Laplace transform of the input signal is

$$\mathbf{F}(s) = \frac{\Delta F}{s}$$

Using the expression for the transfer function, the Laplace transform of the output signal is found

$$\mathbf{X}(s) = \mathbf{F}(s)\mathbf{T}(s) = \frac{\Delta F}{s(ms^2 + bs + k)}$$

To obtain the solution in the time domain, two cases must be considered. First, assume that both roots of the quadratic term in denominator of $\mathbf{X}(s)$ are real and distinct, that is

$$ms^2 + bs + k = m(s + a)(s + b)$$

where a and b are real numbers and $a \neq$ b. The inverse transform of $\mathbf{X}(s)$ is this case is

$$x(t) = \mathcal{L}^{-1}\{\mathbf{X}(s)\} = \frac{\Delta F}{m}\left(\frac{1 + \dfrac{be^{-bt} - ae^{-at}}{a - b}}{ab}\right)$$

Now, assume that the roots of the quadratic term in denominator of $\mathbf{X}(s)$ are complex conjugate, that is

$$ms^2 + bs + k = m(s^2 + 2\zeta\omega_n s + \omega_n^2)$$

The inverse Laplace transform in this case is

$$x(t) = \mathcal{L}^{-1}\{\mathbf{X}(s)\} = \frac{\Delta F}{k}\left[1 + \left(\frac{1}{\sqrt{1 - \zeta^2}}\right) e^{-\zeta\omega_n t} \sin\left(\omega_n\sqrt{1 - \zeta^2}\, t + \phi\right)\right]$$

where

$$\phi = \tan^{-1}\left(\frac{\sqrt{1 - \zeta^2}}{\zeta}\right)$$

■

It should be noted that the initial conditions needed when using Laplace transforms, as well as for computer simulations, are at $t = 0^-$; whereas solutions using the classical methods of Chapter 5 need initial conditions at $t = 0^+$.

APPENDIX 3

Digital Simulation Program

This appendix describes a simple program in the BASIC language for use on small personal computers to simulate linear dynamic system models. It is intended to provide the reader with an opportunity to gain first-hand experience with the use of numerical integration techniques for dynamic simulation. Assuming that only 64K of random access memory might be available, it was decided to provide only for simulation of systems up to third order, using about 300 uniform time steps for each simulation. This program incorporates a routine that computes an initial estimated time step based on the gains of inherently closed loops within the system. This default value of time step, together with default values for number of initial waiting period steps, frequency of output, and number of simulation steps may be over-ridden by computer prompts. Also initial values for inputs, state-variables, and outputs, all at time $t = 0^-$, are prompted by the computer.

This simulation program provides for the numerical integration of a set of three state-variable equations with provision for generating two inputs and it provides for the use of output equations for two outputs. A linear system format is employed for the state-variable equations (which the user may wish to alter to accommodate nonlinear effects). This set of equations appears in the following form

$$\frac{dq_1}{dt} = a_{11}q_1 + a_{12}q_2 + a_{13}q_3 + b_{11}u_1 + b_{12}u_2$$

$$\frac{dq_2}{dt} = a_{21}q_1 + a_{22}q_2 + a_{23}q_3 + b_{21}u_1 + b_{22}u_2$$

$$\frac{dq_3}{dt} = a_{31}q_1 + a_{32}q_2 + a_{33}q_3 + b_{31}u_1 + b_{32}u_2$$

where the constant coefficients $a_{11} \ldots a_{33}$ and $b_{11} \ldots b_{32}$ reside in the $\mathbb{A}$ and $\mathbb{B}$ matrices

$$\mathbb{A} = \begin{bmatrix} a_{11} & a_{12} & a_{13} \\ a_{21} & a_{22} & a_{23} \\ a_{31} & a_{32} & a_{33} \end{bmatrix} \qquad \mathbb{B} = \begin{bmatrix} b_{11} & b_{12} \\ b_{21} & b_{22} \\ b_{31} & b_{32} \end{bmatrix}$$

The structure provided for the two output equations is of the form

$$y_1 = c_{11}q_1 + c_{12}q_2 + c_{13}q_3 + d_{11}u_1 + d_{12}u_2$$

$$y_2 = c_{21}q_1 + c_{22}q_2 + c_{23}q_3 + d_{21}u_1 + d_{22}u_2$$

where the coefficients $c_{11} \ldots c_{23}$ reside in the $\mathbb{C}$ matrix and the coefficients $d_{11} \ldots d_{23}$ residing in the $\mathbb{D}$ matrix include d_{13} and d_{23} which may be needed in generating u_1 or u_2 or for generating nonlinear terms in the state-variable equations.

$$\mathbb{C} = \begin{bmatrix} c_{11} & c_{12} & c_{13} \\ c_{21} & c_{22} & c_{23} \end{bmatrix} \qquad \mathbb{D} = \begin{bmatrix} d_{11} & d_{12} & d_{13} \\ d_{21} & d_{22} & d_{23} \end{bmatrix}$$

For the option which uses an attached printer to list the system data and line-by-line computed data, only one of the two inputs is activated, and only one of the two potential outputs is used in order to be able to keep the output within the width of 9 and 1/2 inch wide printer paper.

A second option may be invoked which causes the input and output to be plotted versus time on the monitor screen. Additional statements may be inserted by the user who wishes to plot more inputs and/or outputs on the screen.

The numerical integration routine is based on the "exponential Euler method" described in Chapter 6 which uses an exponential extrapolation for each state variable in the interval from t to $t + \Delta t$. Because of the need to incorporate sufficiently up to date "feedback" from the other state variables into the right-hand-side terms of each state-variable equation, an "average" or "half-way" value of each of the other state variables is employed in producing the full step extrapolation for each state variable in its turn.

To facilitate the exponential extrapolation process the state variable equations are rearranged into a set of first-order input-output equations with the right-hand side terms representing "inputs" for each equation

$$\frac{dq_1}{dt} - a_{11}q_1 = \mathrm{sum}_1$$

$$\frac{dq_2}{dt} - a_{22}q_2 = \mathrm{sum}_2$$

$$\frac{dq_3}{dt} - a_{33}q_3 = \mathrm{sum}_3$$

where

$$\mathrm{sum}_1 = a_{12}q_2 + a_{13}q_3 + b_{11}u_1 + b_{12}u_2$$

$$\mathrm{sum}_2 = a_{21}q_1 + a_{23}q_3 + b_{21}u_1 + b_{22}u_2$$

$$\mathrm{sum}_3 = a_{31}q_1 + a_{32}q_2 + b_{31}u_1 + b_{32}u_2$$

So that the exponential extrapolation solutions take the form

$$q_1(t + \Delta t) = q_1(t)e^{a_{11}\Delta t} + [(\text{sum}_1)_{\text{ave}}](1 - e^{a_{11}\Delta t})/a_{11}$$

$$q_2(t + \Delta t) = q_2(t)e^{a_{22}\Delta t} + [(\text{sum}_2)_{\text{ave}}](1 - e^{a_{22}\Delta t})/a_{22}$$

$$q_3(t + \Delta t) = q_3(t)e^{a_{33}\Delta t} + [(\text{sum}_3)_{\text{ave}}](1 - e^{a_{33}\Delta t})/a_{33}$$

Each sum_i term, representing feedback from the other newly-computed state-variable equation solutions, needs to be an average over the interval from t to $t + \Delta t$. These "averages" are provided here by using the set of q's computed at the half-way point, $t + \Delta t/2$. For instance

$$(\text{sum}_1)_{\text{ave}} = a_{12}q_2(t + \Delta t/2) + a_{13}q_3(t + \Delta t/2) + b_{11}u_1(t) + b_{12}u_2(t)$$

and so on.

For the cases when $a_{ii} = 0$, the routine reverts to the linear extrapolation of the "improved Euler method", using the "half-way" computed values of the q_i's instead of the average of the beginning and ending values of the q_i's.

To facilitate the process of assessing the quality of the digital simulation produced by using this program and to see how the results are affected by the choice of the time step Δt, a linear third-order undamped system model has been provided for the user to make test runs that can be compared to a known analytic solution obtained by classical methods.

Using a system model having a characteristic equation with a pair of imaginary roots creates a very critical test when the amplitude of the simulated sinusoidal response is examined carefully after the first-order time constant portion of the response has decayed to a very small value. Ideally the simulated sinusoid should maintain the correct amplitude over a period of several cycles.

A simulation block diagram for the third-order undamped system model to be used here is shown in Figure A3.1.

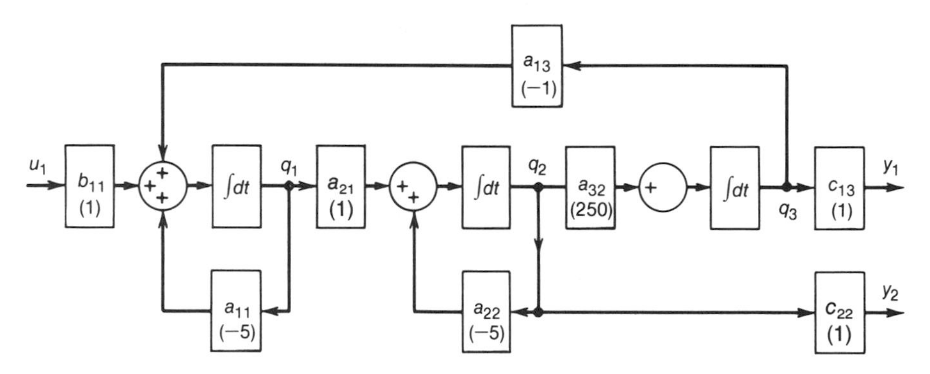

FIGURE A3.1. Simulation block diagram of third-order undamped model for test runs.

The values of the matrix coefficients are:

For the row 1 a's:	$a_{11} = -5$	$a_{12} = 0$	$a_{13} = -1$
For the row 2 a's:	$a_{21} = 1$	$a_{22} = -5$	$a_{23} = 0$
For the row 3 a's:	$a_{31} = 0$	$a_{32} = 250$	$a_{33} = 0$
For the row 1 b's:	$b_{11} = 1$	$b_{12} = 0$	
For the row 2 b's:	$b_{21} = 0$	$b_{22} = 0$	
For the row 3 b's:	$b_{31} = 0$	$b_{32} = 0$	
For the row 1 c's:	$c_{11} = 0$	$c_{12} = 0$	$c_{13} = 1$
For the row 2 c's:	$c_{21} = 0$	$c_{22} = 1$	$c_{23} = 0$
For the row 1 d's:	$d_{11} = 0$	$d_{12} = 0$	$d_{13} = 0$
For the row 2 d's:	$d_{21} = 0$	$d_{22} = 0$	$d_{23} = 0$

The system input-output differential equation for the test model is

$$\frac{d^3y}{dt^3} + \frac{10d^2y}{dt^2} + \frac{25dy}{dt} + 250y = 250u_1$$

The characteristic equation is

$$s^3 + 10s^2 + 25s + 250 = 0$$

The roots (poles) of the characteristic equation are

$$p_1 = -10$$

$$p_2 = +j5$$

$$p_3 = -j5$$

With all initial values of u's, q's and y's equal to zero, the classical solution for a unit step change in u_1 at $t = 0$ is given by

$$y = 1.0 - .2e^{-10t} - .4\sqrt{5}\sin(5t + .5205)$$

The period of the sine wave is

$$T_p = \frac{2\pi}{\omega_n} = \frac{6.28}{5} = 1.256 \text{ sec.}$$

The phase shift time is

$$T_{ph} = \frac{\psi}{\omega_n} = \frac{.4205}{5} = .0841 \text{ sec.}$$

A copy of the screen plot showing the unit step input, $u_1(t)$, and the output $y_1(t)$ when this program was run with $\Delta t = .015$ sec is shown in Figure A3.2.

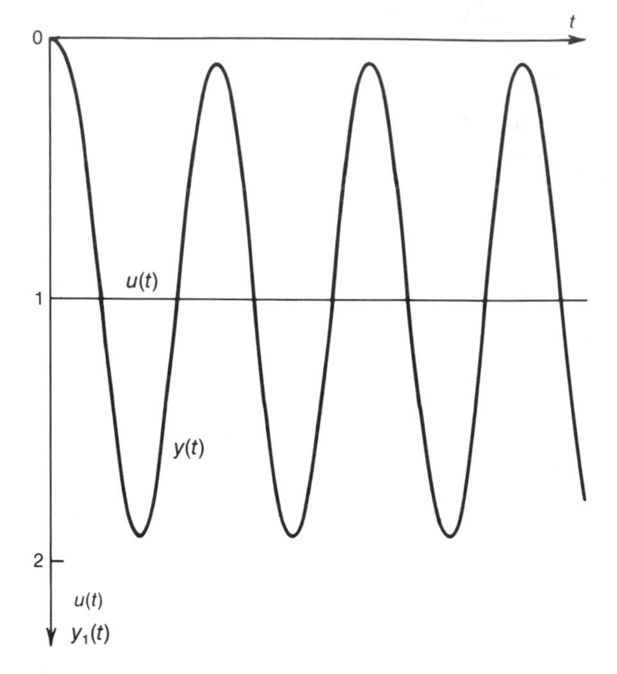

FIGURE A3.2 Copy of screen plot showing response of $y_1(t)$ to a unit step input $u_1(t)$ using t = .015sec.

The computer program that was used to produce this screen plot is listed below and is available on a 5¼ inch diskette for use with an IBM-compatible personal computer.

```
5 REM THIS VERSION SAVED AT 1100 HRS. ON 19 DEC 1989
10 REM  THIS IS "A3PROB" REVISED FROM "A3PROG", TEXT
VERSION OF
30 REM  FILE "SIM3T/BAS" FOR USE IN APPENDIX 3 OF D.M.C.E.S.
40 REM
50 REM  PROGRAM FOR DIGITAL SIMULATION OF 3RD ORDER LINEAR SYSTEMS
60 REM   'STRIPPED-DOWN' FOR ONLY ONE INPUT AND ONE OUTPUT IN ORDER FOR
70 REM   OUTPUT COLUMNS TO FIT ON 9-1/2 INCH WIDE FAN-FOLD PRINTER PAPER.
80 REM
90 REM THIS PROGRAM MAY BE USED TO
100 REM   A) PRINT COMPUTED OUTPUT DATA AT CONNECTED PRINTER,
110 REM OR
120 REM   B) PLOT INPUT AND OUTPUT VERSUS TIME ON MONITOR SCREEN.
130 REM
140 REM DEFAULT IS TO PRINT INPUT AND OUTPUT DATA AT PRINTER, TO BE OVER-RIDDEN
150 REM  AT PROMPT IF INPUT AND OUTPUT ARE TO BE PLOTTED ON SCREEN INSTEAD.
160 REM
170 REM NOTE!  PRINTER MUST BE CONNECTED IF OPTION A) TO PRINT IS USED.
180 REM
190 REM SYSTEM CAN HAVE TWO INPUTS, THREE STATE VARIABLES, AND TWO OUTPUTS.
200 REM
210 REM INITIAL VALUES OF INPUTS U1 AND U2 ARE STORED AS UI(1) AND UI(2).
220 REM INITIAL VALUES OF STATE-VARIABLES Q1 . . Q3 ARE STORED AS QI(1) . . QI(3).
230 REM
240 REM STATE-VARIABLE COEFFICIENTS ARE STORED IN MATRICES:
250 REM A-MATRIX COEFFS. RESIDE IN THREE 3-ELEM. ROWS AS A(1,1) . . . A(3,3)
260 REM B-MATRIX COEFFICIENTS RESIDE IN THREE 2-ELEM. ROWS AS B(1,1) . . B(3,2).
270 REM C-MATRIX COEFFICIENTS RESIDE IN TWO 3-ELEMENT ROWS AS C(1,1) . . C(2,3).
280 REM D-MATRIX COEFFICIENTS RESIDE IN TWO 3-ELEMENT ROWS AS D(1,1) . . D(2,3).
290 REM
```

```
300 INPUT "TITLE:"; NA$
310 DIM UI(2),QI(3),YI(2),Q(3,250),T(250),U(2,250),A(3,3),B(3,2),C(2,3),D(2,3),S(3)
315 REM
320 REM SET TIMER (IF SET-TIMER FEATURE IS AVAILABLE, ACTIVATE NEXT LINE).
330 REM TIMER = 0
340 PRINT "THE STARTING VALUE OF TIMER IS:";TIMER
350 REM
360 REM NEXT ARE THE STATEMENTS TO PROMPT THE NECESSARY INPUT DATA.
370 PRINT "RESPOND TO THE FOLLOWING PROMPTS WITH NUMERICAL VALUES FOR"
380 PRINT "RUN NUMBER AND INITIAL CONDITIONS AT T = 0-"
385 PRINT " "
390 INPUT "RUN NUMBER, NR="; NR
400 INPUT "UI(1) AND UI(2):";UI(1),UI(2)
410 INPUT "QI(1), QI(2) AND QI(3):";QI(1),QI(2),QI(3)
420 REM
430 REM  NEXT THE PROMPTS FOR THE MATRIX ROWS:
440 PRINT "RESPOND TO ALL THE FOLLOWING PROMPTS FOR MATRIX COEFFICIENTS"
445 PRINT " "
450 INPUT "A(1,1), A(1,2), AND  A(1,3):"; A(1,1), A(1,2), A(1,3)
460 INPUT "A(2,1), A(2,2), AND A(2,3):"; A(2,1), A(2,2), A(2,3)
470 INPUT "A(3,1), A(3,2), AND A(3,3):"; A(3,1), A(3,2), A(3,3)
480 INPUT "B(1,1) AND B(1,2):"; B(1,1), B(1,2)
490 INPUT "B(2,1) AND B(2,2):"; B(2,1), B(2,2)
500 INPUT "B(3,1) AND B(3,2):"; B(3,1), B(3,2)
510 INPUT "C(1,1), C(1,2) AND C(1,3):"; C(1,1), C(1,2), C(1,3)
520 INPUT "C(2,1), C(2,2) AND C(2,3):"; C(2,1), C(2,2), C(2,3)
530 INPUT "D(1,1), D(1,2) AND D(1,3):"; D(1,1), D(1,2), D(1,3)
540 INPUT "D(2,1), D(2,2) AND D(2,3):"; D(2,1), D(2,2), D(2,3)
550 REM
560 REM COMPUTE INITIAL VALUE OF TIMESTEP DELT AND PROVIDE OVER-RIDE FOR
570 REM  ROUND-OFF OR TO CHANGE QUALITY OF SIMULATION.
580 REM   METHOD INVOLVES SUCCESSIVE ELIMINATION OF ESTIMATES BASED ON LOOP
590 REM   GAINS, HENCE RESPONSE RATES.
600 REM   IT IS BASED ON THE NOTION THAT THE INITIAL DELT SHOULD BE ABOUT
610 REM   ONE-FOURTH THE SMALLEST "TIME-CONSTANT" IN THE SYSTEM, WHERE
620 REM   "TIME-CONSTANT" IS TAKEN TO BE THE INVERSE OF THE "BREAK-FREQUENCY"
630 REM   OF A GIVEN LOOP.
640 REM
650 REM FIRST SET UP ALL THE LOOP "BREAK-FREQUENCIES*4".
660 REM
670 REM FOR THE SINGLE INTEGRATOR LOOPS:
680 W1=4*ABS(A(1,1)) : W2=4*ABS(A(2,2)) : W3=4*ABS(A(3,3))
690 REM
700 REM FOR THE TWO-INTEGRATOR LOOPS:
710 W4=4*SQR(ABS(A(1,2)*A(2,1)))
720 W5=4*SQR(ABS(A(2,3)*A(3,2)))
730 W6=4*SQR(ABS(A(1,3)*A(3,1)))
740 REM
750 REM FOR THE THREE-INTEGRATOR LOOP:
760 W7=4*(ABS(A(2,1)*A(3,2)*A(1,3)))^.333
770 REM
780 REM NOW FIND THE LARGEST, CALLING IT WB.
790 WB=W1
800 IF W2>WB THEN 820
810 GOTO 830
820 WB=W2
830 IF W3>WB THEN 850
840 GOTO 860
850 WB=W3
860 IF W4>WB THEN 880
870 GOTO 890
880 WB=W4
890 IF W5>WB THEN 910
900 GOTO 920
910 WB=W5
920 IF W6>WB THEN 940
930 GOTO 950
940 WB=W6
950 IF W7>WB THEN 970
960 GOTO 980
970 WB=W7
980 DT= 1.0/WB
990 REM
1000 REM PRINT ON SCREEN THE COMPUTED ESTIMATE OF DELT:
1010 PRINT "COMPUTED DELT="; DT
1020 REM
1030 REM PROVIDE MEANS TO OVER-RIDE COMPUTED ESTIMATE OF DELT.
1040 INPUT "OVER-RIDE DELT--YES (ALL CAPS) OR NO?"; DT$
1050 IF DT$ = "YES" THEN 1070
```

```
1060 GOTO 1110
1070 INPUT "OVER-RIDE VALUE OF DELT IS:";DT
1080 REM
1090 REM  SET UP DEFAULT VALUES AND MEANS TO OVERRIDE FOR NP, NW, NSIM
1100 REM
1110 REM SET DEFAULT VALUES OF NP (PRINT ONLY NPTH DATA SET, NW (TIME INCREMENTS
1120 REM  INCREMENTS FOR INITIAL WAITING PERIOD), NSIM (NUMBER OF TIME
1130 REM  INCREMENTS FOR SIMULATION).
1140 NP=2 : NW=2*NP : NS=230
1150 REM
1160 PRINT "RESPOND TO THE FOLLOWING PROMPTS FOR POSSIBLE OVER-RIDE VALUES"
1170 PRINT "FOR NPER (PRINTING FREQUENCY), NWAIT (WAITING PERIOD STEPS), AND"
1180 PRINT "NSIM (SIMULATION STEPS)"
1185 PRINT " "
1190 REM PRINT DEFAULT VALUE OF NPER ON SCREEN.
1200 PRINT "NPER=";NP
1210 REM
1220 REM MEANS TO OVERRIDE VALUE OF NPER:
1230 INPUT "OVER-RIDE NPER -YES OR NO?";N1$
1240 IF N1$="YES" THEN 1260
1250 GO TO 1280
1260 INPUT "OVER-RIDE VALUE OF NPER (MAX IS 6):";NP
1270 REM
1280 REM PRINT DEFAULT VALUE OF NWAIT ON SCREEN.
1290 PRINT "NWAIT=";NW
1300 REM
1310 REM MEANS TO OVERRIDE VALUE OF NWAIT:
1320 INPUT "OVER-RIDE NWAIT--YES OR NO?";N2$
1330 IF N2$="YES" THEN 1350
1340 GO TO 1370
1350 INPUT "OVER-RIDE VALUE OF NWAIT (MAX IS 5):";NW
1360 REM
1370 REM PRINT DEFAULT VALUE OF NSIM ON SCREEN.
1380 PRINT "NSIM=";NS
1390 REM
1400 REM MEANS TO OVERRIDE VALUE OF NSIM:
1410 INPUT "OVER-RIDE NSIM--YES OR NO?";N3$
1420 IF N3$="YES" THEN 1440
1430 GO TO 1460
1440 INPUT "OVER-RIDE VALUE OF NSIM IS (MAX IS 240-NWAIT-1):"; NS
1450 REM
1460 REM COMPUTE TIME PERIOD PARAMETERS N1, N2, NM
1470 N1=NW : N2=N1+1 : NM=NS+N2
1480 REM
1490 REM SET UP "WAITING-PERIOD" VALUES OF U'S, Q'S, AND Y'S:
1500 T(0)=-N1*DT
1510 FOR N=1 TO N1
1520 U1=UI(1)
1530 U(1,N)=U1
1540 U2=UI(2)
1550 U(2,N)=U2
1560 Q(1,N)=QI(1)
1570 Q(2,N)=QI(2)
1580 Q(3,N)=QI(3)
1590 T(N)=T(N-1)+DT
1600 NEXT N
1610 REM
1620 REM PRINT MESSAGE ON SCREEN WHEN SIMULATION BEGINS (OPTIONAL).
1630 PRINT "SIMULATION STARTS NOW, TIME="; TIMER
1640 REM
1650 REM NEXT, THE STATEMENTS TO COMPUTE THE U'S, Q'S, AND Y'S FOR N=N2 TO NM.
1660 FOR N = N2 TO NM
1670 REM COMPUTE THE VALUE OF T(N):
1680 T(N) = T(N-1)+DT
1690 REM FIRST GENERATE A UNIT STEP CHANGE IN U1, AND KEEP U2 ZERO.
1700 REM  (HERE A SUBROUTINE COULD BE CALLED TO GENERATE THESE FUNCTIONS).
1710 U1=1.0
1720 U(1,N)=U1
1730 U2=0.0
1740 U(2,N)=U2
1750 REM NEXT, SET UP RIGHT HAND SIDE SUMS S(1). . .(S(3) BASED ON NEW U'S
1760 REM  AND PREVIOUS Q'S.
1770 S(1)=B(1,1)*U1+B(1,2)*U2+A(1,2)*Q(2,N-1)+A(1,3)*Q(3,N-1)
1780 S(2)=B(2,1)*U1+B(2,2)*U2+A(2,1)*Q(1,N-1)+A(2,3)*Q(3,N-1)
1790 S(3)=B(3,1)*U1+B(3,2)*U2+A(3,1)*Q(1,N-1)+A(3,2)*Q(2,N-1)
1800 REM
1810 REM  NOW THE STATE-VARIABLE EQUATION SOLUTIONS AT DELT/2, INCLUDING THE
1820 REM  CASES WHEN A(I,I)=0.
1830 REM
1840 FOR I=1 TO 3
```

```
1850 IF A(I,I)=0 THEN GOTO 1880
1860 Q(I,N)=((Q(I,N-1)*A(I,I)+S(I))*EXP(A(I,I)*DT/2)-S(I))/A(I,I)
1870 GOTO 1890
1880 Q(I,N)=Q(I,N-1)+S(I)*DT/2
1890 NEXT I
1900 REM
1910 REM  NOW IMPROVE THE SUMS S(1)...S(3)
1920 REM  WITH SAME U'S.
1930 S(1)=B(1,1)*U1+B(1,2)*U2+A(1,2)*Q(2,N)+A(1,3)*Q(3,N)
1940 S(2)=B(2,1)*U1+B(2,2)*U2+A(2,1)*Q(1,N)+A(2,3)*Q(3,N)
1950 S(3)=B(3,1)*U1+B(3,2)*U2+A(3,1)*Q(1,N)+A(3,2)*Q(2,N)
1960 REM
1970 REM  AND NOW THE FULL-STEP EXTRAPOLATIONS, USING THE 'HALF-STEP' S'S.
1980 FOR I=1 TO 3
1990 IF A(I,I)=0 GOTO 2020
2000 Q(I,N)=((Q(I,N-1)*A(I,I)+S(I))*EXP(A(I,I)*DT)-S(I))/A(I,I)
2010 GOTO 2030
2020 Q(I,N)=Q(I,N-1)+S(I)*DT
2030 NEXT I
2040 REM PRINT ON SCREEN WHEN "MILEPOSTS" ARE PASSED (ASSUMING TIMER IS AVAILABLE).
2050 IF N = INT(N2+NS/4) THEN PRINT "SIMULATION IS AT 1/4 POINT, T=";TIMER
2060 IF N = INT(N2+NS/2) THEN PRINT "SIMULATION IS AT MIDPOINT, T=";TIMER
2070 IF N = INT(N2+3*NS/4) THEN PRINT "SIMULATION IS AT 3/4 POINT, T=";TIMER
2080 IF N = NM THEN PRINT "SIMULATION IS COMPLETE, T=";TIMER
2090 NEXT N
2100 REM SET UP PROMPT TO OVERRIDE DEFAULT THAT PRINTS DATA OUTPUT.
2110 PRINT "THE DEFAULT VERSION OF THIS PROGRAM LISTS INPUT AND OUTPUT DATA"
2120 PRINT "AT PRINTER, WHICH MUST BE CONNECTED TO PROCEED WITH DEFAULT VERSION"
2125 PRINT " "
2130 INPUT "DO YOU WISH SCREEN PLOT OF INPUT AND OUTPUT INSTEAD--YES OR NO?";PR$
2140 IF PR$="YES" THEN 2630
2150 REM
2160 REM PRINT INPUT DATA AND OUTPUT AT PRINTER.
2170 REM
2180 LPRINT "TITLE:";NA$
2190 LPRINT "NR="; NR, "DELT="; DT, "NPER=";NP, "NWAIT=";NW
2195 LPRINT " "
2200 LPRINT "NSIM=";NS, "N2=";N2, "NMAX=";NM
2210 LPRINT "UI'S="; UI(1), UI(2)
2220 LPRINT "QI'S="; QI(1), QI(2), QI(3)
2230 LPRINT "YI'S="; YI(1), YI(2)
2240 LPRINT "ROW1 A'S="; A(1,1), A(1,2), A(1,3)
2250 LPRINT "ROW2 A'S="; A(2,1), A(2,2), A(2,3)
2260 LPRINT "ROW3 A'S="; A(3,1), A(3,2), A(3,3)
2270 LPRINT "ROW1 B'S="; B(1,1), B(1,2)
2280 LPRINT "ROW2 B'S="; B(2,1), B(2,2)
2290 LPRINT "ROW3 B'S="; B(3,1), B(3,2)
2300 LPRINT "ROW1 C'S="; C(1,1), C(1,2), C(1,3)
2310 LPRINT "ROW2 C'S="; C(2,1), C(2,2), C(2,3)
2320 LPRINT "ROW1 D'S="; D(1,1), D(1,2), D(1,3)
2330 LPRINT "ROW2 D'S="; D(2,1), D(2,2), D(2,3)
2340 REM
2350 REM PRINT COLUMN HEADS AT PRINTER FOR T, U1, Q'S, AND Y1 (IF THIS NEXT
2360 REM  STATEMENT IS MODIFIED TO INCLUDE U2 AND Y2, WIDER PRINTING PAPER
2370 REM  IS NEEDED).
2380 LPRINT "T(N)", "U(1,N)", "Q(1,N)", "Q(2,N)", "Q(3,N)", "Y1(N)"
2390 REM
2400 REM INITIALIZE INTEGER VARIABLE X NEEDED TO SELECT EVERY NP'TH DATA
2410 REM  SET FOR PRINTING,LETTING Z DESIGNATE THE PAST VALUE OF X.
2420 X=0
2430 Z=X
2440 REM NOW THE (NM/NP) ITEMS IN EACH COLUMN:
2450 FOR N = 1 TO NM
2460 X = INT(N/NP)
2470 IF (X-Z)=1. THEN GOTO 2490
2480 GO TO 2550
2490 REM  COMPUTE Y1 ;AND Y2 FOR POSSIBLE PRINTING.
2500 Y1=C(1,1)*Q(1,N)+C(1,2)*Q(2,N)+C(1,3)*Q(3,N)+D(1,1)*U(1,N)+D(1,2)*U(2,N)
2510 Y2=C(2,1)*Q(1,N)+C(2,2)*Q(2,N)+C(2,3)*Q(3,N)+D(2,1)*U(1,N)+D(2,2)*U(2,N)
2520 REM  PRINT T, U1, Q'S, Y1 AT PRINTER.
2530 LPRINT T(N), U(1,N), Q(1,N), Q(2,N), Q(3,N), Y1
2540 Z=X
2550 NEXT N
2560 REM PRINT TOTAL TIME ON SCREEN AND AT PRINTER-(OPTIONAL).
2570 REM  DELETE NEXT 4 STATEMENTS IF YOUR COMPUTER DOES NOT HAVE "TIMER,MEM"
2580 REM  OR "TIMER".
2590 PRINT "TOTAL TIME,T="; TIMER
2600 LPRINT "TOTAL TIME, T="; TIMER
2610 LPRINT "REMAINING MEMORY=";MEM
2620 GOTO 3210
```

```
2630 REM PICK UP HERE FOR OVERRIDE TO PLOT ON SCREEN INSTEAD OF PRINT.
2640 REM
2650 REM SET UP DEFAULT VALUES FOR INORM (NI), AND ONORM (NO), NORMALIZING
2660 REM  FACTORS TO SCALE THE INPUT AND OUTPUT PLOTS.
2670 NI=1    :    NO=1
2680 REM
2690 REM PROVIDE FOR MEANS TO OVERRIDE DEFAULT VALUES OF NI AND NO
2700 PRINT "YOU MAY OVERRIDE THE DEFAULT VALUES OF NORMALIZING FACTORS"
2710 PRINT "INORM AND ONORM (USED TO SCALE INPUT AND OUTPUT PLOTS) BY"
2720 PRINT "RESPONDING TO THE FOLLOWING PROMPTS"
2725 PRINT " "
2730 PRINT "DEFAULT VALUE OF INORM=";NI
2740 INPUT "OVER-RIDE VALUE FOR INORM--YES (ALL CAPS) OR NO?";NI$
2750 IF NI$="YES" THEN 2770
2760 GOTO 2780
2770 INPUT "OVER-RIDE VALUE FOR INORM=";NI
2780 PRINT "DEFAULT VALUE OF ONORM IS:";NO
2790 INPUT "OVER-RIDE VALUE OF ONORM--YES OR NO?";NO$
2800 IF NO$="YES" THEN 2820
2810 GOTO 2830
2820 INPUT "OVER-RIDE VALUE FOR ONORM=";NO
2830 REM
2840 REM SET UP FOR PLOTTING INPUT AND OUTPUT VS. TIME ON MONITOR SCREEN.
2850 SCREEN 1
2860 CLS
2870 REM
2880 REM DRAW BORDER 'BOX'
2890 LINE (10,10)-(246,152),,B
2900 REM
2910 REM DRAW X (HORIZ.) AND Y (VERT.) AXES.
2920 LINE(20,15)-(20,150)
2930 LINE(15,20)-(240,20)
2940 REM
2950 REM SET UP THE H AND V EQNS. BASED ON
2960 REM  FOR THE TIME AXIS:
2970 REM   H=16+N*220/NM
2980 REM  FOR THE INPUT:
2990 REM   V=20+120*IN(N)/(2*NI)
2995 REM  FOR THE OUTPUT:
3000 REM   V=20+120*OU(N)/(2*NO)
3005 REM
3010 REM FIRST THE LOOP TO PLOT THE INPUT CURVE.
3020 FOR N=1 TO NM
3030 H=16+N*220/NM
3040 V=20+120*U(1,N)/(2*NI)
3050 PSET (H,V),1
3060 NEXT N
3070 REM
3080 REM THEN THE LOOP FOR THE OUTPUT CURVE.
3090 FOR N=1 TO NM
3100 H=16+N*220/NM
3110 REM
3120 REM COMPUTE Y1 FOR PLOT.
3130 Y1=C(1,3)*Q(3,N)
3140 V=20+120*Y1/(2*NO)
3150 PSET(H,V)
3160 NEXT N
3170 REM CREATE DUMMY LOOP TO HOLD PLOT ON SCREEN FOR A FEW SECONDS.
3180 FOR N=1 TO 6000
3190 X=N+1
3200 NEXT N
3210 END
```

Glossary

		English Units	SI Units
$\mathbb{A}$	state matrix		
A	area	in.^{2}	m^2
a	acceleration $\mathbb{A}$-matrix coefficient	in./sec^2	m/sec^2
$\mathbb{B}$	input matrix		
B	rotational damping constant	lb-in.-sec/rad	N-m-sec/rad
b	$\mathbb{B}$-matrix coefficient translational damping constant	 lb-sec/in.	 N-sec/in.
$\mathbb{C}$	output matrix		
C	capacitance, electric capacitance, hydraulic capacitance, pneumatic capacitance, thermal	Farad or a-sec/v in.^{5}/lb in.^{2} Btu/degF	a-sec/v m^5/N m^2 kgcal/degC
c	$\mathbb{C}$-matrix coefficient coefficient specific heat	 in^2/sec^2-degF	 m^2/sec^2-degC
$\mathbb{D}$	direct transmission matrix differential operator, d/dt displacement per unit motion	 1/sec in.^{3}/rad or in.^{2}	 1/sec m^3/rad or m^2
d	$\mathbb{D}$-matrix coefficient diameter	 in.	 m
db	decibel, $20 \times \log_{10}$		
$\mathscr{E}$	energy	lb-in.	N-m or Joule
e	electric potential system error	v	v

		ENGLISH UNITS	SI UNITS
$\mathscr{F}$	Fourier transform		
F	force	lb	N
f	function		
	cycle frequency	cyc/sec	cyc/sec
	(subscript) for fluid		
G	transfer function		
g	acceleration of gravity	in./sec^2	m/sec^2
	coefficient		
H	transfer function		
h	coefficient		
	(subscript) for heat		
Im	imaginary part		
I	inertance, hydraulic	lb-sec^2/in^5	N-sec^2/m^5
	inertance, pneumatic	sec^2/in.^{2}	sec^2/m^2
i	electric current, coulomb/sec	amp	amp
	integer		
	(subscript) for input		
J	rotational inertia	lb-in.-sec^2/rad	N-m-sec^2/rad
j	integer		
	square-root of -1		
K	torsional spring stiffness	lb-in./rad	N-m/rad
k	translational 1 spring stiffness	lb/in.	n/m
	integer		
	specific heat ratio, c_p/c_v		
	Laplace transform		
L	electrical inductance	Henry or v-sec/a	v-sec/a
ℓ	length		
	integer		
	(subscript) for load		
m	mass	lb-sec^2/in.	kg = N-sec^2/m
	integer		
N	integer		
n	integer		
o	(subscript) for output		
$\mathscr{P}$	power	hp or lb-in./sec	watt = N-m/sec
P	pressure	lb/in.^{2}	Pascal = N/m^2
p	pole (root)	1/sec	1/sec
Q	flowrate, heat	Btu/sec	kgcal/sec or watt
	flowrate, volume	in.^{3}/sec	m^3/sec
	flowrate, weight	lb/sec	n/sec
q	state variable		
	electric charge	coulomb = a-sec	coulomb
Re	real part		

		English Units	SI Units
R	resistance, electric	ohm or v/a	v/a
	resistance, hydraulic	lb-sec/in.^{5}	N-sec/m^5
	resistance, pneumatic	sec/in.^{2}	sec/m^2
	resistance, thermal	degF-sec/Btu	degC/watt
R	ideal gas constant	in.^{2}/sec^2-degR	m^2/sec^2-degK
r	radius	in.	m
S	slope		
s	complex variable	1/sec	1/sec
T	transfer function		
T	torque	lb-in.	N-m
	time parameter	sec	sec
	temperature	degF	degC
	temperature, absolute	degR	degK
t	time	sec	sec
u	input variable		
	chamber volume	in.^{3}	m^3
V	volume displaced, $\int Q dt$	in.^{3}	m^3
v	translational velocity	in./sec	m/sec
	work	lb-in.	N-m
W	electric power	watt = v-a	v-a
w	weighting function		
	bilinear transformation variable		
x	displacement (motion)	in.	m
	auxiliary state variable		
y	output variable		
$\mathfrak{Z}$	z-transform		
z	complex variable for discrete systems		
α	coupling coefficient		
β	bulk modulus	lb/in.^{2}	N/m^2
γ	weight density	lb/in.^{3}	N/m^3
Δ	gravity-displaced reference	in.	m
	difference or change		
ϕ	phase angle	rad	rad
λ	flux linkage, $\int e dt$	v-sec	v-sec
ψ	angle	rad	rad
σ	real part of s		
	Stefan-Boltzmann constant		
θ	angle (motion)	rad	rad
τ	time constant	sec	sec
ζ	damping ratio		
Ω	angular velocity	rad/sec	rad/sec
ω	radian frequency	rad/sec	rad/sec

Index